Synchronous Ethernet and IEEE 1588 in Telecoms

Synchronous Ethernet and IEEE 1588 in Telecoms

Next Generation Synchronization Networks

Jean-Loup Ferrant, Mike Gilson,
Sébastien Jobert, Michael Mayer,
Laurent Montini, Michel Ouellette,
Silvana Rodrigues, Stefano Ruffini

Series Editor
Pierre-Noël Favennec

WILEY

First published 2013 in Great Britain and the United States by ISTE Ltd and John Wiley & Sons, Inc.

Apart from any fair dealing for the purposes of research or private study, or criticism or review, as permitted under the Copyright, Designs and Patents Act 1988, this publication may only be reproduced, stored or transmitted, in any form or by any means, with the prior permission in writing of the publishers, or in the case of reprographic reproduction in accordance with the terms and licenses issued by the CLA. Enquiries concerning reproduction outside these terms should be sent to the publishers at the undermentioned address:

ISTE Ltd
27-37 St George's Road
London SW19 4EU
UK
www.iste.co.uk

John Wiley & Sons, Inc.
111 River Street
Hoboken, NJ 07030
USA
www.wiley.com

© ISTE Ltd 2013

The rights of Jean-Loup Ferrant, Mike Gilson, Sébastien Jobert, Michael Mayer, Laurent Montini, Michel Ouellette, Silvana Rodrigues, Stefano Ruffini to be identified as the authors of this work have been asserted by them in accordance with the Copyright, Designs and Patents Act 1988.

Library of Congress Control Number: 2013933689

British Library Cataloguing-in-Publication Data
A CIP record for this book is available from the British Library
ISBN: 978-1-84821-443-9

Printed and bound in Great Britain by CPI Group (UK) Ltd., Croydon, Surrey CR0 4YY

Table of Contents

Foreword xi

Abbreviations and Acronyms xv

Acknowledgments xxvii

Introduction xxxiii

Chapter 1. Network Evolutions, Applications and Their Synchronization Requirements 1

   1.1. Introduction 1
   1.2. Evolution from plesiochronous digital hierarchy to optical transport networks 3
      1.2.1. Plesiochronous digital hierarchy and public switch telephone networks 3
      1.2.2. Evolution toward SDH and synchronous optical network 7
      1.2.3. Introduction of optical transport network in transport networks . . 11
   1.3. Migration and evolution in the next-generation networks: from time division multiplexing to packet networks 12
      1.3.1. Circuit emulation synchronization requirements 14
   1.4. Mobile networks and mobile backhaul 17
      1.4.1. Synchronization requirements in mobile networks 22
   1.5. Synchronization requirements in other applications 27
   1.6. The need to define new synchronization technologies 28
   1.7. Bibliography 30

Chapter 2. Synchronization Technologies 33

   2.1. Fundamental aspects related to network synchronization 33
   2.2. Timing transport via the physical layer 42
      2.2.1. Synchronous Ethernet 42

2.3. Packet timing.. 47
    2.3.1. Packet timing using traffic data........................ 47
    2.3.2. Packet-based methods................................... 52
2.4. IEEE 1588 and its Precision Time Protocol..................... 55
    2.4.1. Some essentials of IEEE 1588........................... 56
    2.4.2. IEEE 1588-2002: origin and limitations................. 64
    2.4.3. IEEE 1588-2008 and PTPv2............................... 68
2.5. The concept of "profiles"..................................... 75
    2.5.1. Frequency profile....................................... 77
    2.5.2. Phase and time profile (ITU-T G.8275.1)................ 81
2.6. Other packet-based protocols.................................. 82
    2.6.1. Packet-based timing: starting with CES................. 82
    2.6.2. Dedicated timing TDM PW................................ 86
    2.6.3. NTP.. 87
    2.6.4. Summary and comparison................................. 91
2.7. GNSS and other radio clock sources............................ 94
    2.7.1. Global and regional space-based timing system.......... 94
    2.7.2. Regional terrestrial systems........................... 102
    2.7.3. Comparison... 104
2.8. Summary.. 105
2.9. Bibliography... 107

Chapter 3. Synchronization Network Architectures in Packet Networks.. 111

3.1. The network synchronization layer............................. 111
    3.1.1. Network layers and abstraction......................... 111
    3.1.2. The synchronization layer.............................. 116
3.2. Functional modeling... 117
3.3. Frequency synchronization topologies and redundancy schemes using SyncE... 119
    3.3.1. Introduction... 119
    3.3.2. Network topologies..................................... 120
    3.3.3. Redundancy and source traceability..................... 121
    3.3.4. Use of SSM in real networks............................ 122
    3.3.5. Networks involving SSUs................................ 130
    3.3.6. Classical errors during SSM configuration.............. 131
    3.3.7. Conclusion on synchronization topologies............... 133
3.4. The IEEE 1588 standard and its applicability in telecommunication networks.. 133
3.5. Frequency synchronization topologies and redundancy schemes using IEEE 1588... 134
    3.5.1. Redundancy schemes using IEEE 1588..................... 136

    3.6. Time synchronization topologies and redundancy schemes 139
        3.6.1. Locally distributed PRTC 140
        3.6.2. Packet-based method 141
        3.6.3. Resiliency and redundancy schemes 142
    3.7. Bibliography 146

Chapter 4. Synchronization Design and Deployments 149

    4.1. High-level principles 149
        4.1.1. Network evolution 152
        4.1.2. Typical mobile networks requirements and evolutions 158
    4.2. MAKE or BUY network synchronization strategies 161
        4.2.1. Relationships between MAKE or BUY strategies for network
        connectivity and Synchronization 162
        4.2.2. MAKE or BUY network synchronization source strategies 167
        4.2.3. Fixed/mobile network scenarios 170
    4.3. Deployment of timing solutions for frequency
    synchronization needs 182
        4.3.1. Overview of synchronization solutions for frequency needs 183
        4.3.2. Synchronous Ethernet deployments 184
        4.3.3. IEEE 1588 end-to-end deployments 196
    4.4. Deployment of timing solutions for accurate phase/time
    synchronization needs 220
        4.4.1. GNSS deployments and associated issues 221
        4.4.2. IEEE 1588 full timing support deployments 225
        4.4.3. Possible migration paths toward IEEE 1588
        phase/time profile 236
    4.5. Bibliography 237

Chapter 5. Management and Monitoring of Synchronization
Networks 239

    5.1. Introduction 239
    5.2. Network management systems and the telecommunications
    management network (TMN) 240
    5.3. Synchronization Network management: the synchronization
    plan and protection 242
    5.4. Provisioning and setup: manual versus automatic 245
    5.5. Monitoring functions 246
    5.6. Management issues in wireless backhaul 249
    5.7. Network OS integration: M.3000 versus SNMP 250
    5.8. Bibliography 252

Chapter 6. Security Aspects Impacting Synchronization. 255

6.1. Security and synchronization 255
   6.1.1. Terminology used in security. 257
   6.1.2. Synchronization in network security ensemble 259
6.2. Security of the timing source 261
   6.2.1. Access security to device 262
   6.2.2. GNSS signal vulnerability. 263
   6.2.3. Protecting and mitigating from compromise signal 270
6.3. Security of synchronization distribution 274
   6.3.1. Security aspects of physical timing transmission 275
   6.3.2. Security aspects of packet-based timing transmission 277
6.4. Synchronization risk management 282
6.5. Bibliography 284

Chapter 7. Test and Measurement Aspects of Packet Synchronization Networks. 287

7.1. Introduction. 287
7.2. Traditional metrics 287
7.3. Equipment configuration. 292
7.4. Reference signals, cables and connectors 293
7.5. Testing Synchronous Ethernet. 293
   7.5.1. Testing the performance of SyncE EEC 294
   7.5.2. Testing the ESMC protocol 304
7.6. Testing the IEEE 1588 end-to-end telecom profile 308
   7.6.1. Testing the telecom profile – protocol. 308
   7.6.2. Testing the telecom profile – performance of packet networks . . . 316
   7.6.3. Testing the telecom profile – performance of a PTP
   packet slave clock 319
7.7. Bibliography 326

Appendix 1. Standards in Telecom Packet Networks Using Synchronous Ethernet and/or IEEE 1588 329

A1.1. Introduction 329
A1.2. General content of ITU-T standards 330
   A1.2.1. Network level 330
   A1.2.2. Equipment level. 331
   A1.2.3. Use of network and equipment specification. 332
A1.3. Summary of standards 332
   A1.3.1. Standards related to SyncE 333
   A1.3.2. Standards related to IEEE 1588 end-to-end telecom
   profile for frequency. 335

 A1.3.3. Standards related to IEEE 1588 full timing support
telecom profile for phase and time transport.   337
 A1.4. Bibliography   339

**Appendix 2. Jitter Estimation by Statistical Study (JESS)
Metric Definition**   341

 A2.1. Mathematical definition of JESS   341
 A2.2. Mathematical definition of JESS-w   342

Permissions and Credits   345

Biography   349

Index   353

Foreword

Synchronization is the bedrock of the telecommunication highways. Much like the pavement of a highway, synchronization is often taken for granted, since it is effectively invisible when it is working. But it is an enabler of many aspects of the transfer of voice and data. Some services directly require synchronization. More specifically, synchronization can optimize the use of a given bandwidth, increasing available throughput in a fixed band.

Time-and-frequency issues are rich and complex enough that they form their own discipline. Unfortunately, there are few places of study that have this unique focus as an institute unto itself. Hence, the importance of this book. There is much confusion about principles of time and frequency, in part because our human experience of time is so intriguing. Time is a major focus in human culture, in art, philosophy, and song. Yet time in science, engineering and metrology is a different thing.

Scientific time and frequency start with a clock, a device that realizes a theoretical principle in a physical way. The underlying principle in any clock is always a law of physics that predicts circumstances in which the states of a system will repeat at a constant rate. A clock physically realizes this theoretical principle and produces this rate, or frequency, with some level of accuracy and stability. The underlying principle is a theory of physics that forces the theoretical rate of the clock to be constant by definition. Time from a clock comes by counting the states as they repeat themselves, just as counting days produces the calendar. Hence, a clock fundamentally produces a frequency, with time optionally produced by using a counter. The standard frequency is a physical quantity, equivalent to an energy. The time standard, however, is a man-made artifact. A standard frequency signal can be produced by a single device, a cesium standard. To get standard time from a clock, as opposed to time intervals, the clock must first be set, or synchronized, against a reference. Then, since the clock is at best a frequency device with some white noise on the signal, any two clocks will wander off from each other in time without bounds if they are not re-synchronized periodically, whereas the best clocks may be bounded in their native frequency differences.

So where does the time reference come from? Metrologically, the practical time standard is a weighted average of clock times from all over the world. This is produced by the International Bureau of Weights and Measures (Bureau International des Poids et Mesures – BIPM) in the forms of International Atomic Time (TAI) and Universal Coordinated Time (UTC). These time scales are produced only after the fact. Any real-time time signal can be only a prediction of what the correct time will be when it is defined later.

The result of these facts is that a system that requires only frequency can have a stand-alone device that produces the signal. But a system that requires time must compare the count of time on its local device to an external reference. This has broad implications in telecommunications systems, and in the many other systems that require some level of accurate time coordination. Not only must clocks be chosen and implemented to run properly, but their signals must be transported and measured properly.

This book describes the needs for synchronization in telecommunications networks and the current evolution of methods and standards that enable it. The challenge of supplying needed synchronization to telecommunications systems is primarily an engineering problem, not a theoretical or scientific one. Because the requisite frequency devices can be expensive, and because the necessary time synchronization must be transported, there is a need for a synchronization network. The requirements of a communications network fundamentally conflict with those of a synchronization network. The communications network ideally separates functional layers, so devices interact with other devices only one layer up or down. Synchronization requires direct access to the lowest layer, the physical layer, since synchronization, unlike data, requires a physical signal. Applications that are required to consume some form of synchronization signal can be in any layer of the communications network. Hence they must break or tunnel through the isolation of layers to get access to the synchronization signal, in violation of the layer principles of a communications network.

Alternatively, a synchronization signal can be supplied from a source external to the communications network. Receivers of Global Navigation Satellite Systems (GNSS) such as the U.S. Global Positioning System (GPS) are commonly used to provide both time and frequency synchronization. These cannot be used everywhere, however, both because of expense and because of difficulties in getting the synchronization signals where they are needed. In addition, GNSS signals are vulnerable to interference – both intentional and unintentional interference. Thus, even with GNSS signals available, a synchronization network remains essential.

As I write in 2013, synchronization in communication systems is in the midst of evolving from the role of primarily frequency synchronization to the role of precise time synchronization. The metrology community separates these two types of synchronization by calling "synchronization" in frequency the name "syntonization", though the telecommunications community uses the word synchronization for both. The transport of networks in the late 80 s and 90 s was itself synchronous in frequency, or syntonous. With the advent of packet networks, the transport no longer needed syntonization, yet many applications and services still required various forms of either syntonization or synchronization or both.

Among other things, synchronization optimizes available bandwidth, enabling a more efficient use of the spectrum. Today, wireless networking is becoming more ubiquitous. Time synchronization is becoming essential to allow high data rates in the limited wireless spectrum. This becomes a complex engineering problem, as many different scenarios require different types of synchronization. Traditional synchronous networks still remain in use and require syntonization, while packet networks dominate all new roll-outs. Further, many size-scales of wireless networks are being deployed, from macro-cells over cities to femto-cells in a small interior of a building. These hybrid networks challenge operators to supply needed synchronization to all the requisite applications and services.

Within the context of these circumstances, this book is timely. As synchronization becomes both more complex and more necessary, there still remains a dearth of training and learning options. This book is a comprehensive effort by experts who have been developing standards and engineering devices, and employing these in real networks. It should help fill the void for those trying to negotiate the diverse and complex world of time and frequency issues in communications systems.

This book is a collaboration of major figures in the creation and use of synchronization. I will not repeat the information in the biographic, but I want to mention my appreciation and respect for this team of authors and the current effort. The authors are a mixture of standards experts, equipment building and testing experts, and operators who must implement and maintain synchronization. This book is a work of significant magnitude, involving many hours of development and coordination.

Synchronization in telecommunications is a fascinating field. It involves complex concepts and difficult engineering efforts. Concomitantly, the field creates great benefits for many users, facilitating an increasing ease for human social communications underpinned by large data transfers. In addition, synchronization facilitates large industries representing many billions of dollars. This book can take the user a long way along this river. Enjoy!

Marc WEISS
Mathematical Physicist, GPS and Telecom Sync Expert at NIST

Abbreviations and Acronyms

This section lists the abbreviations and acronyms used in this book.

3GPP	Third-Generation Partnership Project
3GPP2	Third-Generation Partnership Project 2
AAA	Authentication, Authorization and Accounting
AAL1	ATM Adaptation Layer 1
ACMA	Australian Communications Authority
ACR	Adaptive Clock Recovery
ADEV	Allan Deviation
ADM	Add Drop Multiplexer
ADPCM	Adaptive Differential Pulse Code Modulation
ANSI	American National Standards Institute
ATIS	Alliance for Telecommunications Industry Solutions
ATM	Asynchronous Transfer Mode
AVB	Audio Video Bridging
BC	Boundary Clock
BIPM	International Bureau of Weights and Measures
BITS	Building Integrated Timing System
BMC	Best Master Clock

BMCA	Best Master Clock Algorithm
BNC	Bayonet Neill-Concelman (connector)
BPM	People's Republic of China's National Time Signal Service
BS	Base Station
BSC	Base Station Controller
BSS	Base Station Subsystem
BTS	Base Transceiver Station
CBR	Constant Bit Rate
CDMA	Code Division Multiple Access
CDR	Clock Data Recovery
CEM	(SONET/SDH) Circuit Emulation Service over MPLS (RFC5143 – obsoleted by CEP)
CEP	(SONET/SDH) Circuit Emulation over Packet (RFC 4842 – obsoletes CEM)
CERN	European Organization for Nuclear Research
CES	Circuit Emulation Service
CESoPSN	Circuit Emulation Service over Packet-Switched Network (RFC5086)
CF	Correction Field
CI	Characteristic Information (ITU-T Rec. G.805)
CLI	Command Line Interface
CNSS	Compass Navigation Satellite System
CoMP	Coordinated Multipoint
CPE	Customer Premise Equipment
CPRI	Common Public Radio Interface
CSG	Cell Site Gateway
D/A	Digital to Analog
DCF	Dispersion Compensating Fibers
DCN	Data Communications Network
DCR	Differential Clock Recovery
DCF77	Radio Time Service in Germany
DNU	Do Not Use (QL value interpretation)
DoS	Denial of Service

DPLL	Digital Phase Lock Loop
DS1	Digital Signal 1 (1.544 Mbit/s)
DSL	Digital Subscriber Line
DSLAM	Digital Subscriber Line Access Multiplexer
DTI	DOCSIS Timing Interface
DUS	Don't Use (QL value interpretation – equivalent to DNU)
DUT	Device Under Test
DVB-T/H	Digital Video Broadcast – Terrestrial/Handheld
E1	Digital signal (2.048 Mbit/s)
E2E	End-to-End
EEC	(Synchronous) Ethernet Equipment Clock
eICIC	Enhanced ICIC (Inter-cell Interference Coordination)
eLORAN	Enhanced LORAN
EPC	Evolved Packet Core (LTE)
EPL	Ethernet Private Line
ESI	External Sync Interface
ESMC	Ethernet Synchronization Messaging Channel
ETH	Ethernet MAC Layer Network (IU-T)
ETSI	European Telecommunications Standards Institute
ETY	Ethernet PHY Layer Network (ITU-T)
FCC	Federal Communications Commission
FCS	Frame Check Sequence
FCAPS	Fault, Accounting, Configuration, Performance and Security Management
FDD	Frequency Division Duplexing
FLL	Frequency Lock Loop
FPP	Floor Packet Percentage
GAARDIAN	GNSS Availability Accuracy Reliability and Integrity Assessment for Timing and Navigation
GBAS	Ground-Based Augmentation System
GE	Gigabit Ethernet
GFP-F	Generic Framing Procedure-Framed

GLONASS	*Globalnaya Navigatsionnaya Sputnikovaya Sistema* (Global Navigation Satellite System)
GM	Grand Master
GMP	Generic Mapping Procedure
GNSS	Global Navigation Satellite System
GPS	Global Positioning System
GRI	Group Repetition Interval
GSM	Global System for Mobile communications
GUI	Graphical User Interface
HL	Hop Limit
HLR	Home Location Register
HOL	Head Of Line
HOLB	Head Of Line blocking
HRM	Hypothetical Reference Model
HRX	Hypothetical Reference Connection
HSPA	High-Speed Packet Access
HSS	Home Subscriber Server
IANA	Internet Assigned Numbers Authority
ICIC	Inter-cell Interference Coordination
ID	Identifier or Identity Description
IED	Improvised Explosive Device
IEEE	Institute of Electrical and Electronics Engineers
IETF	Internet Engineering Task Force
IMA	Inverse Multiplex for ATM
IP	Internet Protocol
IP FRR	IP FastReRoute
IPDV	Inter-Packet Delay Variation or IP Packet Delay Variation
IRIG	Inter-Range Instrumentation Group
IRNSS	Indian Regional Navigational Satellite System
ITSF	International Telecom Sync Forum
ITU	International Telecommunication Union
ITU-T	International Telecommunication Union – Telecom

Iu	Interconnection point between an RNC or a BSC and a 3G Core Network
Iub	Interface between an RNC and a Node B
IWF	Interworking Function
JESS	Jitter Estimation by Statistical Study
JLOC	Jammer Location
LACP	Link Aggregation Control Protocol
LAN	Local Area Network
LBAS	Local Based Augmentation System
LORAN	Long Range Aid to Navigation
LSP	Label Switched Path
LTE	Long-Term Evolution
LTE-A	LTE Advanced
Lx	Layer x
M-CMTS	Modular Cable Modem Termination System
MAC	Media Access Control
MAFE	Maximum Averaged Frequency Error
MATIE	Maximum Averaged Time Interval Error
MB(M)S	Multicast Broadcast (Multimedia) Services
MBSFN	Multicast Broadcast Single Frequency Network
MDEV	Modified Allan Deviation
MEF	Metro Ethernet Forum
MI	Management Information
MIB	Management Information Base
MinTDEV	Minimum TDEV
MME	Mobility Management Entity
MNO	Mobile Network Operator
MPLS	Multi-Protocol Label Switching
MRTIE	Maximum Relative Time Interval Error
MS	Mobile System
MSAN	Multi-Service Access Node
MSC	Mobile Switching Center

MSF	UK Low frequency time signal and standard frequency radio station based on the NPL time scale UTC(NPL)
MTIE	Maximum Time Interval Error
MTOSI	Multi-Technology Operations System Interface
MTU	Maximum Transmission Unit
MW	Microwave
NE	Network Element
NGN	Next-Generation Network
NIST	National Institute of Standards and Technology (USA)
NMS	Network Management System
NPU	Network Processor Unit
NS	Network Synchronization (ITU-T)
NTP	Network Time Protocol
NTR	Network Timing Reference
OAM	Operations, Administration, Maintenance
OAM&P	Operations, Administration, Maintenance and Provisioning
OC	Ordinary Clock
OCXO	Oven-Controlled Crystal Oscillator
ODUk	Optical data unit of level k
OFDM	Orthogonal Frequency-Division Multiplexing
OLT	Optical Line Terminal (PON)
ONU	Optical Network Unit
OPEX	Operational Expense
OS	Operating System
OSC	Optical Supervisory Channel (OTN)
OSI	Open Systems Interconnection
OSPF	Open Shortest Path First
OSS	Support System
OSSP	Organization-Specific Slow Protocol
OTDR	Optical Time-Domain Reflectometer
OTN	Optical Transport Network
OTT	Over-The-Top

OUI	Organization Unique Identifier
P2P	Peer to Peer
PABX	Private Automatic Branch Exchange
PAR	Project Authorization Request
PCM	Pulse Code Modulation
PCP	Port Control Protocol
PDF	Probability Density Function
PDH	Plesiochronous Distribution Hierarchy
PDN-GW	Packet Data Network-GateWay
PDU	Protocol Data Unit
PDV	Packet Delay Variation
PHY	Physical
PLL	Phase-Locked Loop
PM	Packet Master
PMC	Packet Master Clock
PNT	Position, Navigation and Timing
PON	Passive Optical Network
POS	Packet over SONET (or SDH)
ppb	Part per billion
ppm	Part per million
PPS	Pulse per second
pps	Packets per second
PRC	Primary Reference Clock
PRS	Primary Reference Source
PRTC	Primary Reference Time Clock
PS	Packet Slave
PSN	Packet-Switched Network
PSTN	Public-Switched Telephone Network
PTP	Precision Time Protocol
PTSF	Packet Timing Signal Failure
PW	Pseudo-Wire

PWS	Pseudo-Wire Service
QL	Quality Level
QZSS	Quasi-Zenith Satellite System
RAIM	Receiver Autonomous Integrity Monitoring
RAN	Radio Access Network
RBAS	Regional Based Augmentation System
RF	Radio Frequency
RNC	Radio Network Controller
RNSS	Radio Navigation Satellite Service
RT	Residence Time
RTP	Real-Time Protocol
SA	Selective Availability
SAE-GW	System Architecture Evolution-GateWay
SASE	Stand Alone Synchronization Equipment
SATop	Structure-Agnostic TDM over Packet (IETF)
SBAS	Satellite-Based Augmentation System
SD	Synchronization Distribution (ITU-T)
SDCM	System for Differential Corrections and Monitoring (GLONASS)
SDH	Synchronous Digital Hierarchy
SDO	Standardization Development Organizations
SDSL	Symmetric Digital Subscriber Line
SEC	SDH Equipment Clock
SETG	Synchronous Equipment Timing Generator
SETS	Synchronous Equipment Timing Source
SFN	Single Frequency Network
SFP	Small Form Factor Pluggable
SGSN	Serving GPRS Support Node
SG15	Study Group 15 (ITU-T)
SLA	Service Level Agreement
SMA	SubMiniature Version A (connector)
SMB	SubMiniature Version B (connector)

SMC	SubMiniature Version C (connector)
SNMP	Simple Network Management Protocol
SNTP	Simple Network Time Protocol
SOF	Start Of Frame
SONET	Synchronous Optical Network
SOOC	Slave Only Ordinary Clock
SP	Service Provider
SRTS	Synchronous Residual Time Stamps
SSH	Secure SHell
SSM	Synchronization Status Message
SSU	Synchronization Supply Unit
ST3	QL Value for Stratum3
STM-N	Synchronous Transport Module (level N)
SyncE	ITU-T Synchronous Ethernet
S1	Interface between an eNB and an EPC
T-BC	Telecom-BC (boundary clock)
T-GM	Telecom-GM (grandmaster)
T-SC	Telecom-Slave Clock
T-TC	Telecom-Transparent Clock
T-TSC	Telecom-Time Slave Clock
TAI	*Temps Atomique International* (International Atomic Time)
TASI	Time Assignment Speech Interpolation
TC	Transparent Clock
TCO	Total Cost of Ownership
TD-SCDMA	Time Division-Synchronous CDMA
TDD	Time Division Duplexing
TDEV	Time Deviation
TDF	*TéléDiffusion de France* (radio time service broadcasted by TDF)
TDM	Time Division Multiplexing
TDMA	Time Division Multiplexing Access
TICTOC	Timing over IP Connection and Transfer of Clock (IETF WG)

TIE	Time Interval Error
TKS	Time Keeping System
TLV	Type Length Value
TMF	Telemanagement Forum
TNC	Threaded Neill-Concelman (connector)
TNM	Telecommunications Management Network
ToD	Time of Day
TSG	Timing Signal Generator
TTL	Time To Live
TTT	Timing Transparent Transcoding
TWSTFT	Two-Way Satellite Time and Frequency Transfer
TWTT	Two-Way Time Transfer (protocol)
UDP	User Datagram Protocol
UE	User Equipment
UI	Unit Interval
UMTS	Universal Mobile Telecommunications System
US	United States
USNO	US Naval Observatory
UTC	Coordinated Universal Time
UTP	Unshielded Twisted Pair (cable)
UTRAN	UMTS Transport Radio Access Network
Uu	Radio Interface between the UE and the NodeB
VCO	Voltage Controlled Oscillator
VDSL	Very High Speed Digital Subscriber Line
VLAN	Virtual LAN (Local Area Network)
VoIP	Voice over Internet Protocol
VPN	Virtual Private Network
WAAS	Wide Area Augmentation System
WAN	Wide Area Network
WCDMA	Wideband CDMA

WDM	Wavelength Division Multiplexing
WG	Working Group
WiMAX	Worldwide Interoperability for Microwave Access
WS	Work Station
WSTS	Workshop on Synchronization in Telecommunication Systems
WWWF	Radio Time Service from NIST (US)
xDSL	x (any type of) Digital Subscriber Line

Acknowledgments

Jean-Loup Ferrant

I would like to thank Alcatel, now Alcatel-Lucent, and specially Bernard Point and Bernard Sales, who supported my work on synchronization and standardization in ITU, ETSI, IEEE and IETF.

I would like to thank Tommy Cook, CEO of Calnex Solutions, who has been sponsoring my activity in ITU-T Q13 after I retired from Alcatel-Lucent.

I want to thank all the participants of ITU-T Q13 for their work during the last decade, which allowed Q13 to address the new issues raised by the transport of synchronization in packet networks.

I want also to thank my family who supported me during my work on this book.

Mike Gilson

I would like to thank BT and specifically both Tony Flavin and Glenn Whalley for their support of my work on this book and, in the wider context, their ongoing support for the synchronization subject in general. I would also like to thank the team of professionals I work with, Greg Mason, Sean Taylor and Trevor Marwick; both Sean Taylor and Trevor Marwick have worked on the evolving SyncE/1588 technology and provided me with considerable support for which I am indebted to them.

It has been a privilege to work on the development of synchronization standards and see their evolution from concepts to finished standards and then follow their adoption into deployed networks for the benefit of all. In the 1990s, during my first term in telecoms standardization, I met many inspirational people – some are still around, Dr Ghani Abbas being one of them who taught me much. During my second term from the early 2000s, I have had the pleasure to work with many new people on this subject. I feel fortunate to have met them all and worked with many, although too many to name it is primarily the collective group in ITU-T SG15Q13, ITSF and WSTS.

I would like to thank my family who have put up with the many weeks away over the years that have been due to my standards participation. Special thanks go to my partner Jan Longthorp, who has dropped me off and collected me from many different airports and who has also put up with many broken weekends resulting from both standards participation and the subsequent writing of this book. On behalf of all the authors, I would also like to thank Jan for all the help she gave us in the final proof reading.

Finally, I dedicate this book to my parents.

Sébastien Jobert

I would like to very much thank my company, France Télécom Orange, for supporting this work, and all the studies that we initiated with my colleagues in Lannion. I thank in particular Jean-Paul Cornec, my predecessor representing France Télécom in ITU-T SG15 Q13, for having kindly shared his expertise when I joined the team. I also thank Pierre-Noël Favennec for his kind help in finding the publisher of this book.

I would like to acknowledge the excellent technical work that has been done over the years in ITU-T with participants of other companies that are not part of this project (but who could have been for sure without any problem), in particular, Geoffrey Garner and Kenneth Hann (there are many others, but the list would be too long …). The 7–8 year period preceding the publishing of this book has, indeed, been very fruitful in standards and is very likely one my most exciting experiences.

I obviously thank my family very much, my lovely wife Stéphanie, daughter Jessica and son Romain, for kindly supporting the hard work and the long days behind the computer writing this book or attending standardization meetings.

Finally, I dedicate this book to my father.

Michael Mayer

First and foremost, I would like to thank my wife Zsuzsanna for her support and patience during the many months involved with the preparation of this book. Words are not enough to express my gratitude to her for accommodating my absences during the many standards meetings where much of the content of this book was honed and refined. My children Fanni, David and Ben also deserve thanks for putting up with my travels.

Much of the content of this book has been developed over the many years of collaboration with my other colleagues in various companies and standards bodies, particularly the ITU-T and COAST-SYNC. I am very thankful for the privilege to work with so many bright and talented people.

I would like to also thank my coauthors for the opportunity to participate in this truly collaborative effort.

Finally, I would like to dedicate this book to the memory of my parents.

Laurent Montini

I would like to thank the Q13/15 attendees, most coauthors, who, in 2005, welcomed the first "packet guy", and to Marc Weiss who provided me with extended help for this book.

I would like to thank Cisco as a company for promoting thought leadership and providing me with the opportunity to meet and collaborate with so many great talents and individuals. I also thank all my managers (Axel Clauberg, Cedrik Neike, Art Feather, Jane Butler, Chip Sharp and Russ Guyrek) who trusted me in supporting this work.

I would like to particularly recognize Stewart Bryant as an early and staunch supporter and Leonid Goldin as a faithful partner on synchronization for years. I must express my profound gratitude to my very first mentor Jean Guylane, long before my tenure at Cisco.

I would like to thank my four beloved sons and daughters who have, through their patience, strongly contributed to this book. Special thanks go to my wife Marie for her love and support over the years.

I dedicate this book to my parents.

Michel Ouellette

I would like to thank my lovely wife Kim, beautiful daughters Haley and Oliva, and my parents Francois and Monique for their support and looking after me during those long evenings and weekends while I was spending time with my other wife "the Computer".

Special thanks go to Dr James Aweya, an outstanding mentor who taught me so many things. I also thank Bob Mandeville from Iometrix for the opportunity he provided me, as well my previous colleagues at Huawei Technologies and to this great university that was once called Nortel.

Silvana Rodrigues

I would like to thank Integrated Device Technology (IDT), my coworkers, specially my former manager Jim Holbrook, and my current manager Louise Gaulin for their support of my work on this book. I would also like to thank IDT for its support of the standards activities.

It has been a pleasure to work with several colleagues from different companies at the ITU-T and IEEE standards meetings, and ITSF, ISPCS and WSTS workshops. Over the years, several colleagues became very good friends. I feel very fortunate to work with such a great group of people. I would also like to thank John Eidson who provided me with advice while writing this book.

Special thanks go to my loving family, my husband Claudio and sons Nicholas and Thomas. Their support throughout my career was fundamental for my development, with so many weeks away from home participating in standards meetings, and many weekends writing this book. I also would like to thank my sisters Celi and Ivani, and my brother Fabio for their support. I dedicate this book to my father (in memory) and to my mother.

Stefano Ruffini

I wish to thank all the people who have supported me working on this book. They include my company, Ericsson, which has given me the opportunity to take part in this project and has provided with great support over the many years in my participation in the standardization activities, and in particular my current manager, Roberto Sabella, and Ghani Abbas whose experience in the standardization activities has often been a great help; all colleagues, especially those I have met during the ITU-T Q13/15 meetings, and who during these years have inspired me and with whom I often have established extraordinary relationships; and my loving family – my son Dante, daughter Emma Vittorina and wife Elin – who have supported me in this project.

Moreover, I wish to remember the Fondazione Ugo Bordoni and, in particular, Domenico De Seta, with whom more than 20 years ago I started to learn about synchronization in SDH systems.

Finally, I would like to dedicate this book to the memory of my parents, Giovanni and Vittorina.

Introduction

I.1. The importance of synchronization in future telecommunications networks

Exchanging digital data has always required some level of synchronization: the receiver of a telecommunication system must correctly acquire the "rhythm" or frequency of the bits sent by the transmitter in order to recover the data correctly. But synchronization within a telecommunications network has a much wider scope than between a transmitter and receiver on a local link: the delivery or distribution of a common timing reference across a telecommunications network is required for various network applications to ensure proper network operation. This timing distribution effectively becomes a network within a network. In many cases, this distribution follows the same path across a network as the data. However, in some cases the data may flow across the network but the timing will flow from a central point to the edge and only follow the same path for part of the distribution, typically at the edge. Ideally engineering the synchronization network should take place at the same time as engineering the network to carry data. However, this is not always the case.

Synchronization is often thought of for time division multiplexing (TDM) based fixed line voice and data-based infrastructure networks. However, few realize the importance that synchronization plays in allowing various applications to work correctly, one such application that is gathering pace is the increase in mobile telecommunications through long-term evolution (LTE). For instance, in mobile telephony, the synchronization requirements of the air interface are critical. How would wireless mobile (based on technologies such as global system for mobile communications (GSM), *code division multiple access* (CDMA) and LTE) communications work without synchronization? Clearly it would not and Quality-of-Service issues would arise if it were not considered.

Quality-of-Service and synchronization have always had close links. This has been true in the TDM world where accurate synchronization is required to limit the occurrence of slips.

Synchronization is still needed today and in future for mobile applications and networks:

– to stabilize the radio frequencies used by the mobile base station;

– to allow efficient spectrum usage;

– to avoid radio interference between neighboring cells;

– to allow seamless hand over between cells.

Poor synchronization within a telecommunications network may have important impacts on the end user:

– The communication can degrade (voice communication can become inaudible).

– The throughput of data connections in the networks can reduce.

– The network's connections (in the case of the internet) might even be totally lost.

– In the case of mobile communications, hand over between cells could fail and quality of experience degrade.

Ensuring a proper design for a synchronization network should therefore not be underestimated when considering these potential impacts.

I.2. Purpose of this book

One of the problems faced by the network engineers responsible for building suitable synchronization architectures is that synchronization is not necessarily a well-known or even a well-understood topic within telecommunications. Many engineers may have a limited knowledge about the subject and may feel uncomfortable with existing synchronization technology. Equally, there are also many engineers who have no knowledge on the subject at all and no basis on which to develop their understanding of new synchronization technology in the packet world.

Now, with the evolution toward new packet-based technology and consequently new synchronization technologies, a review of some of the key principles and concepts of synchronization and timing is useful. This is especially useful when applying synchronization to new packet-based technologies to understand how they work and where they apply, how they can be tested and what challenges may exist from a network design or operational management perspective.

Dissemination of such knowledge is critical for subject areas such as synchronization that are not common or well understood, but even more so when these new technologies are being considered in new network architectures. Like any subject area, spreading a proper understanding takes time but has wider and longer-term benefits.

Synchronization design and the vagaries around the subject is often a specific discipline that is practiced by a relatively few individuals on a day-to-day basis. These individuals tend to be experts that sit in a wide range of companies within the telecoms and associated industries. For example:

– Large network operators will often have a few individuals who understand the issues across their deployed technologies and scale of operations. When these operations cover many different types of voice and data applications and span tens or hundreds of thousands of elements, supporting many millions or even billions of dollars in revenue, it becomes apparent that the scale of this challenge can be large and the risk under failure conditions is high.

– Systems vendors will have experts in designing and integrating synchronization capabilities within their products and may well have expertise in designing these products into some aspects of the synchronization network.

– Silicon vendors will have experts in designing and integrating their components into the various systems.

– Other organizations may well have a small group of experts, for example these could be test houses or consultancies. A few specialist companies also make it their business to design specific synchronization-related products or provide expert consultancy or both.

However, it is worth knowing that the number of people worldwide with a fair knowledge in the industry is probably in the order of a few thousand (and could actually be below a thousand) and this drops to only a couple of hundred who have real knowledge of how synchronization is designed into real networks and how the industry is evolving. At events such as International Telecom Synchronization

Forum (ITSF) and Workshop on Synchronization in Telecommunication Systems (WSTS) through the late 2000s, typically between 70 and 120 experts attended each year. The experts actively involved in standardization results in an even lower number: in 2003, the expert question in the International Telecommunications Union (ITU) dealing with synchronization had less than 10 experts dealing with the subject and even in 2013 at most it will be 45.

Some of the disciplines that are involved in synchronization require the engineer to:

– have a good understanding of the overall network architecture and the different technologies that make this architecture;

– understand how these technologies work (certainly at a high level) but in some cases to some depth;

– have an appreciation of digital networking;

– have a detailed understanding of analogue technology and some of the factors affecting oscillator performance and other clock components;

– have an understanding of the services carried and the performance requirements. Also how they may be degraded and what may degrade them to sort out potential synchronization problems from the normal service problems;

– have an understanding of how to test for synchronization problems in live networks and for lab-based evaluation. The engineer also needs to understand how the various pieces of test equipment work and what may influence the results, for example a badly tuned reference oscillator;

– understand the standards and what can and cannot be achieved in terms of architecture, technology and performance;

– take the network architecture that has often been developed to carry services without any thought to synchronization and work out how to add synchronization for the services that often have a demanding commercial criteria;

– have an understanding of packet networks and packet technology in the world of NGNs.

There are probably more, but hopefully the reader can see that the synchronization engineer while being a specialist in the field of synchronization also needs, to a certain degree, to be a "jack of all trades". That is they have a wide base of knowledge that can practically be applied. In many cases, although synchronization knowledge can be acquired through study, it will often be learnt over many years of dealing with practical problems of design, problem resolution and testing.

One thing that should be clearly stated is that the content within this book is the work of many experts over many years. Much of this work has become codified within standards and has obviously been based on contributions to standards bodies such as ITU-T throughout the years, for example public switched telephone network (PSTN) synchronization in the 1980s, then synchronous digital hierarchy (SDH) technology and its respective synchronization in the 1990s. On the more recent technologies, these contributions have been from an increasing group of experts that now regularly attend ITU-T. The authors of this book have attempted to distill this collective knowledge and translate it into an up-to-date body of work. Some of this work is the authors' own work and other aspects are based very much on the work of others distilled into a consensus view by the ITU-T process with an attempt by the authors to translate this into a useful reference source.

There may be many reasons why a reader may find this book useful:

– It brings interested engineers very much up to date with the latest technology on this topic.

– It provides some useful guidance based on the latest standardization.

– It provides a handy reference for expert engineers who need to check technical aspects on the latest technology.

This book aims to provide some clarity to the subject and highlight the importance of synchronization so that the subject is not forgotten, or considered only at a very late stage, or poorly designed inside an overall network architecture, especially in the packet network world.

As discussed earlier, experts in the subject of synchronization tend to be fairly rare, with the non-expert in the subject split between those that know the requirement for synchronization and timing exists, and those that do not. This book will have something for the non-expert readers as well and should clarify and build on their existing telecoms knowledge or provide a base on which to explore the subject further. This should help to demystify some aspects of synchronization.

Any engineer involved in synchronization will be familiar with the challenges faced when service is failing or Quality-of-Service metrics are declining and the first response is that "it must be the synchronization". It is true synchronization is sometimes to blame. However, there are many other issues that can cause problems that look as though they are synchronization related. Certainly unfamiliarity with the topic does not help. However, the subject of synchronization essentially revolves around the simple concept of distribution of accurate frequency, time or phase from point A to point B. Some of the detail behind this is complex, but in reality it can be divided into areas that can be explained with simple clarity and intuitive examples which hopefully this book will achieve. The danger is that oversimplification misses certain aspects of the topic or creates yet further misunderstanding.

All synchronization designers have seen "The Good, The Bad and The Ugly" – to use the title from the Sergio Leone spaghetti western film starring Clint Eastwood – in terms of network synchronization designs. Sometimes bad and ugly designs are created through necessity or inherited from previous work; they may have resulted through network migration or are determined through architecture or are sometimes driven by expedient commercial needs. This book cannot comment on those reasons, but what it can do is talk about the approach taken in the development of synchronization standards, which is the key starting point in creating good reusable synchronization designs that have a solid technical foundations and are proven to interwork correctly and stand up commercially (i.e. in both capital investment terms and operation costs). Standardization provides a key coordinating framework around which equipment and their respective interfaces can be specified, network limits can be appropriately designed and how architectures can be created to meet the various performance requirements using an agreed set of design rules. This book, written by a group of people deeply involved in the standardization process of recent synchronization technologies developed over the last decade, attempts to clarify these results.

The authors have all been in various standards bodies and contributed to the development of these new synchronization technologies developed over the last decade by the ITU Telecommunication Standardization Sector (ITU-T), such as *Synchronous Ethernet* or *Precision Time Protocol version 2* (PTPv2) telecom profiles based on IEEE Standard 1588-2008.

As indicated, one of the objectives of this book is to describe the state-of-the-art of these technologies and what can and cannot be achieved with them. It also aims to show how the standards that have been developed should be used and understood. It further discusses the evolving needs for synchronization in a 21st Century telecoms environment and illustrates some of the challenges related to synchronization and its evolution.

Another important goal of this book is to help dispel a few myths sometimes stated or answer questions sometimes asked in the telecom industry, such as:

– synchronization is only for legacy networks;

– synchronization is not needed when migrating toward Internet Protocol (IP) networks;

– Global Positioning System (GPS) is sufficient when synchronization is needed;

– Precise Time Protocol (PTP)/IEEE 1588 is the best solution for synchronization over packet networks;

– PTP is more precise than Network Time Protocol (NTP);

– why synchronize the Ethernet physical layer?

This book, as the title indicates, addresses Synchronous Ethernet and IEEE 1588 as it is used in telecom networks. This book helps to understand how and why Synchronous Ethernet technology was developed, and how it can interwork with existing SDH-based synchronization networks.

Likewise with IEEE Standard 1588-2008 ("1588v2", or "PTPv2"), this book will also provide up-to-date information about this new technology that has been developed, when it can be applied to telecoms networks, and explain the concept of "profiles" and the different PTP telecom profiles developed or under development at the ITU-T. It will explain how to use these standards, the limitations and possible combinations. It will also show how this can be linked with Synchronous Ethernet technology when moving from simple end-to-end (E2E) frequency-based transport to very high precision time and phase transport.

The synchronization world has – like any technology area – many concepts or terms that can be used and are also misused or even used interchangeably. For example, where only frequency is concerned, the oscillators and associated filtering within equipment are collectively called a clock, but these clocks traditionally do not tell the time. However, in the context of next-generation synchronization, clocks will have some concept of a time base or transfer time. Hence timing can now also mean phase and time. Similarly, the term synchronization is often called timing but tends to mean frequency synchronization. However, again in the next-generation synchronization network, timing can now also mean phase and time. Similarly, new terminology previously not associated with traditional synchronization networks will also be introduced.

An attempt will be made to clarify some of these terms and put them into propercontext within the packet-based world these technologies will exist. For example, this book will describe the concepts of packet delay variation (PDV), a term that is often called jitter in the data world. Although both can be measured over time, the jitter in packet networks tends to represent a "maximum value" when PDV is associated to a "variation over time", which is much more important to understand for timing recovery and can be analyzed and metrics developed to quantify. Some discussion will also take place on PDV metrics and their importance in the context of timing recovery in packet-based systems. Jitter, as a term when used in this book, is the strict definition of jitter when applied to synchronization.

Every synchronization specialist has probably felt at least once in their working career that the telecoms world is split into three groups, those who have no idea that the subject of synchronization exists, those who do have a vague idea that it exists – but know little about it, prefer to ignore or have a dangerously vague understanding of the topic and those who are involved in the subject. The synchronization world tends to be a small technical world that does not have a large body of written technical information published and books on the subject are relatively rare (see [BRE 02, SHE 09]) compared to some specializations. This book aims to help to fill the gap. Regardless of the reasons for reading this book, be it for information, for reference, to aid learning of a new subject or refreshing and enhancing knowledge on the subject and bringing oneself up to date with new concepts and technology, the authors hope that this book will be of value.

I.3. Differences between frequency and phase/time

The word synchronization has multiple meanings. Even only in telecommunication systems, it can correspond to very different notions: for example, data synchronization (e.g. synchronizing the data from a smartphone to a computer or a network, or synchronizing data between databases or between two entities of a redundant network device). These notions are all based on information that, in many respects, is a totally different notion from frequency, phase and time synchronization as used in this book. However, if you think of frequency or time as information the analogy is not so different, that is you are maintaining certain information between two points within certain limits one of the limits being time itself.

Dr Marc A. Weiss, National Institute of Standards and Technology (NIST) mathematician and GPS expert, has helpfully tried to encapsulate some of the key aspects and concepts and challenges of frequency, phase and time synchronization into the following paragraphs written during the development of this book:

> One fundamental issue is the differences among frequency, phase and time synchronization. Since a clock is a frequency device, one can have a stand-alone accurate frequency standard, such as a commercial cesium clock. Note, however that there are major differences among the frequency accuracy of different atomic clocks. Many atomic frequency standards are built for stability, not accuracy. Rubidium standards and hydrogen masers are extremely stable. Frequency accuracy is relative to the definition of the second based on the cesium atom. The most accurate devices are built and maintained in laboratories, and are accurate to better than 15 decimal places. These are not available as commercial devices. Note that chip-scale atomic clocks, while they may be based on the cesium atom, have little

accuracy and stability. Their advantage is producing a signal more stable than the best Quartz oscillators with minimal size, weight and power.

Time synchronization brings in new problems and concepts not considered for frequency synchronization (frequency synchronization is properly called syntonization). Time accuracy, unlike frequency accuracy, in principle requires transfer from a source of UTC. Time accuracy requires a method for transfer. Most commonly, this is provided from a GNSS receiver. Even with initially synchronized clocks, any two clocks will walk off from each other in time without bound. Perfectly accurate clocks will have some white noise in frequency, which will produce a random walk in time. Time synchronization accuracy means agreement with UTC, which in turn means traceability to a lab that produces a real-time estimate of UTC. In some cases, phase accuracy is the requirement instead of true accuracy to UTC. For phase accuracy, transfer among devices is still required. An additional complication is that UTC is coordinated among all the industrialized nations, and is postprocessed. Any real-time UTC signal is a prediction of the true UTC, which is available at least 1 month after any real-time signal. Thus, different sources of UTC differ in the time they are generating. The best sources disagree by about 10–20 ns.

This book will explain and use these concepts in the chapters that follow. Some simple illustrations are used to further explain certain concepts and these, together with the information contained within the book, will provide the reader with a useful reference. With the material in the book, a greater understanding of the topic will be achieved hopefully placing the paragraphs from Dr Weiss into full context. The reader is then encouraged to revisit those words.

Starting with some simple examples, Figures I.1 and I.2 show the concept of frequency, phase and time synchronization using clock faces.

Figure I.1 is an example where both clocks are synchronized in frequency. However, they have been set with a 10 min difference between them; they are running with exactly the same frequency (same time base), so in theory, they will always display a 10 min difference between them.

Figure I.2 is an example where both clocks are synchronized in frequency, phase and time. They have been set to the same time at 03:00, as they are well synchronized, they will keep on running with the same frequency and, theoretically, will always display the same time.

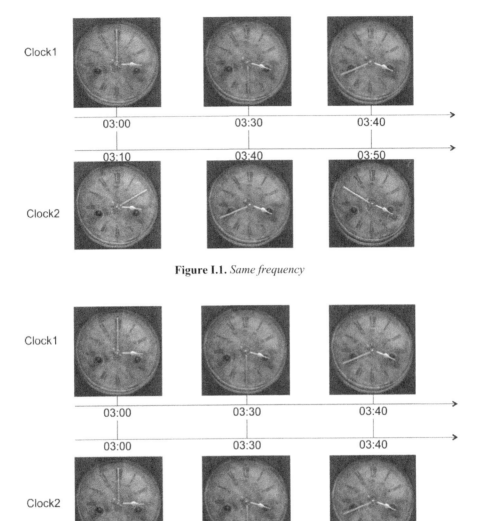

Figure I.1. *Same frequency*

Figure I.2. *Same frequency, phase and time*

Getting into more technical terms, this book will focus on the following types of synchronization, using the ITU definitions. Definitions are based on ITU-T Recommendations G.8260 [G.8260] and G.810 [G.810]:

– *Frequency synchronization (also called "syntonization")*: distribution of reference timing signals experiencing similar frequencies (within the relevant frequency accuracy and stability requirements) to a set of network nodes. However, the reference signals are not necessarily phase-aligned. Two different types of frequency synchronization are distinguished:

– *Frequency-locked systems:* two systems are considered frequency-locked if their frequency difference remains bounded by a given value. If frequency-locked systems are not phase-locked systems, then the maximum phase error accumulation is unbounded. Frequency-locked systems may be based on frequency-locked loops (FLL).

– *Phase-locked systems:* two systems are considered phase-locked if their phase difference remains constant in the long term (they might experience fluctuation in the short term). A fixed phase offset is allowed to be arbitrary and unknown. Phase-locked systems are generally based on phase-locked loops (PLL). Note that the notion of phase-locked systems is different from phase-aligned systems.

– *Phase synchronization:* distribution of reference timing signals whose significant events occur at the same instant (within the relevant phase accuracy requirement) to a set of network nodes. Phase synchronization includes compensation for the propagation delay of the reference signals. Phase synchronization is equivalent to the notion of "phase alignment". Note this concept should not be confused with the concept of "phase-locking" described earlier, which is more related to frequency synchronization.

– *Time synchronization:* distribution of a time reference to a set of network nodes. All the associated nodes have access to information about time (in other words, each period of the reference timing signal is marked and dated) and share a common time scale and related epoch (within the relevant time accuracy requirement). Time synchronization and phase synchronization are very close concepts: time synchronization simply consists of naming the significant instants delivered with phase synchronization (e.g. with time-of-day information).

Time synchronization implies phase synchronization and phase synchronization implies frequency synchronization. In other words, if two reference timing signals are time synchronized, they are also phase synchronized, and if two reference timing signals are phase synchronized, they are also frequency synchronized.

I.4. From traditional TDM synchronization to new mobile applications

The requirement for frequency synchronization was originally developed to support the needs of early digital networks using TDM. For example, the voice network was based on the use of 64 kbit/sec switching with buffering to take account of network variations. Without synchronization and all the switches in the E2E path being synchronized, data would be clocked in and out of different switches at different rates. This would result in data loss and hence an impact on the service. Of course, buffers can be made bigger, but this results in delays on the data and, in the case of voice, it will generate echo and other voice-based impairments. Synchronizing the switches allows much smaller buffering to be used – effectively just enough to soak up timing variations. These switches would be connected together using PDH technology providing an ideal means to transparently transport synchronization at the physical layer. With the development of mobile telecommunications systems, essentially the mobile switching center nodes and base stations, in very simple terms, become an extension to this system.

With the development of SDH transmission to replace PDH transmission this also required synchronization. Essentially the SDH systems could become the synchronization transport medium that the switches were connected to. In both the PDH and SDH cases, the frequency accuracy of the source of synchronization is based on atomic standards such as cesium and is hence very high.

Concerning time, many of the early requirements were based on the need to provide a time stamp to computer-based networks for file management, database management and to indicate most recent or current file. In the telecommunications environment, time stamps were used to provide a means to time stamp alarm events and correlate alarms on different equipment for management purposes. Billing devices also required time stamps. Early on, these time stamps may have been sourced at each specific equipment in turn from low-frequency radio transmitters. As computers and packet-based networks became more prevalent, the use of packet-based protocols such as NTP prevailed. However, the requirements for many of these applications were quite crude and time stamps in the order of ± 0.5 sec only were required.

All these applications and methods to transport frequency and time are effectively now very mature. However, with the move from deterministic TDM-based networks to packet-based networks typically based on Ethernet/IP, the situation for frequency and time transfer changes. Also, the requirements at the application have changed. Mobile telecommunications are a classic example.

Mobile networks typically used PDH backhaul to connect the base station to the mobile switch site. This allowed frequency synchronization to be provided minimizing buffer slips and also provided a means to synchronize the "air interface", that is radio side of the system. Introducing SDH did not really change this model. The use of SDH allowed much greater bandwidths and simplified the backhaul network, but still allowed the same synchronization designs to be used all primarily built around voice being the main application, with text and some data. However, with the emergence of newer more capable mobile phones and a consumer desire to use more data, the use of PDH/SDH-based backhaul no longer cost effectively provides the required data rates or backhaul/transport efficiency.

Ethernet and increasingly IP provide the new backhaul. Although this drives the cost down and in some respects simplifies the network, in other respects this provides complications and challenges. Synchronization is one such area. Packet-based networks were natively not planned for transporting or technically capable of distributing frequency synchronization. In recent years, this has driven developments such as Synchronous Ethernet and PTP to attempt to overcome these weaknesses. Other challenges have also emerged. With the consumer demand for data and hence capacity from mobile, the demand for spectrum and the efficient use of spectrum at the base station has increased. New radio frequency techniques and the desire to precisely position frequency spectrum has resulted in the emergence of new stringent frequency and phase/time requirements. These requirements allow the radio spectrum to not only be used more efficiently but also in slightly different ways.

Although the fundamental frequency requirements have essentially remained the same since the early 1980s that are traceable to a reference clock based on an atomic source, the phase/time-based requirements have now changed from the typical ± 0.5 sec requirement for alarms in the 1990s to typically the 500–1,500 ns region less than 20 years later. Both Synchronous Ethernet and PTP have been designed to deliver high-precision frequency transport and form part of the solution to allow the phase/time-based information to be transported supporting these new requirements.

I.5. Structure of the book

The book is divided into a number of chapters and a quick summary of each is provided below to allow a clear understanding of the book structure.

Chapter 1 presents an overview of the evolution of the telecom networks and of the related synchronization needs, for instance when moving from TDM to packet technologies. The increased need of synchronization in mobile networks (including phase synchronization) is also mentioned as one of the main drivers for the definition of the new synchronization technologies.

Chapter 2 introduces new technologies, such as Synchronous Ethernet and packet-based timing (in particular using IEEE 1588) that are applicable to packet networks. Basic principles of synchronization in telecom networks are summarized in order to provide some background and to put these new technologies into context. The use of Global Navigation Satellite Systems (GNSS) (e.g. GPS) and other radio techniques completes the picture in terms of synchronization techniques. Indeed, GNSS will play a fundamental role especially when it comes to distribute a common time synchronization reference.

Chapter 3 introduces some key architectural concepts to allow better understanding of the role of the synchronization networks as part of the overall telecom network architecture. In particular, the layered approach is used to better explain what the consequences are of carrying timing over packets networks. Architectural aspects for both frequency and time synchronization are introduced including some basic aspects related to redundancy and restoration schemes.

Chapter 4 introduces the evolution of TDM to packet-based networks and discusses some of the issues and considerations related to the deployment of the Synchronous Ethernet and IEEE 1588 technology within the synchronization network. In particular, this chapter addresses some important aspects that an operator (both mobile and transport) has to take into consideration when deploying these techniques, either when the mobile and transport are the same operator or in the case when one operator may be carrying timing traffic of another operator. It covers aspects of the E2E frequency transport and the impact of network stress, issues when using Synchronous Ethernet and the advantages of interconnecting into existing synchronization architectures based around SDH, and introduces aspects of time and phase transport at both the transport and access level. Some of the considerations made in this chapter are based on the experience derived from some of the network operators who actively contributed to the ITU-T work during the 2005–2008 and 2009–2012 study periods. Readers of this chapter are encouraged to look at the developing standards and, if designing networks, to carefully assess the latest standards information with the body of knowledge within this chapter and indeed this book. Operators are advised to look at their own networks and adapt the assumptions made in this chapter according to their specific network needs.

Chapter 5 discusses some of the aspects related to network management and monitoring of the synchronization network. Synchronization distribution with Synchronous Ethernet and IEEE 1588 may need consideration when planning support from network management systems. Ongoing network monitoring, specific issues for managing wireless backhaul and to the introduction of these new technologies into existing network management systems is also briefly discussed.

Chapter 6 introduces possible security threats that can specifically impair synchronization services. Security is clearly a very broad subject covering multiple areas that can interact and is well beyond the scope of this book. However, synchronization can be a subset of various security aspects and with the move toward packet network is a topic that deserves some consideration. This chapter attempts to introduce some aspects of this such as attacks and security techniques and risk management in the context of synchronization distribution.

Chapter 7 provides aspects related to the testing of packet synchronization networks and technologies, primarily Synchronous Ethernet and IEEE 1588. It

Introduction   xlvii

provides guidance on some of the terms and metrics used when performing testing, measuring and configuring synchronization-related equipment. The test scenarios are certainly not exhaustive but provide an understanding of the baseline tests that can be conducted.

A number of appendices provide additional information that cannot easily be covered within the chapters and are provided for additional reference.

Appendix 1 provides an overview of standards in packet networks using SyncE and/or IEEE 1588.

Appendix 2 provides an overview of Jitter Estimation by Statistical Study (JESS) metric definition.

I.6. Standardization

Before the reader goes through the detail of the book, it is worth discussing the standardization process. The ITU [ITU] is one of the key standardization bodies worldwide and is governed by international treaty and has a long history dating back to 1865. The ITU-T is one of three sectors of the ITU. The ITU-T coordinates standards for telecommunications. This body is split into a number of study groups that study a range of topics. Study Group 15 (SG15) is the group that develops all the transport standardization. ITU-T SG15, to give SG15 its full title, is responsible for the optical transport networks and the various copper- and fiber-based access technologies, and many other aspects related to transport networks. As part of its work this study group is also responsible for the various aspects of synchronization and time. It should be noted that the ITU's work plan is based on 4 year study periods and although the works of a study group remain stable during a study period they can change in a new study period. The current study period runs from 2013 to 2016 and the preceding study period was from 2009 to 2012. To put some context around the study periods and standards development, much of the early packet synchronization analysis was carried out from the early 2000s, Synchronous Ethernet was effectively standardized in the 2005–2008 period and IEEE 1588 E2E carrying frequency in the 2009–2012 period. Toward the end of the last study period and into the 2013–2016 study period, phase/time transport hasbecome the key area under development in ITU standards.

To break down responsibility to a manageable level to efficiently and effectively carry out the standardization tasks, the ITU-T sets up a number of working parties and within these working parties specific working groups or "Questions" study these different topics. Question 13 (Q13), "the timing question" is the expert group that is tasked with the development of standards that encompass all aspects of frequency, time, phase, synchronization and syntonization. It is responsible for the clocks, the network limits, the architecture and the rule set as well as studying and developing new metrics and measurement techniques. Development of the work is carried out by the participants bringing their input to the meetings in the form of contributions. These contributions are then discussed and developed into agreed text that forms the basis for future standards or revision of existing standards. Development is very much based on a consensus approach and obviously requires both contribution into meetings and participation.

To allow this work to progress and align with other standards, Q13 works with other questions within SG15, other study groups within ITU and also other Standards Development Organizations (SDO's) to advise them of synchronization aspects and to take input in return. As discussed earlier, it should be noted that Q13 is the valid question in the current study period and it may well change after 2016 as could any other question. Likewise, it should be noted that historically during the 1990s the "timing question" was in ITU-T SG13 and was called Question 18.

Other standards bodies Q13 works with frequently are:

– Institute of Electrical and Electronics Engineers (IEEE) is responsible for Ethernet technology on which much of the new synchronization networks are designed. IEEE, for example, developed the PTP protocol (specified in IEEE 1588) that is profiled by ITU for use in telecoms networks. IEEE also helped ITU to specify the communications channel used by Synchronous Ethernet and is providing support on aspects of Ethernet technology layering when related to PTP.

– Internet Engineering Task Force (IETF) whose mission is to define the protocols and mechanisms governing the Internet. IETF is responsible for IP and Multi-Protocol Label Switching (MPLS) aspects and, from a synchronization perspective, has been developing Management Information Base (MIB's) for the use of IEEE 1588, looking at security aspects of packet timing as well as the transmission of PTP using IP and Ethernet mappings over MPLS. IETF also has responsibility for NTP.

– Metro Ethernet Forum (MEF) that develops implementation agreements and promotes interoperability to aid development of Carrier Ethernet networks. MEF is building on ITU-T work to help them to develop service definitions for Ethernet synchronization.

– Third-Generation Partnership Project (3GPP) that unites a number of telecommunications standard development organizations (including Alliance for Telecommunications Industry Solutions (ATIS) and European Telecommunications Standards Institute (ETSI)). 3GPP is responsible for a range of aspects covering mobile technology including the radio access networks (RAN). The RAN would be seen in Q13 terminology as the application (e.g. the mobile base station).

– Regional standards bodies also exist. For example, in the North American region the ATIS is a technical planning and standards development organization accredited by the American National Standards Institute (ANSI). In the European region, this role is assumed by the ETSI. Both these bodies either feed into ITU or may take their lead from ITU. Other bodies also exist that cover other regions or are country specific.

Within SG15 Question 2 (Q2) and Question 4 (Q4), respectively, are the experts on various aspects of fiber and copper access technology. This is important as one of the aspects that Q13 has to consider (especially in the transport of phase/time) is the budgeting of this capability from where it is generated (i.e. the reference) to where it is required (i.e. the application). In both the cases of Q2 and Q4, they are studying and developing methods to transport synchronization through their respective technologies cognizant of the requirements Q13 has specified based on application requirements and the methods and architectures it has been developing.

A couple of very simple hypothetical reference connections (HRX) are given in Figure I.3 with blocks covering Reference-Transport (e.g. core/metro)-Access-Application sections of network. This HRX designed to explain the issues of budgeting was originally developed as a contribution [GIL 11] into ITU and then developed further within Q13 for joint discussion between Q2, Q4 and Q13. For this book, Figure I.3 has been added to and modified further to illustrate the issue of budgeting with ITU standards and between standards bodies.

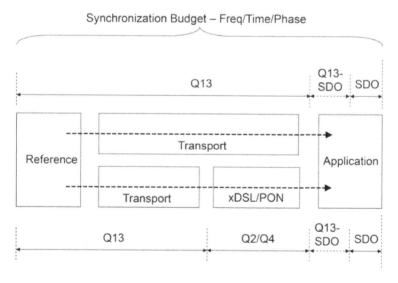

Figure I.3. *Timing flow through network domains showing ITU-T SG15 synchronization budgeting in relation to application*

Within Q13, the synchronization architecture and performance is being developed to meet the requirements of a range of applications. In some cases, the transport architecture may stretch straight from the reference to the application, that is the upper section of Figure I.3. Note the work of other questions within SG15 which develop the optical transport architecture is not shown for clarity. But in other cases, the transport aspect forms only a part of the connection, and part of the specification is defined by the access technology, that is the lower section of Figure I.3. In this example, in both upper and lower cases, the mobile sector would control and be responsible for the application, so in this example the SDO would be 3GPP. If all the aspects of the application are fully defined, the budgeting responsibility would be determined by the SDO and Q13 would have a clear view on the requirement at the input to the application. However, in practical terms this is not always the case and Q13 has to consider the aspects such as the tolerance characteristics of the application.

In both Q2 and Q4, optical and copper based technologies, respectively, are being developed and a key aspect of this work is the wider Q13 objective to ensure that they will connect E2E to deliver the required performance for the end application. The interface types and the performance through these systems are therefore critical. Note that the actual interface type may well influence the budget. Another aspect that Q13 is trying to resolve is the appropriate interface type on which to transfer the timing information. A critical issue is the amount of budget that is taken in any one portion of the HRX as it will have to be considered or budgeted for in another part of the HRX. It is critical that one portion of the HRX does not take too much of the budget and, if anything, uses inherent capability within the technology to best advantage. A classic case in passive optical network (PON) systems would be to use the inherent physical layer frequency transport and the ranging capability. While Q13 is not interested in the detail within each access technology type, it is interested in the amount of delay (both minimum and maximum), variability of delay, asymmetry aspects, etc. It is also interested in ensuring that the properties of how the access technology will work to maintain the E2E timing architecture and operation so that both Synchronous Ethernet and PTP capabilities can be maintained.

So, it can be seen that the task of developing the next generation of synchronization especially phase/time is a complex one requiring interaction and interworking across a range of questions within the same standards bodies and between standards bodies across all the layers.

As a final note, readers are reminded that the topics in this book can be complex. The authors advise caution in dealing with live networks. As the work in standards is continually evolving, readers are advised to consult the latest published standards as this book can only provide a snapshot in time.

I.7. Bibliography

[BRE 02] BREGNI S., *Synchronization of Digital Telecommunications Networks*, John Wiley and Sons, June 2002.

[G.810] ITU-T Recommendation. G.810, Definitions and terminology for synchronization networks, 1996.

[G.8260] ITU-T Recommendation. G.8260, Definitions and terminology for synchronization in packet networks, 2012.

[GIL 11] GILSON M., Synchronization Budget, COM 15 – C1391 – E, BT contribution to ITU-T SG15Q13 Plenary Meeting, Geneva, February 2011.

[ITU] International Telecommunications Union – Telecommunications (ITU-T site). Available at http://www.itu.int/en/ITU-T/Pages/default.aspx

[SHE 09] SHENOI K., *Synchronization and Timing in Telecommunications*, BookSurge Publishing, September 2009.

Chapter 1

Network Evolutions, Applications and Their Synchronization Requirements

1.1. Introduction

The subject matter contained within this book relates to synchronization in telecommunication networks. Synchronization is widely studied and plays a significant role in how the telecommunication network allows us to communicate. As a topic, the rationale behind some aspects of synchronization often appears to be shrouded behind a veil of secrecy, understood only by a few practitioners.

Synchronization is a widely used term in language that has the connotation of alignment in time. As a term, databases can be synchronized, watches are said to be synchronized and networks can be synchronized. Within this book, we limit the term synchronization to pertain to the ability to transfer either time, frequency or phase from one system, or clock, to another system. On a network-wide scale, network synchronization means that all network elements may be configured to share a common frequency, phase or time relationship.

In most telecommunication networks, the need to distribute the information necessary to allow all network elements to be "synchronized" results in the creation of a dedicated synchronization network. This network provides a path from the network clock (e.g. Primary Reference Clock (PRC)) to the individual network elements requiring timing, and may involve the use of other dedicated clocks (e.g. a Synchronization Supply Unit (SSU)). In many cases, the service transport technology has been designed to provide the capability to carry network timing via dedicated synchronization interfaces. For example, Synchronous Digital Hierarchy (SDH) has special input and output timing specific interfaces. This results in a tight linkage between the synchronization network and the transport network.

The requirements for the synchronization network are based on the needs of the services carried over the network. Services offered over the telecommunication network are rapidly changing and evolving. In a sense, the requirements may be evolving, but the capabilities of the network may be set by a specific technology or service.

As the network evolves, it may be found that the synchronization provided by the network is not sufficient for new services. To understand the current capabilities of the synchronization network, it is useful to understand how the telecoms network has evolved to the modern data communication network it is today.

Telecommunications began with telegraphy, which was simply a very rudimentary, low bit rate data service. The ability to telecommunicate by natural voice was, once demonstrated, something highly desired by the cognoscenti. This then spurred the rapid development of technology, first based on manual analog connections, then later based on user-initiated connections in automatic exchange switches. The rapid increase in popularity of telephony and increase in traffic drove the development of transmission systems capable of carrying multiple voice channels simultaneously and miraculously on a single wire. As technology continued to evolve, digital technology was gradually introduced in the mid-20th Century with the development of Pulse Code Modulation (PCM) where the analog voice channel was sampled and converted to a digital bit stream. This technology allowed replacement of analog carrier systems, greatly reducing the cost of sending information by wire. During the 1970s, the first all digital voice switches were developed and represented a quantum leap in technology. Service quality increased dramatically, as the effect of analog impairments was drastically reduced. The technologies used to offer data services were also evolving during this time. Data transmission relied on the use of the analog voice channel to transmit data. Initial data rates in the order of 300 bits/s very gradually increased as modulation techniques improved. At that apex of technology, analog voice band modems were capable of providing up to 56 kbit/s over a nominal 4 kHz channel. Digital data, the carriage of data directly without modulation, became possible with the Integrated Services Digital Network (ISDN) that resulted from a combination of digital voice switches, coupled with digital PCM transmission systems. Many volumes have been written on these technologies.

During the 1970s, the first packet-switched data networks were being placed into service, again leading to dramatic technical development in network technology, notably Asynchronous Transfer Mode (ATM), which transmits and switches packet of fixed length (cells). Packet or cell switching was understood to provide advantages due to the statistical nature of data transmission, although these were primarily for computer-to-computer or terminal-to-computer communications. ATM standards, however, added aspects that allowed its use as a general purpose transport layer, including support for constant bit rate (CBR) services.

The ability for users to access the network has changed dramatically. The traditional telephone channel was restricted to a 4 kHz analog channel. The concept of wider access bandwidths had always been an objective. Digital Subscriber Lines (DSL) was an advanced modem technology that boosted access rates to the low megabit per second range at relatively low cost. Communications were between the subscriber and special equipment (digital subscriber line access modules). Similarly, cable modem technology was also evolving to support a bi-directional channel over the coaxial infrastructure that was common for television cable systems. These wireline access technologies resulted in the transformation of the network from a voice-based network to a network that now predominantly carries data.

However, perhaps the most dramatic swing was toward the use of wireless technology as a network access infrastructure. During the early 1980s, cellular telephone systems provided mobile access to the telecommunication network. Initial systems were analog systems, with voice as the primary service. Later generations of technology migrated to digital transmission and included data communication channels. Following the example of wireline telephony, wireless telephony access grew to the point where the majority of traffic is data, rather than the initially intended voice telephony.

1.2. Evolution from plesiochronous digital hierarchy to optical transport networks

1.2.1. *Plesiochronous digital hierarchy and public switch telephone networks*

Synchronization became a very important aspect of telecom networks with the introduction of the plesiochronous digital hierarchy (PDH) in transport networks and digital switches in public switch telephone networks (PSTN).

The PDH hierarchy is based on two basic rates, 2.048 Mbit/s (E1) for the European networks based on the European Telecommunications Standards Institute (ETSI) standards and 1.544 Mbit/s (DS1) for the North American networks based on the American National Standard Institute (ANSI) standards. They could be multiplexed into the higher rates defined by the PDH hierarchy, that is 8.448, 34.368 and 139.264 Mbit/s for ETSI and 6.312 and 44.736 Mbit/s for ANSI. It is common to call the multiplexed input signal a "tributary" and the resultant signal an "aggregate".

PDH has two main advantages. First, it is able to transport timing since the multiplex principle defined in the PDH hierarchies is based on a 1 bit justification process, allowing appropriate jitter and wander performance on the egress 2.048 and 1.544 Mbit/s to be compliant with the synchronization interface specification defined in International Telecommunication Union Telecommunication Standardization Sector (ITU-T) Recommendations G.823 [G.823] and G.824 [G.824]. The second advantage of PDH is that it does not require any synchronization for itself since all PDH rates can work within a relatively large rate of frequency.

One of the characteristics of these multiplexers is that the full aggregate signal needs to be demultiplexed to extract a single 2.048 or 1.544 Mbit/s tributary signal. As an example, extraction of a single 2.048 Mbit/s from a 139.264 Mbit/s requires the following operations: demultiplexing the 139.264 Mbit/s into its four 34.368 Mbit/s, then demultiplexing one of the 34.368 Mbit/s into four 8.448 Mbit/s and then extracting the 2.048 Mbit/s from one of the 8.448 Mbit/s.

Digital switches were specified to switch the 64 kbit/s time slots transported by the PDH signals; these time slots carry either voice channels or data channels. The bandwidth of these data channels was either 64 kbit/s for ETSI or 56 kbit/s for ANSI, depending on the type of signaling channel. These switches are the key elements of the PSTN.

These switches multiplex and demultiplex 30 time slots of the 2.048 Mbit/s, 24 time slots of the 1.544 Mbit/s and either decode the voice or transmit the data to the 64 kbit/s data interfaces for transmission to the access network.

On its input ports, a switch stores the incoming time slots of a 2.048 Mbit/s, or 1.544 Mbit/s, in a buffer at the rate of the input signal and on the output ports, it generates 2.048 Mbit/s, or 1.544 Mbit/s, at a rate given by its internal system clock. Buffers, or other mechanisms, are designed to allow some short-term variation in the incoming rates to be absorbed.

However, in cases where a fixed frequency offset may be present, the distance, in bits, between the read and write pointers will continually increase, or decrease. At some point the read and write pointers will collide and data will be corrupted. This situation is often referred to as buffer spill, or slip, and will occur with a periodicity based on the frequency offset and the size of the buffer used.

Of course, a very large buffer would result in rare data loss, but buffer size will ultimately impact latency and cost. Initial voice switches had buffers that allowed processing of a full PCM frame. For 1.544 Mbit/s signals buffers sizes were therefore based on 193 bits as this was a convenient size to accommodate 1 bit and 24 bytes, representing 24 voice channels of 64 kHz. By choosing the buffer size to be a multiple of 24 voice channels, a slip resulted in the loss of only one PCM voice sample, which in many cases only minimally impacted the service quality, provided that slips were relatively infrequent. For 2.048 Mbit/s signals, buffer size was based on 256 bits, representing one frame. If the PCM sample represented voice traffic, the slip was largely tolerable. However, with the growing amount of data, carried either as voice band data (or even fax) or later digital data, the impact of a slip could vary. To control performance, international standards, for example ITU-T Recommendation G.822 [G.822], defined acceptable slip rates.

The slip rate could be controlled if all switches were timed from a common source. This required a dedicated network solely for the distribution of a common clock frequency; PDH signals were commonly used to synchronize the switches. Network elements within a network would receive timing from this network. To accommodate the real possibility of failures (e.g. the so-called "Back-Hoe Fade"), intermediate clocks were developed that would minimize the drift rate during these failure conditions. Holdover performance, for example, was directly defined by the slip rate objectives necessary to enable acceptable performance of voice switches. This aspect of the synchronization networks can therefore be attributed to the introduction of digital switching.

The introduction of PSTN has created a need for the transport of frequency. Digital switches have to be synchronized in order to switch 64 kbit/s time slots transported by 2.048 and 1.544 Mbit/s signals without generating slips.

The desynchronization of a single switch in a chain of switches causes two periodic generations of slips of opposite nature, a byte canceled or doubled, in two consecutive switches, as shown in Figure 1.1, without any compensation between the two opposite slips.

The digital switches, switching 2.048 or 1.544 Mbit/s, have to be locked on a reference frequency to prevent the generation of slips. This is achieved by transporting the reference frequency in the 2.048/1.544 Mbit/s PDH signals connecting the digital switches. For this reason, it is essential that all switches of the PSTN are locked to a common reference frequency; this resulted in the development of a first synchronization network.

At this time the synchronization network was composed of a Primary Reference Clock, the PRC defined in ITU-T Recommendation G.811 [ITU-T G.811] with its well known 1.10^{-11} long term accuracy (with regard to the Coordinated Universal Time (UTC) as defined in G.811) allowing to limit the periodicity of slips to less than one per 70 days. The maximum drift between two switches locked to two different PRCs is twice 10^{-11}; this causes a phase error of twice 824 ns per day or

115 µs per 70 days. A slip is generated when the drift reaches 125 µs, the period of a 2.048 or 1.544 Mbit/s frame.

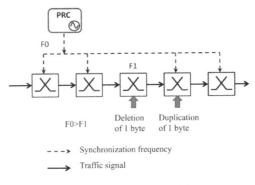

Figure 1.1. *Generation of slips*

The delivery of this frequency reference to equipment generating a 2.048 Mbit/s signal is done via a 2.048 MHz, or even a 10 MHz, signal. It is also very frequent that the frequency reference is provided via a 2.048 or 1.544 Mbit/s signal.

The 2.048 Mbit/s or 1.544 Mbit/s signal can be multiplexed within a higher PDH rate and transported to another PSTN switch where the 2.048 Mbit/s is extracted for the PDH signal and used as a synchronization input by the clock of the switch.

Note that some switches, handling only higher PDH rates, do not need to be synchronized, due to the specific justification process used. This enables these switches to provide timing transparency to the 2.048 Mbit/s signals.

There may be many switches in a PSTN, causing noise accumulation along the chain of 2.048/1.544 Mbit/s; for this reason, a new type of high-quality clock called SSU has been defined in ITU-T Recommendation G.812 [G.812] with a very narrow bandwidth in the mHz range in order to filter the noise outside this bandwidth. This SSU ensures also other important features such as holdover to maintain the quality of the clock in case the incoming 2.048 or 1.544 Mbit/s disappears due to any failure in the network; it also distributes the reference frequency to other equipment within the node where it is installed. These clocks might be embedded in a switch directly or might be in stand-alone equipment.

To ensure robustness of the synchronization network, a switch must be able to receive the reference frequency from at least two independent inputs. This is an important task for the network operator to verify that in any failure condition, a timing loop will not be generated. As defined in section 3.3.1, a timing loop is a

Network Evolutions, Applications and Requirements 7

situation where equipment is synchronized on a timing signal it has generated. Careful network planning is required.

The transport of 2.048 or 1.544 Mbit/s signals between digital switches is mainly done by multiplexing those signals within higher PDH rates. The PDH justification process allows this multiplexing into a higher rate PDH signal without the need for the different rates to be synchronized to a common reference frequency, but it is simply needed that all these rates are within a certain frequency range specified in the ITU-T Recommendation G.703 [G.703]. This allows the transport network to remain asynchronous from the digital switches requiring reference frequency, while being the medium used to propagate the synchronization between these digital switches.

1.2.2. *Evolution toward SDH and synchronous optical network*

The increased demand in bandwidth and the introduction of the optical fiber and its connecting technology pushed the standard bodies, ANSI, ETSI and ITU-T, to develop a new transport hierarchy allowing an increase to the possible rates and hence the transported bandwidth. ETSI defined SDH, and ANSI defined Synchronous Optical Network (SONET).

Prior to the introduction of SDH, multiplexing systems were based on PDH multiplexing technology. Bit stuffing justification mechanisms were used to rate-adapt a lower bit rate signal for multiplexing into a higher bit rate "carrier".

The new SDH/SONET hierarchy was defined so that it made it possible to extract and insert individual tributary signals without being forced to terminate all the tributaries, as is the case in PDH. This feature allowed the development of a new type of equipment, the Add-Drop Multiplexer (ADM), which will be the basic element for a new telecoms architecture with the deployment of rings, rather than linear lines connecting switches centers in the previous PDH and PSTN networks.

SDH was defined and standardized to alleviate some of the issues with PDH-based networking. PDH multiplex systems were generally limited to a specific type of client interface and the addition of new clients or rates usually required considerable network and equipment redesign. With the subsequent development of data networks, initially ATM, the need for flexibility was also seen in terms of the types of signals and associated bandwidth that could be carried on a uniform transport system.

SDH, while seemingly offering the same functionality as PDH systems, differed in that it offered two levels of rate justification. This was key to the flexibility required to carry various types of signals in a ubiquitous and flexible network infrastructure. Client signals were mapped into one or more "containers". For traditional PDH client signals, this could involve a bit stuffing mechanism similar to what is used for PDH. Multiple containers were then synchronously mapped together to form a higher rate signal. The advantage of this is that it allowed containers to be "switched" by simply byte swapping in a manner very similar to that which was used for the voice switch discussed earlier. Recovery of the client signal only required a desynchronizer function at the final edge of the network where the client signal was then handled off. The client signal would still "float" within the container, but the container could be visible. Higher bit rate systems were simply made by adding containers, rather than remapping. These containers were then visible at the higher levels.

The second level of justification was provided to allow containers to be rate adjusted. This was done by the pointer processor. The pointer contained within a special overhead was allocated so that it could be used to occasionally carry data when the bit rates required adjustment. This, conceptually, is not significantly different from the justification methods used in PDH systems. In this case, the pointer adjustments were made with a multiple of 8 bits and were relatively coarse compared to the bit stuffing used in PDH. These were also of relatively low frequency and would therefore be seen as a phase change on the client signal. While PDH used positive justification, SDH used a +/0/– justification mechanism, which had the potential of adding low-frequency wander, which was difficult to remove. These differences, however, were considered in the specification of SDH desynchronizers.

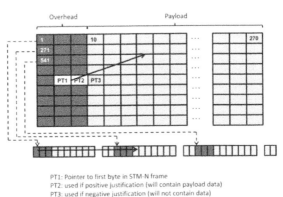

PT1: Pointer to first byte in STM-N frame
PT2: used if positive justification (will contain payload data)
PT3: used if negative justification (will not contain data)

Figure 1.2. *SDH frame and pointer*

The control of the pointers was critical as this was directly related to the jitter that would be imparted on the client signal. A key point was to synchronize all SDH network elements. This was generally not an issue, since a synchronization network already existed. A second consideration was that failure of the synchronization network could occur resulting in an offset between the network equipment clocks and thus continuous pointer generation. Similarly, transients due to synchronization switching could also occur. Pointer generation due to these processes also needed to be considered in the specification of the desynchronizers, in order to limit the jitter that is imparted on the client signal exiting the SDH network.

The introduction of SDH network elements provided great flexibility and value to the network as witnessed by the long lifespan of the technology. The pointer processor mechanism, which was responsible for much of the flexibility, also had considerable impact on the existing synchronization network. Prior to the introduction of SDH, PDH signals were used to distribute timing between various points in the network. The pointer mechanism, due to its 8 bit granularity, could produce a transient close to 4 μs on a 2.048 Mbit/s signal that could exceed the tolerance of some PDH equipment or network clocks. Since the synchronization links were carried as client signals to PDH transport equipment, simple replacement of PDH equipment with SDH had a significant, negative impact on the pre-SDH synchronization network.

To address this, SDH provided functionality that allowed direct access to the Synchronous Transport Module (level N) (STM-N) line frequency rate, which could be used to generate a 2.048 Mbit/s rate signal and then passed to external equipment, for the specific purpose of transporting network timing. Provided that the transmitting SDH equipment was timed from a network clock, SDH now had capability to transfer network timing over the optical carrier signal itself avoiding the pointer processor entirely. The timing specific interfaces were based on the existing interfaces (e.g. 2.048 Mbit/s and 2.048 MHz) used within the existing synchronization network allowing SDH replacement of PDH to have minimum impact on the synchronization network. The direct support for network timing resulted in SDH becoming a fully integrated component of the synchronization network.

The transport of timing via STM-N now implied that there will be a chain of SDH equipment transporting the reference frequency. This means that the length of the chain and the control of the jitter and wander accumulation, the characteristics of the clock in the SDH equipment and the architectural aspects needed to be specified.

The clock of the SDH equipment, called SDH Equipment Clock (SEC) was defined and specified in ITU-T Recommendation G.813 [G.813]. Two types have been specified, one for clock based on ETSI requirements and one for clock based on ANSI requirements, to reflect the different regional requirements.

The most important parameters specified for SDH clocks are the operating range of frequency, the bandwidth and the holdover performance. Proper specification of these parameters is necessary for overall network synchronization performance. To control the accumulation of jitter and wander along the chain, an additional clock of good quality may be required in order to further reduce accumulated noise: this is the role of the SSU, which can be positioned after up to 20 SECs, and has a bandwidth in the range of 1 mHz.

A synchronization reference chain is defined in ITU-T Recommendation G.803 [G.803]; there must not be more than a total of 60 SECs from the PRC to the last SEC of the chain and not more than 10 SSUs. There must be no more than 20 SECs in tandem.

To enhance overall network reliability, all clocks should have at least two independent inputs as a protection against equipment and network failures. To allow secure protection in linear chains and rings, the Synchronization Status Message (SSM) has been defined and its use is specified in ITU-T Recommendation G.781 [G.781]. A detailed description of the protection mechanism using SSM is presented in Chapter 3 for the Synchronous Ethernet case, which is similar to the SDH/SONET case.

As discussed above, the transient produced by a pointer could exceed the tolerance of some PDH interfaces. To avoid problems in the implementation of the synchronization network due to pointer adjustments, two types of 2.048 Mbit/s interface have been specified in ITU-T G.823: the traffic (ITU-T Figure 1/G.823) and the synchronization (ITU-T Figure 10/G.823) interfaces.

The traffic interface has been specified with a relaxed MTIE mask, compatible with the effect of the VC12 pointer justification event; it is defined with a 9 µs plateau in order to be compatible with two VC12 pointers of same polarity, and a wander value of 18 µs for observation intervals larger than 64 sec in order to be compatible with some switches defined prior to the introduction of SDH.

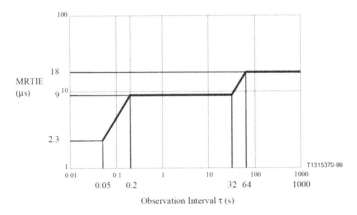

Figure 1.3. *2.048 Mbit/s interface output wander limit (ITU-T Figure 1/G.823)*

A 2.048 Mbit/s synchronization interface is specified with a much smaller amount of noise since it transports a reference timing signal used for network synchronization.

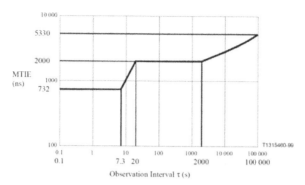

Figure 1.4. *Network limit for wander (MTIE) at PDH synchronization interface (ITU-T Figure 10/G.823)*

1.2.3. Introduction of optical transport network in transport networks

Optical Transport Network (OTN) was defined in 2000 initially to provide standardized wavelength division multiplexing (WDM) networking. At the time of its initial development SDH was seen as the primary tributary signal to be carried. However, OTN was defined to carry different types of clients including, for example, Ethernet. During the initial discussions, it was decided to define OTN as an asynchronous system that would not require external synchronization. This maintained SDH as the primary signal for carrying timing within the network. As a result, the main synchronization requirement for OTN was to be transparent to the timing transported by SDH. The OTN, as an asynchronous network, has network elements with free running clocks, each within a frequency range of ±20 ppm from the nominal frequency. Transparent mapping means that the SDH client signal mapped and demapped from the OTN Optical data unit of level k (ODUk) has to comply with the relevant ITU-T Recommendation G.825 [G.825] "SDH jitter and wander requirements" in order to effectively carry SDH clients.

Initial mappings for Ethernet within ODUk only required mapping of the Ethernet media access control (MAC) frames and not the entire Ethernet signals. The mapping was not transparent to timing but this was not considered an issue, since, at that time, Ethernet signals did not transport synchronization in their physical layer. Since the introduction of Synchronous Ethernet to recognize the growing need to transport timing over data networks, this statement had to be revised. A new timing transparent mapping for Synchronous Ethernet was specified in 2010. This evolution was also considered important in order to follow the evolution of transport networks because the assumption that SDH tributaries will always transport network timing is no longer valid since timing can now be transported by Synchronous Ethernet.

Given this timing transparency characteristic of OTN for both SDH and Synchronous Ethernet tributaries, OTN does not play any direct role in the synchronization network, as long as the number of mapping/demapping operations complies with the requirements provided in ITU-T Recommendation G.8251 [G.8251].

1.3. Migration and evolution in the next-generation networks: from time division multiplexing to packet networks

Section 1.1 outlined how the various technologies used in the telecommunication networks have changed over time. Of key importance is the transition from the connection-oriented network optimized for voice traffic to a connectionless network optimized for data.

The voice network was based on circuit switching voice connections. As discussed earlier, PCM multiplexing systems carried either 24 or 30 PCM-encoded voice channels, each channel requiring 64 kbit/s. This figure was based on the need to sample and replicate the 4 kHz analog voice channel. The 8 bit PCM encoding allowed the development of Time Division Multiplexing (TDM) circuit switching but the type of PCM encoding used originally provided only minimum amplitude compression of the analog voice channel via either the Mu-law (North America) or A-law (Europe) encoding to achieve a desired signal-to-noise ratio when 8 bits encoding was used.

Methods for achieving greater levels of compression for transmission systems were later developed such as Adaptive Differential Pulse Code Modulation (ADPCM) and Time Assignment Speech Interpolation (TASI), but due to their negative impact on voice band data and fax – both rapidly growing services – were not widely deployed. Voice switches continued to switch full rate 64 kb/s channels.

The digital transmission systems developed for digitizing and carrying voice traffic formed the base technology for the emergence of full digital voice switching technology. Now that voice channels were encoded, the functional equivalent of an analog switch could be realized simply by extracting a PCM channel from one multiplex stream and reinserting it into a second stream. This, while seemingly simple, revolutionized the telecommunication network by allowing the development of all digital voice switches. From the user's perspective, the quality of the telecommunication network was now dramatically increased in terms of connection setup time and voice quality, since the connection noise no longer depended on distance. From the perspective of the operator, digital voice switches occupied less space and reduced overall operating costs.

Soon after the first voice switches were deployed in the late 1970s, the industry began to recognize the benefits of the "all digital" networks. While designed primarily for voice, the potential for a clear 64 kbit/s channel across a network provided an opportunity to drastically increase the capacity of a data "call" that could be placed over the network. It is important to recognize that about the time the first digital voice switches were deployed, voice band modems – the only means of carrying data at that time – were extremely slow, typically 1.200 kbit/s, and the existence of a 64 kbit/s channel across a switched voice network was seen as offering huge potential for expanding data service offerings.

Recognizing that switched services could evolve, voice-switching technology also evolved to provide the capability of supporting different service types using the ISDN as a potential to define a unified network for all services. The requirements for voice networks defined the requirements for the synchronization network by the definition of slip rate objectives. Access technology to this network was still an issue.

While digital voice switches were being deployed, data services were also being rolled out in increasing numbers. Services such as X.25 packet were some of the first network services to offer data connectivity using packets or data grams. These offered increased network utilization through the use of statistical multiplexing in contrast to the time domain multiplexing used in the voice network. X.25 was one of the first packet network technologies developed. Subsequent technologies for deployment in public networks included Frame Relay and ATM. Packet technologies initially developed for Local Area Network (LAN) application include Ethernet and Internet Protocol (IP), both now seen in the public network. Although X.25 predated IP by several decades some of the protocol aspects found in IP are directly traceable to X.25.

In parallel with the growth of the data traffic, new access technologies such as xDSL provided higher available bit rates for data services and thus also increased the relative proportion of data to voice carried in the network as a whole. However, the traffic patterns for data "calls" were significantly different from those of the traditional voice network. The holding time of data calls was much longer than the typical duration of a voice call and the increases in the amount of data traffic in the integrated voice network was seen as a concern. xDSL, however, provided carriers with the opportunity to "offload" data from the voice network. DLS terminals would connect voice channels to the TDM-based voice switches, while data would be routed to the packet network.

14   Synchronous Ethernet and IEEE 1588 in Telecommunications

With the migration of data off the voice network, it was now possible not only to provide compression at the encoding level (e.g. ADPCM), but also to consider packetizing voice (i.e. voice over IP) to get additional gains from statistical multiplexing. Voice is now delivered as data and the desire to maintain a single service infrastructure still remained. Rather than the TDM-based infrastructure being defined as the unifying infrastructure for services, the advent of services such as voice over IP resulted in the use of IP as a unifying service layer. Below the IP layer, Multi-Protocol Label Switching (MPLS) or Ethernet is seen as the supporting network.

However, since voice and some services are sensitive to delay, retransmission due to dropped or lost packets will impact perceived service quality. To maintain a high level of performance for voice services, carriers will often run a dedicated packet network to support their voice switches rather than using the public Internet. At a minimum, some form of Quality of Service (QoS) (e.g. DiffServ) is needed to maintain acceptable performance for voice traffic.

In most carrier networks, full replacement of an entire network technology at one time is not possible. Migration to a packet environment may take place over an extended period of time. Alternatively, there may also be cases where the carrier has migrated to a packet infrastructure, but still needs to provide TDM interfaces. Examples of where this may occur include continuing to support DS-1/E1 private line services or support of specific types of legacy equipment (e.g. voice switches or base stations (BSs)). Circuit Emulation Services (CESs) is the primary method to support existing 2.048/1.544 Mbit/s services over a packet infrastructure as described below.

1.3.1. *Circuit emulation synchronization requirements*

The term "circuit emulation" has been used in relation to packet-switched technologies that emulate either the characteristics of a circuit-switched network or of a PDH/SDH transport network, in order to carry CBR services (e.g. 2.048 Mbit/s) (see [G.8261]). Figure 1.5 presents an example of a CES architecture.

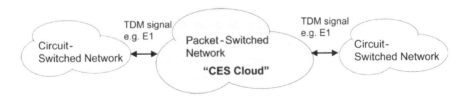

Figure 1.5. *CES Architecture*

The transport of CBR clients (e.g. those used in TDM networks such as a 2.048 Mbit/s signal) over packet networks as such does not introduce new specific requirements. The timing of the client at the output of the packet network will meet the traditional jitter and wander requirements as described in section 1.2.2 (e.g. [G.823], [G.824]). Moreover, the long-term frequency must meet certain requirements in order to control the slip rate (with objectives defined in [G.822]). This means that either the original timing of the TDM client is transparently carried across the packet network, or assuming a synchronous set up (i.e. a network where all clocks have the same long-term accuracy, within some tolerance margin, as generally applicable in the case of a single operator), the timing of the outgoing TDM signal (e.g. 2.048 Mbit/s) is generated by a PRC (see [G.811]) traceable signal (i.e. a signal with the same long-term accuracy of the PRC of the synchronization chain, see also Chapter 2).

Several documents have been produced in various Standards Developing Organizations (SDOs) to address the synchronization requirements applicable in case of CES. ITU-T Recommendation G.8261 [G.8261] presents a comprehensive analysis where the general case of a CES segment inserted into a TDM network has been considered. In particular, several deployment cases have been considered.

The first case (deployment case 1) concerns the case of a packet network replacing one SDH island, so that the CES segment is located as an island between the two switches of the reference model defined in G.823, Annex A (see Figure 1.6, which is based on Figure 12/G.8261, [G.8261]). In fact, [G.823] defines a reference network where the 2.048 Mbit/s client is transported between equipment with slip buffers (e.g. digital switch), across a series of SDH islands. G.8261 has generalized the model where one of the SDH islands is replaced by a network segment based on packet-switched technologies, and the client is carried via a CES.

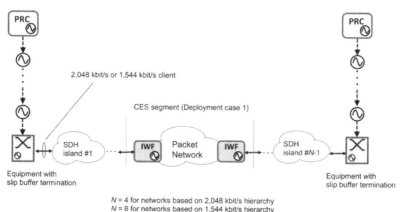

Figure 1.6. *CES segment, deployment case 1*

16 Synchronous Ethernet and IEEE 1588 in Telecommunications

A second scenario (deployment case 2) concerns the case of a packet network directly connected to the end equipment. In this case, the CES segment is located outside the network elements containing the slip buffers (see Figure 1.7, which is also based on Figure 12/G.8261, [G.8261]).

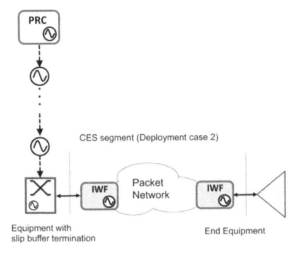

Figure 1.7. *CES segment, deployment case 2*

The study of the various deployment cases has led to the definition of wander budget allocated to the CES segment. As an example, the requirement application to an E1 is described in Figure 1.8 for deployment case 2 in terms of MRTIE (see Chapters 2 and 7 for more information on the synchronization parameters used to characterize a network). Figure 1.8 is based on Figure 11/G.8261, see [G.8261].

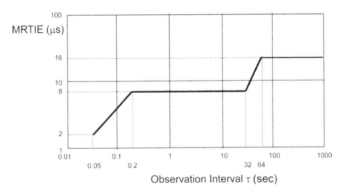

Figure 1.8. *CES network limits: 2.048 Mbit/s interface wander budget for deployment case 2A (ITU-T Figure 11/G.8261)*

Because of the smaller network scenario, a larger portion of the overall wander budget can be allocated to the CES segment in deployment case 2 than that applicable in deployment case 1.

It should be added that [G.8261] defines the two sub-cases for deployment case 2. The requirement described in Figure 1.8 concerns the sub-case indicated as "2A", that is the most general sub-case. Here, multiple interfaces can be connected to the end equipment and therefore it is required to control the differential wander between a generic incoming signal and the wander carried by the interface that is also used to synchronize the system clock of the end equipment.

As already mentioned, in the case of a CES service for a CBR signal, in addition to controlling the wander accumulation, it is also important that the long-term frequency on the recovered signal must be the same as that of the client entering the packet network (or within a certain tolerance) in order to control the slip rate. Note, however, that when the CES signal is used to synchronize BSs, in order to provide a reference timing signal of suitable quality, the long-term accuracy must also be better than 16 ppb; in this way, the BS can generate a signal on the radio interface with accuracy better than 50 ppb (see also section 1.4).

1.4. Mobile networks and mobile backhaul

The evolution of mobile networks and of the related transport network is one key topic impacting synchronization networks, especially when it comes to the synchronization solutions currently defined for packet networks.

Global system for mobile communications (GSM) is currently the most widely deployed mobile technology. This standard has been developed by ETSI, and is also known as the second-generation mobile technology.

The basic architecture of a GSM network is shown in Figure 1.9.

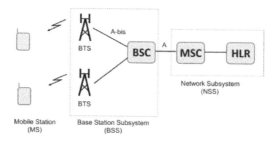

Figure 1.9. *GSM network architecture*

As shown in the figure, the network is composed of the base station subsystem (BSS), which includes the base station controller (BSC) and base station transceiver (BTS), that is the node generating data over the air interface, and the network subsystem, also known as core network, which includes the mobile switching center (MSC), that is the node that controls the network switching subsystem elements. The home location register (HLR) is a central database that contains details of each mobile phone subscriber authorized to use the GSM network.

Similar architectures apply in case of other relevant mobile technologies. Most of these technologies have been defined by the 3rd Generation Partnership Project (3GPP). In particular, 3GPP is the standardization body that has specified the evolution of the mobile technologies moving from GSM to universal mobile telecommunications system (UMTS) and long-term evolution (LTE).

In the case of UMTS (see [TS 25.401]), the BSC role is replaced by the radio network controller (RNC), and the BS is denominated "NodeB" (see Figure 1.10 that is based on Figures 1 and 4 from [TS 25.401]).

Functionally, the UMTS network can be divided into the radio access network (RAN), called UTRAN in UMTS (UMTS Terrestrial RAN), and analogous to the GSM BSS (in fact also referred to as the GSM RAN), that manages the radio-related functionality, the core network handling switching and routing calls and data communication to external networks. Finally, the user equipment (UE) corresponds to the terminal (i.e. the mobile phone) that interfaces with the user.

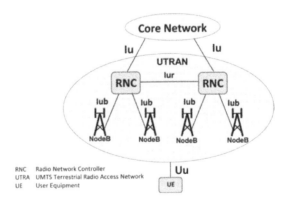

Figure 1.10. *UMTS and UTRAN Architecture*

The UTRAN is the portion of the UMTS architecture that is relevant for mobile backhaul networks. Here, the NodeB converts the data flow from the Iub and the Uu interfaces and participates in the radio resource management. The RNC controls and manages the radio resources in its domain (i.e. the NodeBs connected to it).

The latest generation of mobile technologies is called LTE (see [TS 36.401]). This technology has been introduced with Release 8 of the 3GPP specifications. Release 10 also has added LTE Advanced (LTE-A).

The LTE architecture is shown in Figure 1.11 that is based on Figure 6.1-1 from [TS 36.401].

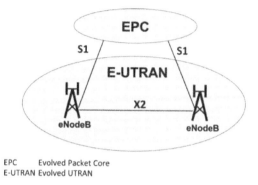

EPC   Evolved Packet Core
E-UTRAN Evolved UTRAN

Figure 1.11. *LTE network architecture*

As shown in the figure, there is only one type of node in the E-UTRAN, the LTE BS (eNodeB). The LTE BSs are connected to the core network (evolved packet core (EPC)) using the core network–RAN interface, S1.

The Core Network includes several fundamental nodes required for the operations in the mobile network (see [TS 23.401]). These include the Serving gateway (Serving-GW) and the Packet Data Network GW (PDN-GW), these take care of the user plane and transport the IP data traffic between the UE and the external networks; the Mobility Management Entity (MME), which deals with the control plane and handles the signaling related to mobility and security for the E-UTRAN access; and finally the Home Subscriber Server (HSS), which is a database that contains user-related and subscriber-related data.

The term System Architecture Evolution GW (SAE GW) is a term sometimes used to include both PDN and Serving GW.

The serving GPRS support node (SGSN) is the node that allows existing GSM and WCDMA/HSPA (i.e. wideband code division multiple access, enhanced with the high-speed packet access technology) systems to be integrated to the evolved system through standardized interfaces.

Synchronization requirements applicable to GSM, UMTS and LTE networks are described in section 1.4.1. Additional requirements may be related to the deployment of new architectures such as in the case of heterogeneous networks.

The term "heterogeneous network" is used to indicate deployments where macro cells (i.e. cells served by high-power BSs and with large coverage) are mixed with low-power BSs, also called pico cells or small cells, having reduced coverage (see [TR 36.814]). The use of pico cell allows enhancing the coverage as well to increase the capacity in the macro area. This generally applies to LTE.

As explained in section 1.4.1, the potential interferences between the macro and pico cells may lead to additional synchronization requirements.

One example of such deployments is shown in Figure 1.12.

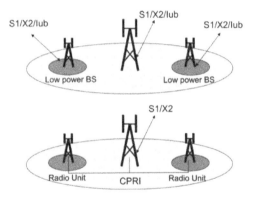

Figure 1.12. *Heterogeneous network example*

In this figure, two main logical architectural configurations are shown. In the first configuration the macro BS covers an area where a number of low-power BSs are also deployed. The macro cell and the pico cells are served by independent backhaul connections and all BSs have their own baseband and radio units.

In the second configuration, a baseband unit is connected to one or more radio units. These radio units may be macro radio units or low-power radio units and are typically connected to the baseband unit via the Common Public Radio Interface (CPRI) (see [CPR]). The radio units may be located at the same site as the baseband unit or connected remotely to the baseband unit. In the latter case, due to the stringent requirements on jitter, symmetric channel and delay between the baseband and the radio unit, they are typically connected via direct links.

The segment of the mobile network that connects BSs to network controllers within a coverage area is called mobile backhaul as shown with an example in Figure 1.13, see [BRI 10]. Mobile backhaul is a fundamental concept that is often mentioned in the context of synchronization.

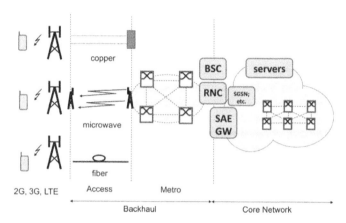

Figure 1.13. *Mobile Backhaul boundaries*

In this architecture, two main parts can be identified: the metro area sometimes also called the aggregation area or high RAN (HRAN) and the access part sometimes also called last mile or low RAN (LRAN).

The access is the cell site access part of the RAN backhaul, which typically includes multiple physical access technologies (e.g. microwave, copper or fiber). The metro is the part of the network that collects, aggregates and concentrates traffic from the access for connecting to the radio controllers. This part of the network is generally based on optical fiber networking infrastructure.

GSM and the first generation of the WCDMA networks connect the BSC/RNC with the radio BSs by means of a 2.048 Mbit/s TDM connection carried over PDH or SDH transport networks. In this case, the 2.048 Mbit/s signal carrying user traffic is also used to deliver a frequency synchronization reference to the BS (the 2.048 Mbit/s is generally synchronous in this context).

The increase in the capacity requirements and the related introduction of packet technologies have brought a significantly more complex architecture as shown in Figure 1.13.

One key aspect is the use of packet technologies. As an example, all the interfaces in the LTE architecture are IP based. In practice, for LTE, Ethernet interfaces carrying IP traffic is the standard.

As described in section 1.4.1, mobile networks have some specific synchronization requirements. Over the years these have led to various considerations in terms of how synchronization could be delivered toward the radio BSs.

In particular, the migration from TDM to packet networks in mobile backhaul, in addition to the introduction of new time and phase synchronization requirements, has been the main driver for much of the work in standards to define the necessary technologies needed to be able to deliver frequency and time synchronization over packet networks. This will be the main topic of this book.

1.4.1. *Synchronization requirements in mobile networks*

The synchronization is one of the key aspects related to mobile technologies operations.

In this case, the needs are mainly related to the correct generation of the radio signal. This means the generation of the carrier frequency as well as the generation of the data on the air interface.

As it will be discussed in section 1.6, the synchronization requirements related to mobile applications have a particular relevance on the recent work in ITU-T related to timing in packet networks; due to that, a list of the most important mobile applications and related synchronization requirements can also be found in the following documents: [G.8261] and ITU-T Recommendation G.8271 [G.8271].

Frequency synchronization is always required in mobile networks, typically in order to allow handover of the UE between cells.

In fact, when moving from one cell to the next cell, the UE must be able to lock in frequency to a different reference timing signal. The effective implementation of the timing recovery circuit in the mobile equipment, for instance in terms of pull-in range, puts some demand on the frequency error that can be allocated to the signal generated by the BS. Some budget must be also allocated to a potential Doppler shift effect (particularly critical in the case of a fast movement), therefore in order to limit the total frequency error that a UE and the connected network must be able to handle (for instance, it may also happen that if the UE is attached to a BS with a poor synchronization reference, the measurements performed on other BSs are impacted and may suffer from inaccuracies, leading sometimes to a late handover procedure being triggered by the network), it was necessary to define some maximum tolerance also to the frequency generated over the radio interface.

A second reason that constrains requirements on the maximum frequency deviation of a BS is to meet the compliance to regulatory aspects, that is generating a signal within the allocated frequency band.

The GSM is one of the first examples of mobile technology with specific synchronization requirements.

With respect to the applicable synchronization requirements, the [TS 145 010] provides the frequency synchronization related requirements applicable to the GSM networks:

> *"The BTS shall use a single frequency source of absolute accuracy better than 0.05 ppm for both RF frequency generation and clocking the timebase. The same source shall be used for all carriers of the BTS.*
>
> *For the pico BTS class the absolute accuracy requirement is relaxed to 0.1 ppm."*

Typically, the reference timing signal used to generate a signal on the air interface and meeting the previous requirements is propagated from the MSC sites down toward the BSC and BTS sites over the 2.048 Mbit/s traffic signals.

As will be described later in this book, the 0.05 ppm (or 50 ppb) synchronization requirement has become one of the main drivers of the recent synchronization studies in the standards (see section 1.6).

A similar frequency synchronization requirement applies to other generations of mobile technologies. For instance, in the case of the third-generation mobile networks (UMTS), the frequency synchronization requirements have been defined in [TS 25.104] for frequency division duplex (FDD) and in [TS 25.105] for time division duplex (TDD) radio technology options.

In either case, the following is specified for the frequency error (i.e. the measure of the difference between the actual BS transmit frequency and the assigned frequency):

> *"The modulated carrier frequency of the BS shall be accurate to within the following accuracy range [...] observed over a period of one timeslot":*

1) Wide area BS: ±0.05 ppm

2) Local area BS: ±0.1 ppm

3) Home BS: ±0.25 ppm

One time slot in this context corresponds to 10/15 ms (about 0.7 ms). This means that the basic requirement is applicable to a very short period. However, the fact that this requirement must always be met also implies that a frequency relationship is required over long observation periods (i.e. requirement on long-term accuracy).

As it can be noted, the requirement in the case of smaller cells is relaxed. One of the main reasons being the reduced budget that has to be allocated to the Doppler shift effect (as it would be more relevant for the UEs in rapid movement in larger cells).

The TDD option of the UMTS standard in addition to a frequency synchronization requirement also requires BSs to have access to a common phase, and therefore requires phase synchronization. This is required for the correct generation of the TDD frame on the radio interface avoiding the interference between the signals generated by adjacent cells.

In fact, as shown in Figure 1.14, the TDD method, in contrast to the FDD method, which requires separate frequency bands for both uplink and downlink, is a Time Division Multiplexing Access (TDMA) system using code division within the time slots and the same frequency band in uplink and downlink with separation on time slots basis.

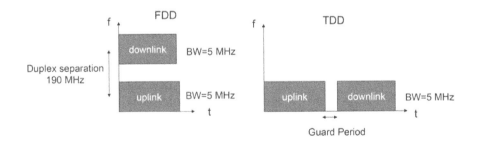

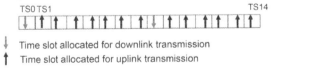

Figure 1.14. *FDD and TDD frame generation*

As shown in this time slot allocation example, at least two slots per frame must be allocated to downlink transmission; the synchronization channel is assigned to these slots.

In this case, in order to achieve "intercell synchronization", interNodeB node synchronization is used with requirements defined in [TS 25.123]:

> "3 μs maximum deviation in frame start times between any pair of cells on the same frequency that have overlapping coverage areas".

In order to achieve this accuracy, 3GPP has provided specifications for the suitable reference timing signals (see [TS25.402]). In particular, the BS (NodeB) must have access to a phase synchronization signal via a port with electrical characteristics as per RS422 and the relative phase difference of the synchronization signals at the input port of any NodeB in the synchronized area should not exceed 2.5 μs in order to provide sufficient margin for meeting 3 μs over the radio interface. In case of LTE for the TDD option, 3GPP specifies the following (see [TS36.133]).

In particular, 3GPP requirements [TS36.133] specify the following:

> "maximum absolute deviation in frame start timing between any pair of cells on the same frequency that have overlapping coverage areas":

– 3 μs for small cell (<3 km radius);

– 10 μs for large cell (>3km radius).

The new generations of mobile networks (in particular, LTE-A) and the related applications and/or architectures are expected to put additional needs for time and phase synchronization, for example related to possible need of coordinating the neighboring BSs in order to avoid interferences. One example is in the case of some of the heterogeneous network configurations described earlier. In such configurations, in order to achieve the maximum capacity (e.g. to increase the capacity for the UEs at the border of the cells), the implementation of specific features generally based on aligning in time or phase pico cells with the macro cell might be required. Coordinated multipoint (CoMP) is an example of such interference coordinating mechanisms.

The applicable synchronization requirements in this case have not been set yet; however, in case of heterogeneous networks, it is expected to be in the same order of magnitude as the one defined for TDD systems. The main difference though is that alignment is only required locally in a cell. That is, there is no need to achieve absolute time synchronization on the air interface as long as the phase deviation between the macro cell signal and the pico cell signal is within the defined level of accuracy.

Different considerations may apply in case of coordination (e.g. CoMP) required between macro cells. In this case, for instance, absolute time synchronization might be required.

Finally, we should be reminded that *code division multiple access* (CDMA) mobile technology was the first technology to introduce a time synchronization requirement. The same requirement has been defined in case of CDMA2000.

In this case, according to the relevant specifications, [C.S0010-B] and [C.S0002-C], the BS will recover the CDMA system time (which is in its turn traceable to Universal Time Coordinated (UTC)) with an accuracy of ±3 µs. In addition, ±10 µs with respect to CDMA system time is allowed in case of failure for a period of at least 8 h (when the external source of CDMA system time is disconnected).

In addition to the previous requirements, it should be noted that the use of some services might also place additional specific requirements on the synchronization. This is the case of multimedia broadcast multicast service (MBMS) that is part of the WCDMA evolution as standardized by 3GPP. This is a technology for broadcast of content over cellular networks to small terminals (handsets), for example for mobile TV.

Ordinary MBMS requires time accuracy in the order of tens of milliseconds. However, when MBMS is based on single-frequency network (SFN) mode (MBSFN), a simulcast transmission technique is realized by transmission of identical waveforms at the same time from multiple cells. These are then combined by the terminal as multipath components of a single cell. The synchronization requirements are driven by the length of the Orthogonal Frequency-Division Multiplexing (OFDM) symbol, and the BS budget should be a portion of this. In particular, in the case of WCDMA, the accuracy will be within 12.8 µs. The requirement applicable to LTE networks is not defined as of Release 10. Values between 1 and 10 µs have been mentioned.

A summary on the most relevant synchronization requirements applicable to mobile networks is provided in Table 1.1. Additional features not included in this table are also discussed where the phase and time synchronization requirement may be significantly more stringent than 1 µs (e.g. positioning).

Duplex mode/feature	Frequency synchronization requirements	Phase synchronization requirements
FDD	50 ppb accuracy (air interface for a macro cell)	None
	100 ppb accuracy (air interface for a micro cell)	
	250 ppb accuracy (air interface for a femto cell)	
TDD	50 ppb accuracy (air interface for a macro cell)	Around microsecond accuracy for the air interface (the exact value depends on the mobile technology, and sometimes on the exact deployment case
	100 ppb accuracy (air interface for a micro cell)	
	250 ppb accuracy (air interface for a femto cell)	
MBSFN (LTE)	As per duplex mode used	Values in the 1–10 μs range have been mentioned (under discussion in 3GPP)
CoMP (LTE-A)	As per duplex mode used	Around microsecond accuracy (under discussion in 3GPP)

Table 1.1. *Overview of the synchronization requirements in mobile networks*

1.5. Synchronization requirements in other applications

Over the years, many other applications, either as part of the telecom industry or related to some other industries, have been shown to require some synchronization support (mainly time synchronization) for the proper operations.

As mentioned earlier in this chapter, transport and switching networks have mainly required some level of frequency synchronization. However, some complementary functions such as management and charging introduced the need to have a time synchronization reference available (time of day).

As an example, the operation and maintenance (OAM) functions are required to time stamp some specific event (e.g. alarm) as to allow the correlation and analysis of significant events in the network (e.g. faults). In this case, the required accuracy is in the order of the sub-second (typically better than half a second).

Charging, or billing, function also puts some "time of day" requirements in the order of sub-second.

The traditional needs for time accuracy in telecommunication networks require relatively low accuracy compared to some of the requirements for wireless systems. These needs are in general addressed using the Network Time Protocol (NTP), presented in Chapter 2.

New emerging applications are increasing the need to deliver time synchronization reference with various accuracy requirements.

One of the first examples is related to packet delay monitoring. In fact, the delivery of data over packet networks has increased the need to monitor the actual latency introduced by the transport network.

Tools to monitor the quality of packet data delivery are, for instance, defined in ITU-T Y.1731. In some of these measurements, the evaluation of the one-way delay is required. This measurement requires that a common time synchronization reference is available at the measurement reference points. If the performance has to be measured in the millisecond range, this leads to distributing a common time reference much better than this requirement, typically in the hundreds of microsecond range.

Accurate time synchronization (in the microsecond range) is required in the power industry, for instance to synchronize phasor measurement units (PMU) used to monitor multiphase power systems to optimize power distribution.

An IEEE 1588 profile has been defined to support this type of application in [C37.238] (common profile for the use of Precision Time Protocol of IEEE Std 1588-2008 in power system protection, control, automation and data communication applications).

Finally, financial and legal time (e.g. to identify the exact instant in time of a financial transaction) and scientific applications (e.g. European Organization for Nuclear Research (CERN) accelerator's control) are further examples of applications where time accuracy in microsecond range (or even more accurate) is required and where the use of IEEE 1588 is being considered as an alternative to GPS.

1.6. The need to define new synchronization technologies

It has been shown that a large number of applications and technologies require various levels of synchronization (frequency only or both phase/time and frequency synchronization).

A detailed analysis of the various options to deliver the necessary reference timing signals in a telecommunication network in order to meet these needs is presented in Chapter 2.

As will be shown, TDM networks (e.g. SDH) have permitted the dissemination of reference timing signals of proper accuracy (frequency synchronization in this case) to all points of the transport and switching network. In fact, the generation of the traffic signals within a network element is controlled by a system clock capable of being frequency locked to the PRC of the network. Thus, in TDM networks, traffic data can also be an accurate frequency synchronization reference.

The migration from TDM to packet networks, in particular with the adoption of an Ethernet physical layer, initially separated data from synchronization distribution, due to the asynchronous operation of Ethernet networks and of the packet switching technologies (Ethernet and IP). In fact, for the correct operation of packet networks, there is no need to synchronize packet switching nodes in frequency, time or phase. The nodes are allowed to work with free running clocks (e.g. in case of Ethernet network, clocks with 100 ppm accuracy are allowed). This means that the physical layer of legacy Ethernet networks does not carry an accurate reference timing signal as was the case for SONET or SDH networks. As noted earlier in this chapter, the development of Synchronous Ethernet now begins to bind delivery of frequency synchronization with packet data.

As it will be shown in this book, various options have then been introduced in the standards in order to address frequency, phase and time distribution. The time synchronization aspects deserve a specific discussion.

For the few examples when accurate time synchronization is required (e.g. CDMA), traditionally GPS receivers have been deployed in the points of the network where frequency, phase and time synchronization is required. When less accurate time synchronization is required (e.g. sub-second accuracy), this can be achieved by using packet protocols such as NTP operating at a software level.

The increased needs of accurate phase and time synchronization (mainly related to emerging mobile applications as discussed in section 1.2) have then driven the definition of alternative synchronization methods, also in this case based on packet timing technology.

Details on the new synchronization technologies are provided in Chapters 2 and 3.

1.7. Bibliography

[802.16] IEEE 802.16-2009, IEEE Standard for Local and metropolitan area networks Part 16: Air Interface for Broadband Wireless Access Systems, 2009.

[BRI 10] BRIGGS P., CHUNDURY R., OLSSON J., "Carrier Ethernet for mobile backhaul", *IEEE Communications Magazine*, vol. 48, no. 10, pp. 94–100, 2010.

[C.S0010-B] 3GPP2 C.S0010-B, Recommended Minimum Performance Standards for CDMA2000 Spread Spectrum Base Stations, 2004.

[C.S0002-C] 3GPP2 C.S0002-C, Physical Layer Standard for CDMA2000 Spread Spectrum Systems, 2002.

[C37.238] IEEE C37.238™-2011, Standard Profile for Use of IEEE Std. 1588 Precision Time Protocol in Power System Applications, 2011.

[CPR] Common Public Radio Interface (CPRI); Interface Specification.

[G.703] ITU-T Recommendation G.703, Physical/electrical characteristics of hierarchical digital interfaces, 1998.

[G.781] ITU-T Recommendation G.781, Synchronization layer functions, 2008.

[G.803] ITU-T Recommendation G.803, Architecture of transport networks based on the synchronous digital hierarchy (SDH), 2000.

[G.811] ITU-T Recommendation G.811, Timing characteristics of primary reference clocks, 1997.

[G.812] ITU-T Recommendation G.812, Timing requirements of slave clocks suitable for use as node clocks in synchronization networks, 1998.

[G.813] ITU-T Recommendation G813, Timing characteristics of SDH equipment slave clocks (SEC), 1996.

[G.822] ITU-T Recommendation G.822, Controlled slip rate objectives on an international digital connection, 1998.

[G.823] ITU-T Recommendation G.823, The control of jitter and wander within digital networks which are based on the 2.048 Mbit/s hierarchy, 2000.

[G.824] ITU-T Recommendation G.824, The control of jitter and wander within digital networks which are based on the 1.544 Mbit/s hierarchy, 2000.

[G.825] ITU-T Recommendation G.825, The control of jitter and wander within didgital networks which are based on the synchronous digital hierarchy (SDH), 2000.

[G.8251] ITU-T Recommendation G.8251, The control of jitter and wander within the optical transport network (OTN), 2010.

[G.8261] ITU-T Recommendation G.8261, Timing and synchronization aspects in packet networks, 2008.

[G.8271] ITU-T Recommendation G.8271, Time and phase synchronization aspects in packet networks, 2012.

[GHE 10] GHEBRETENSAE Z., HARMATOS L., GUSTAFSSON K., "Mobile broadband backhaul network migration from TDM to carrier ethernet", *IEEE Communications Magazine*, vol. 48, no. 10, pp. 102–109, 2010.

[HOL] HOLMA H., TOSKALA A., *WCDMA for UMTS*, John Wiley & Sons, 2004.

[TR 25.836] 3GPP TR 25.836, Node B synchronization for TDD, 2001.

[TR 25.866] 3GPP TR 25.866, 1.28Mcps TDD Home NodeB (HNB) study item technical report, 2009.

[TR 36.814] TR 36.814, Further advancements for E-UTRA physical layer aspects, 2010.

[TR 36.922] 3GPP TR 36.922, TDD Home eNode B (HeNB) Radio Frequency (RF) requirements analysis, 2011.

[TS 23.002] 3GPP TS 23.002, Network Architecture, 2012.

[TS 23.401] 3GPP TS 23.401, General Packet Radio Service (GPRS) enhancements for Evolved Universal Terrestrial Radio Access Network, 2011.

[TS 25 104] 3GPP TS 25.104, Universal Mobile Telecommunication Systems (UMTS), UTRA BS FDD, Radio Transmission and Reception, 2012.

[TS 25 105] 3GPP TS 25.105, Universal Mobile Telecommunication Systems (UMTS), UTRA BS TDD, Radio Transmission and Reception, 2012.

[TS 25.123] 3GPP TS 25.123, Requirements for support of radio resource management (TDD), 2011.

[TS 25.346] 3GPP TS 25.346, Introduction of the Multimedia Broadcast/Multicast Service (MBMS) in the Radio Access Network (RAN); Stage 2, 2011.

[TS 25.401] 3GPP Technical Specification 25.401 UTRAN Overall Description, 2012.

[TS 25 402] 3GPP TS 25.402, Universal Mobile Telecommunications Systems (UMTS); Synchronization in UTRAN Stage 2, 2012.

[TS 36.401] 3GPP TS 36.401, Evolved Universal Terrestrial Radio Access Network (E-UTRAN), Architecture description, 2012.

[TS 36.133] 3GPP TS 36.133, Requirements for support of radio resource management, 2011.

[TS 145 010] ETSI TS 145 010, Radio Subsystem synchronization, 2007.

Chapter 2

Synchronization Technologies

2.1. Fundamental aspects related to network synchronization

This chapter provides some fundamental aspects related to the synchronization of telecommunication networks that also form the basis for the synchronization in packet networks.

A few books can be found in the literature dealing with the synchronization in telecommunication networks (e.g. [BRE 02]). It should be noted that generally the main focus of these is frequency synchronization (syntonization would be the correct term). In fact, the issue of distributing accurate time synchronization, as has been mentioned earlier in this book, is a relatively recent topic.

The key aspects related to the synchronization in telecommunication networks (in particular, time division multiplexing (TDM) based networks) have been described in International Telecommunication Union Telecommunication Standardization Sector (ITU-T) Recommendation G.810 [G.810], a document that was significantly updated during the development of the synchronous digital hierarchy (SDH) synchronization during the second half of the 1990s. More recently, ITU-T Recommendation G.8260 [G.8260] has been developed to also include aspects related to the synchronization in packet networks. Among the main aspects included in [G.8260] is an appendix dealing with the definition of the methodologies to verify if a packet network is suitable for carrying timing via packets (e.g. definition of the packet delay variation (PDV) metrics, see section 7.3).

The key element in the synchronization of telecommunication networks is the clock. As defined in [G.810], a clock is a piece of equipment that provides a timing signal where the timing signal is a nominally periodic signal $s(t)$:

$$s(t) = A\sin[\Phi(t)]$$

where:

A is a constant amplitude coefficient and $\Phi(t)$ is the total instantaneous phase.

As discussed earlier, traditionally the focus of the synchronization networks is the distribution of frequency synchronization. This is achieved by means of the distribution of reference timing signals with a frequency accuracy and phase noise within predefined limits.

The frequency accuracy y is measured as the maximum relative error of the frequency as measured over a defined period of time (e.g. 1 month and 1 year).

$$y = \max(y(t)) = \max((v(t) - v_{nom})/v_{nom})$$

where $v(t)$ is the instantaneous frequency and v_{nom} is the nominal frequency.

Two main architectures have been defined in order to meet this target, see Figure 2.1, which is an adaptation of Figure 1/G.8261 (see [G.8261]):

– Full plesiochronous (or distributed primary reference clock (PRC)).

– Master–slave synchronization.

A third approach, mutual synchronization, was also defined where all clocks are interconnected and there is no underlying hierarchical structure or unique PRC. This approach, however, never became successful mainly due to the complexity of this solution and due to instability issues.

A full plesiochronous synchronization network does not involve any distribution of a reference timing signal in the network and fully relies on the capabilities of the clocks implemented in the nodes of the network.

The target requirement, typically a frequency accuracy of 1×10^{-11}, involves the use of very accurate clocks, based on atomic technology (see [BRE 02]) or Global Navigation Satellite Systems (GNSS) receivers (ultimately synchronized to universal time coordinated (UTC) via atomic clocks on the satellites and predictions uploaded from the ground forming the GNSS system). For more information on GNSS systems, see also section 2.7.

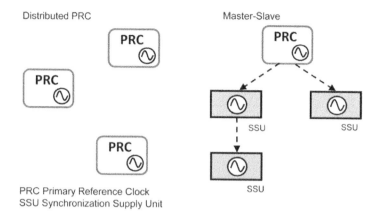

Figure 2.1. *Master–slave and full plesiochronous synchronization networks*

In a master–slave architecture, the reference timing signal is distributed from the master clock of the network to all clocks in the networks via a recursive "master clock-slave clock" relationship where the slave clock is phase-locked to the reference timing signal generated by the connected master clock.

This approach involves a strict hierarchy where the slaves in general can be locked to reference timing signals based on preassigned priorities. When the primary reference is lost, the slave can switch over to a secondary reference. More details on the architectures and protection aspects as typically used in telecommunication networks are provided in Chapter 3.

The key element of a slave clock in a master–slave architecture is the phase-locked loop (PLL) [GAR 05].

A PLL is a control system that generates an output signal whose phase is related to the phase of the incoming reference signal.

A PLL has the following basic blocks (see Figure 2.2):

– a *phase detector*, which provides an output signal proportional to the phase difference between the input and output signal;

– a *low-pass* filter (loop filter), which is required to filter the high-frequency noise at the output of the phase detector;

– a *Voltage Controlled Oscillator (VCO)*, which generates a reference timing signal whose frequency is controlled by the input signal, and that, when no input is present, operates in "free running".

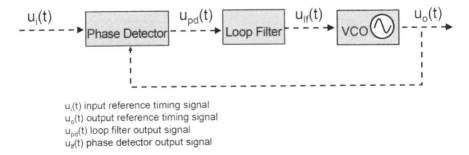

$u_i(t)$ input reference timing signal
$u_o(t)$ output reference timing signal
$u_{pd}(t)$ loop filter output signal
$u_{lf}(t)$ phase detector output signal

Figure 2.2. *PLL model*

A theoretical analysis of the behavior of a PLL is, in general, a complex task. However, with some simplification it is possible to consider a PLL as a linear system with a low-pass transfer function, typically of the second order, that can be characterized by a cutoff frequency (frequency bandwidth). The frequency bandwidth is generally related to the point in frequency where phase noise is attenuated by 3 dB and is indicated as f_{3dB}.

The PLL follows the phase modulations of the input when the frequency components of these modulations are within the frequency bandwidth. This means that assuming an input signal has an ideal constant phase offset, the phase detector in steady state would provide a constant value that controls the VCO to generate a signal with the frequency identical to that of the input signal.

As also described in [G.810], the following main parameters can be identified in a PLL:

– *Hold-in range* is the largest offset between a slave clock's reference frequency and a specified nominal frequency, within which the slave clock maintains lock as the frequency varies arbitrarily slowly over the frequency range.

– *Pull-in range* is the largest offset between a slave clock's reference frequency and a specified nominal frequency, within which the slave clock will achieve locked mode.

– *Pull-out range* is the offset between a slave clock's reference frequency and a specified nominal frequency, within which the slave clock stays in the locked mode and outside of which the slave clock cannot maintain locked mode, irrespective of the rate of the frequency change.

Sometimes the PLL is characterized in terms of its "time constant".

The time constant of a PLL, also known as its characteristic response time, provides an indication of the duration of the effects on the output of the PLL due to a given input. The time constant τ_c is related to the 3 dB bandwidth of the PLL, f_{3dB}, by the following relationship:

$$\tau_c = 1/(2\pi \cdot f_{3dB})$$

As a result of the characteristics of a PLL, a clock can operate in the following modes:

- locked;
- holdover;
- free running.

The free-running mode is an operating condition of a clock, the output signal of which is generated by the internal oscillator and is not controlled by incoming reference. In this mode, the clock has never had a network reference input or the clock has lost external reference and has no access to stored data. Typically, free-run mode would occur when a node powers up either with no provisioned references and/or with references that was lost for an extended period of time prior to the start up.

The holdover mode is an operating condition of a clock that has lost its controlling reference input and is using stored data, acquired while in locked operation, to control its output.

The locked mode is the normal operation of a slave clock. It is an operating condition of a slave clock in which the output signal is controlled by an external input reference and the output signal has the same long-term average frequency as the input reference. In addition, the phase difference between the output clock and reference input is tightly bounded.

Aspects of the synchronization networks based on SDH networks are described in several ITU-T Recommendations: G.803 [G.803], G.781 [G.781], G.811 [G.811], G.812 [G.812], G.813 [G.813], G.825 [G.825], G.823 [G.823] and G.824 [G.824].

The performance objectives for a synchronization network based on SDH or plesiochronous distribution hierarchy (PDH) as well as for the traffic signals carried over PDH and SDH are provided in [G.823] and [G.824].

The synchronization network architecture is based on the use of the following clock types:

– *PRC*: the PRC is the highest accuracy clock used as reference for the network (one or more network segments could be defined, each of them traceable to a PRC). Its requirements are defined in [G.811]. This clock operates always in free running mode.

– *Slave clocks*: slave clocks could be generally defined as two main types: SDH equipment clock (SEC) as defined in [G.813] (two options are defined, option 1 for use in networks optimized for the 2,048 kbit/s hierarchy and option 2 for use in networks optimized for the 1,544 kbit/s hierarchy) and synchronization supply unit (SSU) as defined in [G.812]. Several types of clock are defined in [G.812]: type I for use in networks optimized for the 2,048 kbit/s hierarchy and type II and III for use in networks optimized for the 1,544 kbit/s hierarchy. Other types are also defined but only to describe clocks used in earlier networks. Slave clocks during normal conditions operate in locked mode and may enter holdover mode when the external reference timing signal is lost. The performance of the system when the clock is in holdover is directly related to the performance of the oscillator and circuitry in the system. For example, a system with SSU capability is able to maintain a closer frequency offset to the reference signal compared to a system with SEC capability.

[G.803] and [G.781] provide the main guidelines with respect to the reference architectures including the network reference chain based on clocks defined above (see also Chapter 1).

The synchronization network described in [G.803] requires all clocks to be traceable to a master clock compliant with G.811 (PRC).

Two main types of timing distribution are considered:

– intersites or interstation;

– intrasite or intrastation.

As shown in Figure 2.3 (which is an adaptation of Figure 8-1/G.803, see [G.803]), in the intrasite distribution, the best clock of the site (typically compliant with [G.812]) receives the reference timing signal and is responsible to distribute it according to a star configuration within the site.

Figure 2.4, which is an adaptation of Figure 8-2/G.803 (see [G.803]), describes the intersite timing distribution, where the timing is distributed from the PRC toward all the clocks of the network according to the master–slave architecture.

[G.803] also describes the synchronization network reference chain that provides the basic rules (applicable to the clocks defined for the 2,048 kbit/s networks) that should be followed in order to meet the performance objectives defined in [G.823].

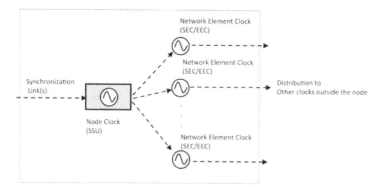

Figure 2.3. *Intrastation synchronization*

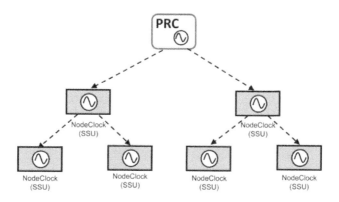

Figure 2.4. *Intersite timing distribution*

According to the network reference chain, after at most 20 SECs, there must be an SSU clock required to filter the wander accumulated over the SEC chain.

The total maximum number of clocks is 60 and the maximum number of SSUs is 10, see Figure 2.5, which is an adaptation of Figure 8-5/G.803 (see [G.803]).

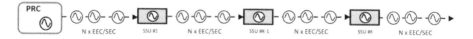

Worst case scenario calculation purposes:
K=10 and N= 20, with limitation that total number of clock is limited to 60

Figure 2.5. *Synchronization network reference chain*

As discussed earlier, the network reference chain applies only to option 1 clocks (i.e. those defined for use in 2,048 kbit/s networks). However, similar considerations can be made in case of option 2 clocks. For this case, some basic rules have been defined in [G.824].

Extension of the SDH-based recommendations to the use of Synchronous Ethernet (SyncE) is described in section 2.2.

To describe the performance of a clock and of a synchronization network, several parameters have been described in [G.810]. The analysis of the performances of a clock and of a synchronization network is based on measurements performed in the time domain. In particular, the measurement is based on the collection of samples that describes a time error function.

The time error of a clock, with respect to a reference signal, is the difference between the time of that clock and the reference signal. Mathematically, the time error function $x(t)$ between a clock generating time $T(t)$ and a reference clock generating time $T_{ref}(t)$ is defined as:

$$x(t) = T(t) - T_{ref}(t)$$

Figure 2.6 shows an example where the error is measured against an ideal reference (i.e. generating the ideal time $T_{ref}(t) = t$).

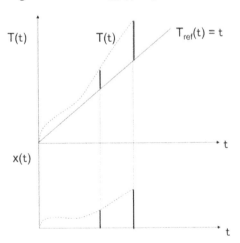

Figure 2.6. *Time error*

The measurement can be done at the output of a clock, for instance, to characterize the performance of a specific clock when it is locked to an ideal input, or can be made at a certain point of a synchronization network, in order to verify the performance of the synchronization network with respect to an ideal reference (see Figure 2.7).

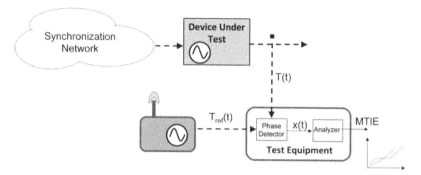

Figure 2.7. *Time error measurement setup*

As shown in Figure 2.7, the $x(t)$ samples are then analyzed by means of metrics specifically defined for the analysis of clocks and networks used in a telecom environment. Maximum time interval error (MTIE) and time deviation (TDEV) are the metrics most commonly used for this task. More details on these metrics and on the measurement techniques are provided in [G.810], [0900003] and in Chapter 7.

As already mentioned, most of the aspects described in literature concern frequency synchronization.

In case of time synchronization, similar concepts can be applied, where the total phase difference from the ideal reference is also relevant. However, time synchronization brings in new problems and concepts not considered for frequency synchronization. Time accuracy, unlike frequency accuracy, *in principle* requires transfer from a source of UTC. Since a clock is a frequency device, we can have a stand-alone accurate frequency standard. Time accuracy requires a method for transfer. In some cases, phase accuracy is the requirement, instead of true accuracy to UTC. For phase accuracy, transfer among devices is still required.

This means, for instance, that the performance must be directly based on time error samples, as metrics such as MTIE and TDEV would not provide information on the actual phase difference from the ideal time. The following sections provide an overview on the methodologies used to distribute reference timing signals in a network. Aspects related to packet timing and packet clocks applicable to both frequency synchronization and time synchronization are provided in section 2.3.

2.2. Timing transport via the physical layer

The transport of timing, that is frequency synchronization, through the physical layer has been used in telecommunication networks for many years; it was used in TDM and synchronous optical network (SONET)/SDH networks.

In telecommunication networks, the synchronization is necessary to maintain the performance of the network. Existing TDM and SONET/SDH networks are based on the distribution of a stable frequency.

For TDM networks, it is important to control the slip performance, see Chapter 1 for more details. For SONET/SDH, it is important to control jitter due to SONET/SDH pointer adjustments.

The physical layer synchronization is based on a very simple concept as shown in Figure 2.8.

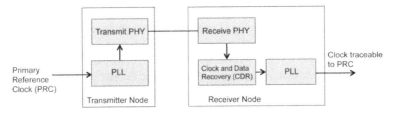

Figure 2.8. *Physical layer synchronization concept*

The constant bit rate (CBR) of the transmit signals carries the transmit clock to the receive node. A clock data recovery (CDR) PLL at the receive node extracts the clock and a digital PLL (DPLL) is used to filter wander and jitter introduced by the network.

2.2.1. *Synchronous Ethernet*

SyncE provides a mechanism to transport frequency over the Ethernet. When it was first introduced, there was a request to make SyncE compatible with synchronization of SDH and SONET networks, so SyncE can be an extension of the SDH and SONET synchronization network. It was also important to have SyncE compliant with the IEEE 802.3 standard.

SyncE specifies the accuracy of the Synchronous Ethernet equipment clock (EEC) to be within ± 4.6 parts per million (ppm), and therefore it is within specifications of IEEE 802.3 that specifies Ethernet clocks to be within ± 100 ppm.

SyncE was first introduced in the first revision of ITU-T Recommendation G.8261 [G.8261] consented in 2006. G.8261 specified SyncE architecture in Annex A.

In 2007, ITU-T Recommendation G.8262 [G.8262] was published. This recommendation defines the clock characteristics and requirements for SyncE. It defines two types of clock: synchronous EEC option 1 and option 2.

EEC-option 1 is based on ITU-T Recommendation G.813 [G.813], SEC option 1, and EEC-option 2 is based on ITU-T Recommendation G.812 [G.812] type IV. SyncE equipment using EEC-option 1 can interwork with equipment designed for synchronization network based on the 2,048 kbit/s hierarchy. SyncE equipment using EEC-option 2 can interwork with equipment designed for synchronization network based on the 1,544 kbit/s hierarchy.

G.8262 defines all the characteristics of the clocks used in SyncE equipments. It defines the frequency accuracy, pull-in/hold-in/pull-out ranges, noise generation, noise tolerance, noise transfer, transient response and holdover performance. Appendix III of G.8262 lists IEEE 802.3 Ethernet interfaces applicable to SyncE. SyncE will be configured according to the network synchronization plan (see section 3.3), but note that the use of 1000BASE-T and 10GBASE-T for SyncE could become incompatible with the network synchronization plan. This could happen due to the configuration of the master–slave relationship as defined by IEEE 802.3. Appendix IV gives considerations related to SyncE over 1000BASE-T and 10GBASE-T. Another key feature of SyncE is the synchronization status messaging (SSM). SSM for SyncE was introduced in the first revision of G.8261 (2006), the details of SSM was part of an appendix, as it was waiting for an organizational unique identifier (OUI) from IEEE. The SSM protocol was further developed and became part of ITU-T Recommendation G.8264 [G.8264] that was published in 2008.

A second revision of G.8261 was published in 2008. It includes detailed network requirements and architecture for SyncE.

2.2.1.1. Synchronization status messaging

SSM is very important to determine the traceability of the clock, as the quality level (QL) of the clock (source of the synchronization trail) is transported through the SSM channel. If there is a failure in the upstream network element (NE), then proper actions are taken based on the clock QL and clock priorities to select a different synchronization feed. One of the main goals of SSM is to help preventing timing loops. However, as it cannot fully eliminate them, proper network design is required in order to avoid the creation of timing loops as described further in section 3.3.4.

ITU-T Recommendation G.781 [G.781] defines the synchronization layer functions and defines the SSM selection algorithm. The QL of the clock, transmitted by SSM, is an indication of the long-term accuracy (holdover) of the NE clock. A new version of G.781 was published in 2008 to include SyncE.

The SSM QL for EEC-option 1 is equivalent to ITU-T G.813 option 1 clock (QL-SEC) and EEC-option 2 is equivalent to ITU-T G.812 type IV clock (QL-ST3); they are defined as follows:

– QL code 1011 – means that the source of the trail is an EEC-option 1;

– QL code 1010 – means that the source of the trail is an EEC-option 2.

These QL values defined in G.8264 and G.781 were added for EEC nodes. All other codes defined in G.781 are still valid for SyncE and they are used as is.

G.8264 specifies the SyncE SSM protocol and formats. For SyncE, the SSM transport channel is called "Ethernet synchronization messaging channel (ESMC)". The structure of the ESMC is based on the organization-specific slow protocol (OSSP) defined by IEEE 802.3, section 57 with Annexes 57A and 57B. The OSSP is a link-by-link protocol and it needs to be processed by each node; therefore, new ESMC frames need to be created for transmission to every SyncE peer nodes.

2.2.1.2. *SSM format and protocol*

G.781 defines processing times for the SSM selection algorithm in order to meet performance requirements for network reconfiguration. To be able to meet the time requirements specified in G.781, two types of message are defined for SyncE: the general message (used principally as "heart-beat") and the event message. The event flag contained in the ESMC protocol data unit (PDU) distinguishes these messages.

The general message is transmitted once a second (hence "heart-beat") and it provides an indication of the QL of the clock. Note that for SDH, the SSM is transmitted in every frame, and therefore every 125 µs. The event message is transmitted immediately if there is a change of the clock QL. If no message is received in a 5 seconds period, then the last received QL value is considered failed.

The event message with a new clock QL is transmitted immediately, and is triggered by a change of the clock QL. The next general ("heart-beat") message is considered as a backup in case the event message is lost.

The ESMC PDU format utilizes the protocol subtype provided by the IEEE for the OSSP and the OUI assigned to ITU-T with values as follows:

– Slow protocol subtype: 0x0A.

– OUI: 00-19-A7.

The ESMC protocol was designed to allow future enhancements and extensions of the protocol by using type length value (TLV). The specific TLV containing the clock QL has a fixed location within the ESMC PDU.

The ESMC PDU format is shown in Table 2.1, which is based on Table 11-3 of G.8264.

Octet number	Size/bits	Field	Field definition
1–6	6 octets	Destination address = 01-80-C2-00-00-02 (hex)	IEEE-defined slow protocol multicast address, as defined in Annex 43B of IEEE 802.3
7–12	6 octets	Source address	MAC address associated with the port which ESMC PDU is transmitted
13–14	2 octets	Slow protocol Ethertype = 88-09 (hex)	Slow protocol Ethertype assigned by IEEE: 88-09 (hex)
15	1 octet	Slow protocol subtype = 0A (hex)	Slow protocol subtype assigned by IEEE: 0A
16–18	3 octets	ITU-OUI = 00-19-A7 (hex)	ITU-OUI assigned by IEEE registration authority
19–20	2 octets	ITU Subtype	ITU subtype was assigned by ITU to be 00-01 for G.8264
21	bits 7:4	Version	Four-bit version number for the ESMC set to 0x1 for version 1 of the protocol. Bit 7 is the most significant bit.
	bit 3	Event flag	A value of 1 indicates an event PDU and a value of 0 indicates an information PDU
	bits 2:0	Reserved	Reserved for future use

Table 2.1. *ESMC PDU format*

22–24	3 octets	Reserved	Reserved for future use
25–1,532	1 octet	8 bits, Type: 0x01	QL TLV
	2 octets	16 bits, Length: 00-04	
	1 octet	Bits 7:4, 0x0 (unused)	
		Bits 3:0, SSM code	
	32–1,490 octets	Data and padding (See point j)	Data (e.g. TLV extension) and padding to achieve the minimum frame size of 64 bytes. It must be an integral number of bytes (octets).
Last 4	4 octets	Frame Check Sequence (FCS)	Frame check sequence as defined in section 4 of IEEE 802.3

Table 2.1. *(Continued) ESMC PDU format*

The ESMC PDU allows for future extensions with the utilization of TLVs as well as the three reserved octets (22–24). The TLV format is defined in Table 2.2.

1 byte	Type
2 bytes	Length (octets)
N bytes	Data plus padding

Table 2.2. *TLV format*

Examples of extensions with new TLVs include:

– a possible extension could be to have a TLV with a field to be incremented as the message traverses an equipment in the network, so the total number of equipment in the network can be calculated;

– the transmission of the source clock identity description (ID);

– the recording of all the clock IDs from the active source clock (current reference of the trail) allowing the detection of a timing loop (receiving node will not find its own clock ID in the TLV).

2.3. Packet timing

As described in section 2.2, the use of the physical layer to distribute a reference timing signal for carrying accurate frequency synchronization has been an ideal option in TDM networks where the traffic data itself needs to be synchronized.

The migration toward packet networks (in particular, based on Ethernet technology) initially made this option impossible.

One main reason is that the packet switching network, because it is based on statistical multiplexing, does not require inherent synchronization. For instance, Ethernet switches have been defined to run with free-running clocks with tolerance of ± 100 ppm [802.3] and the physical layer of packet networks, typically Ethernet, is also not required to be synchronous.

As has been discussed in Chapter 1, the standardization of SyncE has allowed us, in case of new deployments, to use the physical layer of Ethernet networks to deliver frequency synchronization. However, in the early phases of the migration of TDM networks toward packet networks, SyncE has not been an option. In fact, all NEs in the SyncE path must support SyncE, and the wide use of traditional Ethernet equipment not supporting SyncE broke the synchronization chain.

One additional important aspect, mainly relevant in the case of leased lines, is that while the client signals (e.g. E1) transported over SDH or SONET could maintain the original timing, thus allowing a transparent transport of the client timing at the physical layer, this is not the case when data are carried over packet networks. In this case, the client data are carried at the packet layer and the use of the Ethernet physical layer to carry the timing is generally constrained within the administrative domains (one exception being the transport of SyncE over optical transport networks (OTN) as explained in Chapter 4).

Because of the above issues, it was required to define some new methodologies to carry timing references using the packet layer. As will be discussed in this chapter, this can be done either by using the same packets that are also carrying traffic data or by using dedicated packets.

2.3.1. *Packet timing using traffic data*

The use of packets carrying both traffic and timing was the first "packet timing" option applied in telecom (if we exclude the use of Network Time Protocol (NTP) to carry time of day).

This was required in the asynchronous transfer mode (ATM) based networks. In this case, data are carried in cells, as an opposite to preassigned time slot as in case of TDM networks. When CBR services (e.g. E1) is transported over ATM (based over AAL1), it is then required to recover the original CBR timing in order to correctly regenerate the client at the output of the ATM network (see Figure 2.9). In fact, in order to prevent slips, the timing of the CBR service at both ends of the ATM network must be similar. The interworking function (IWF), as described in [Y.1411] is the functional block that translates data from a TDM-based network toward a packet-based network and vice versa. As part of this function, it also has to take care of the correct clock recovery.

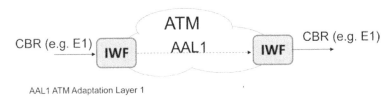

AAL1 ATM Adaptation Layer 1

Figure 2.9. *Transport of CBR services over ATM*

As also described in [TR 101 685], three main timing recovery approaches have been defined in ATM networks. The first approach uses a generic term of adaptive clock recovery (ACR) method.

In this case, the CBR timing is recovered based on the fact that the mean rate of cell arrivals can be used as a measure of the original timing. Due to its characteristics, the cell delay variation impacts the quality of the recovered timing. Analogous technology has been defined to distribute reference timing signal by means of dedicated protocols. The related characteristics are described in more detail in section 2.6.

A second approach that has been defined in the case of ATM networks is the synchronous residual time stamps (SRTS). This method, described in [I.363.1], is applicable when a common frequency reference is available at the entrance and exit points of the ATM networks. In this case, it is possible to encode at the ingress of the ATM network the difference between the CBR frequency and the common reference, and distribute this information toward the egress point of the ATM network where it is possible to recover the original frequency making use of the same common reference. This approach eliminates the dependency of the timing performance from the cell delay variation.

In both cases, the CBR timing is transparently carried across the ATM network.

A third approach is applicable in case the CBR services are synchronous with the ATM network synchronization signal. In this case, it is possible to apply a network synchronous method where the frequency of the outgoing CBR signals is generated by the ATM network synchronization signal.

A generalization of these approaches, applicable to circuit emulation services (CES) in packet networks as detailed in section 2.6, is described in [G.8261]. In particular, Ethernet, Internet Protocol (IP) and Multi-Protocol Label Switching (MPLS) networks are addressed in this reference.

As described in ITU-T Recommendation G.8261 [G.8261], the CES island is a segment of a network based on packet-switched technologies that emulates either the characteristics of a circuit-switched network or of a PDH/SDH transport network, in order to carry CBR services (e.g. E1).

As in case of ATM, a CBR service carried over a CES island (i.e. circuit-emulated TDM signal) requires that the timing of the signal is similar on both ends of the packet network.

Therefore, starting from the ATM experience, the following operating methods are identified in G.8261 that allow the service clock preservation:

– Network synchronous operation.

– Differential methods.

– Adaptive methods.

An additional approach is mentioned in G.8261:

– Reference clock available at the TDM end systems.

These methods are described in the following figures.

Figure 2.10, which is an adaptation of Figure 4/G.8261 (see [G.8261]), describes the network synchronous operation. When the client signal is synchronous with the synchronization network timing (i.e. PRC traceable), it is possible to generate the outgoing signal using this reference.

Figure 2.11, which is an adaptation of Figure 5/G.8261 (see [G.8261]), describes the differential method. This method is a generalization of the SRTS. As for the SRTS, also in this case, one basic assumption is that both ends of the network have access to a reference timing signal traceable to the PRC. By means of this common reference, it is possible to encode the difference between the service timing and the network timing, and use this information on the remote side of the network to recover the original timing.

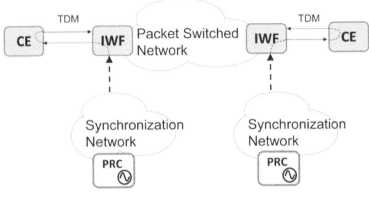

Figure 2.10. *Network synchronous operation*

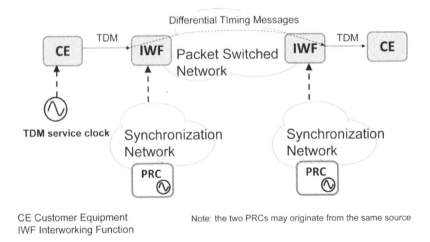

Figure 2.11. *Differential method*

In principle, this methodology could make use of any common reference, however only a common reference traceable to the PRC has been standardized, as this is the only practical approach.

This is a method that allows the transparent transport of the service timing (that in principle, for PDH signals could be in the range of several ppm), with no impact on the performance from the PDV.

Another method that allows the transparent transport of the service timing is the adaptive method. The adaptive method, as already described in the case of the CBR service carried over AAL1, is based on the consideration that the mean packet rate inherently carries the information of the original frequency of the service (see Figure 2.12, which is an adaptation of Figure 6/G.8261, see [G.8261]).

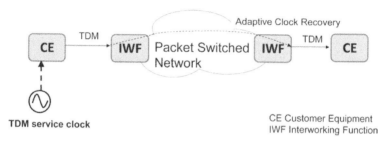

Figure 2.12. *Adaptive methods*

In this case, the timing recovery process can then be based on the arrival time of the packets carrying the CBR data. Because of that the performance is related to the PDV of the network.

If compared with the initial technologies used in the ATM networks, the adaptive algorithms have made significant improvements getting more sophisticated and robust against PDV (although meeting very accurate timing requirements would still be a challenge in some, particularly large, packet networks).

In the original ACR techniques, in most of the cases, all packets were used as input to the clock recovery process. This was, for instance, the case of recovery of the timing by observing the buffer fill.

One typical technology that has been developed in the more recent implementations consists instead of using only those packets that are less impacted by the PDV.

Finally, the arrangement shown in Figure 2.13 has been described in G.8261 (Figure 2.13, in particular, is an adaptation of Figure 7/G.8261 in [G.8261]) to depict possible deployments where the reference clock is available at the end systems. This could be the case of TDM networks segments connected via a packet network segment.

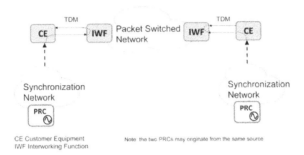

Figure 2.13. *Reference clock available at the TDM end systems*

Assuming that the client service is synchronous with the network timing, it is then possible to loop the timing received from the TDM segments in the generation of the outgoing services.

The deployment cases and related requirements applicable to the CES have also been defined in [G.8261] and a summary has been provided in section 1.3.1.

2.3.2. *Packet-based methods*

The basic principles of packet-based methods, for both the transfer of frequency synchronization and time synchronization, are described in Appendix XII of [G.8261]. As described in Appendix XII, two main protocols used are NTP, see [RFC5905], and Precision Time Protocol version 2 (PTPv2), see [1588-2008].

By means of these protocols, it is possible to exchange timing packets between a master (or server according to the NTP terminology) having access to an accurate reference timing signal and a slave (or client according to the NTP terminology).

The basic principle is shown in Figure 2.14, which is an adaptation of Figure XII.1/G.8261 (see [G.8261]).

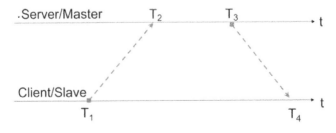

Figure 2.14. *Packet exchange in packet-based methods*

This is an example of two-way time transfer (TWTT). The packet sent at the client's time T_1 arrives at the server/master at the client's time $T_1 + d_{12}$, where d_{12} is the delay time for the packet to travel. Similarly, the packet sent from the server/master at its time T_3 arrives at the client at the server time $T_3 + d_{34}$. Frequency transfer depends largely on the stability of these differential delays. Accurate time transfer depends on these two delays being equal.

In the case of NTP, used in the client-server mode, the client initiates the exchange by sending an NTP packet at time instant T_1. The server receives the packet at time T_2, which sends a second NTP packet at time T_3. This packet includes the three time stamps T_1, T_2 and T_3. The client receives this packet at time T_4.

More details on the NTP protocol and associated mitigation algorithms are provided in section 2.6. A similar exchange is applicable in the case of PTP (see section 2.4).

In both cases, it is possible to recover the frequency of the server (or master in case of PTP) by comparing the sequence of the T_4 (or T_1) time stamps with the related sequence of the T_3 (or T_2) time stamps. In fact, the local clock at the client (or slave) generates the series of T_4 (or T_1), while the server (or master) clock generates the sequence of the T_3 (or T_2). By comparing and properly filtering the difference "$T_4 - T_3$" (or "$T_2 - T_1$"), it is possible to recover the server/master frequency. Because of that, in case only frequency synchronization is required, one-way packet flow is sufficient.

The network between the master and the slave, typically based on packet switching technology, introduces a significant source of error in terms of PDV. Because of that, specific filtering technology is required to optimally estimate the master clock rate at the slave site.

The basic principle of a clock able to handle timing carried via packets is defined in ITU-T Recommendation G.8263 [G.8263]. Figure 2.15, based on Figure I.1/G.8263 (see [G.8263]), shows the logical model of this clock.

If compared with a traditional PLL, it is possible to identify a packet selection block that represents the additional logic that these types of clocks have to implement in order to handle the relatively high noise due to PDV (in the order of tens, hundreds of microseconds or even milliseconds).

Several techniques have been developed for this task. The simplest technique is to select the packets that are transferred across the packet network with the lowest locally measured delay (see Figure 2.16). These packets in most of the cases are those that are less impacted by the packet switch queuing delays and are more suitable to be used by the PLL to recover the frequency synchronization signal.

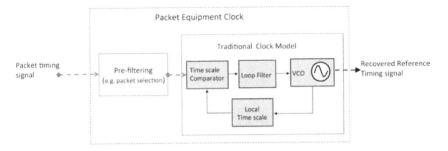

Figure 2.15. *Packet clock model*

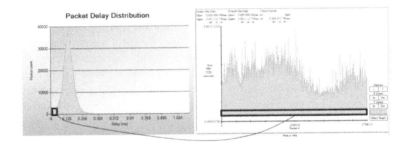

Figure 2.16. *Packet selection principle example*

A detailed discussion on the various methodologies to select packets is presented in [G.8260], when defining various approaches to characterize packet networks. The same methodology can be considered representative of some of the possible implementation of a packet clock.

The use of two-way exchange is important when in addition to frequency it is also required to recover time synchronization. Indeed, both NTP and PTP are protocols that have been created with the purpose of delivering time synchronization.

Having all four time stamps available, the client is also able to evaluate the path delay between the master and the slave.

This is expressed by the following formula:

Path delay = $[(T_4 - T_1) - (T_2 - T_3)]/2$

A basic assumption in this case is that the uplink and downlink paths are symmetric. Any difference in the delay between the uplink and the downlink will generate an error in the recovered time that is half of the overall asymmetry.

Unfortunately, telecommunication networks are generally not symmetric due to several reasons:

– in general, different paths in case of IP or MPLS networks;

– different lengths of the cables when two physical links are used for the two directions of transmission;

– different delays in the buffers of the packet switch (e.g. due to different load on the two directions);

– others.

To overcome some of the main issues mentioned above, the IEEE 1588 standards has defined some additional mechanisms that can be used when accurate time synchronization need to be delivered via packets. This is the objective of section 2.4.

2.4. IEEE 1588 and its Precision Time Protocol

PTP is a TWTT protocol that this section and the following section will introduce with sufficient details in the context of telecommunication networks.

If PTP, the protocol, is a fundamental element of the IEEE 1588 standard, we underline that it is a part but not all of IEEE 1588. In other words, we should reserve "PTP" for discussion of the protocol itself and use "IEEE 1588" when discussing the standard properties. Any expected performance of packet-based frequency and/or time in terms of accuracy, stability or time error cannot be achieved only by a protocol. Hence, if "precise time" is achievable, it highly depends on the engineering:

– of the transport network (set of devices linked together to allow data transmission via a protocol between two points), for example to limit network noise particularly PDV and asymmetry;

– of the timing network (set of devices recovering and/or providing a timing signal, frequency, phase or time of day), for example to limit source noise or filter wander, as will be seen in Chapter 4.

For that reason, to optimize the performance of its protocol, IEEE 1588 also specifies "the node, system and communication properties necessary to support

PTP" (reference: section 6 of [1588-2002] and [1588-2008]) and proposes models of devices, called "clocks", supporting the protocol and aiming to help and enhance the transfer of the packet-based timing reference. In IEEE 1588 terms, a PTP system is a set of IEEE 1588 clocks with specialized functionalities for supporting the protocol, i.e. PTP.

For instance, specifications of an IEEE 1588 clock include some material, such as a port, that is not part of the protocol itself. However, the messages sent and received by each port of an IEEE 1588 clock and the resulting behavior of that clock is part of the protocol. Nonetheless, the standard does not define any performance objective for transport network device or clock. Such a performance target might be specified as part of an IEEE 1588 profile or of a standard accompanying such profile, as discussed in section 3.4.

In the same vein, it is quite worth noting that IEEE 1588:

– does not mandate the data network to be fully formed of "clocks";

– does not define specific implementation either in hardware or in software of those clocks, but suggests clock models;

– does not specify any clock recovery algorithm.

This allows various implementations, systems, behaviors and results that can be adapted to particular needs and environments. This can also create unexpected results due to variable implementations if clock, communication or timing network limits are not specified.

This section highlights the important aspects of IEEE Standard 1588 from the telecommunication operator's standpoint by introducing the main general elements of the standard, looking back on it first release (IEEE Std 1588-2002 [1588-2002] with the PTP version 1 (PTPv1) and its caveats from the telecom standpoint, describing the enhancements of the second and current release [1588-2008] with the version 2 of the protocol (PTPv2), detailing the options specifically defined for the domain of telecommunications and their roots and concluding on items that telecom standard development organizations such as ITU-T had to consider in order to optimize the utilization of IEEE 1588).

2.4.1. *Some essentials of IEEE 1588*

As pointed in the introduction, the IEEE 1588 defines a protocol, PTP, and the properties supporting the protocol. This section focuses on some principal elements of the standard[1] to understand the overall model of distribution. The following sections detail some other important elements introduced by each release of the standard.

IEEE 1588 aims at creating a master–slave timing distribution model across PTP clocks. In IEEE 1588, a clock is a node implementing PTP. In that node, a "PTP port" is a physical, logical or virtual interface supporting PTP and handling PTP communications. To define the state or function of a port, IEEE 1588 defines state machines as part of its so-called best master clock algorithm (BMCA) process.

Actually, BMCA has two key functions in a clock. Based on information received on its PTP port(s), the BMCA of a clock:

– defines the master–slave hierarchy of the PTP distribution;

– selects the best master (i.e. the clock source it will synchronize to).

NOTE.– The distinction between port state and master clock selection in BMCA is an important concept to keep in mind, particularly for telecom approach (e.g. G.8265.1 telecom profile).

Processed for each PTP port of a clock, the port state can end up being in master, slave or passive state as shown in Figure 2.17.

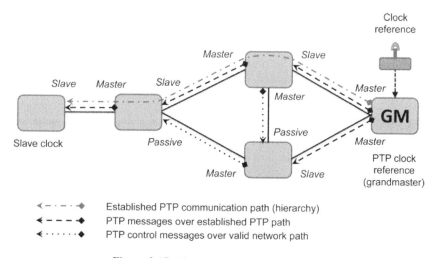

Figure 2.17. *Master, slave and passive ports*

---

1 This book covers only IEEE 1588-2002 and IEEE 1588-2008 and cannot consider evolution of the standard that occur in the beginning of 2013.

Principally, a master port of a clock transmits its packet-based clock signal in the form of PTP messages to a slave port of another clock. Slave ports allow a clock to synchronize with the root of the hierarchy, the ultimate master clock, namely the grandmaster (GM) clock. A passive port prevents timing loop or conflicting master ports in the same hierarchy. The state of a port and the selected clock source (best master or master port) can change over time. In steady network state, the BMC algorithm processed by clocks establishes a hierarchy up to the root, i.e. a GM, and down ultimately to slave clocks, leafs of the hierarchy. The BMCA will avoid creating timing loops through valid transport network paths (i.e. not blocked by transport network protocol, such as Open Shortest Path First (OSPF) for L3 or Spanning Tree for L2 Ethernet). In case of any change in clock (e.g. device failure) or network (e.g. link failure or rearrangement), the BMCA might have to re-establish the hierarchy and/or to reselect a master.

IEEE 1588 defines the notion of PTP domain. This notion can be utilized for different purposes (technical or administrative) but the standard defines the rules. A PTP domain corresponds to one GM delivering its clock signal (or timescale), over the hierarchy established for that domain. In other words, a PTP domain corresponds to a unique hierarchy of two or more clocks synchronized to same GM.

IEEE 1588 allows a clock to support multiple domains but any PTP port of a clock can support only one domain. Logically, because a PTP communication path is established between two PTP ports, a PTP path belongs to one PTP domain.

A communication (transport) network can support multiple PTP domains and a physical network interface can convey multiple PTP paths. Recall, a PTP port is a logical description that can be translated into a physical or virtual interface in a clock implementation. For instance, a PTP port can be assigned to a physical interface, to a Virtual LAN (VLAN) or to a loopback interface.

As Figure 2.18 shows, a network or a clock can support multiple domains on the condition that each domain remains separate. This separation can be physical or logical, as distinct communication networks, VLANs or clocks, and could lead to more complex and expensive design implementation, particularly for maintaining different timescales in one clock.

A PTP domain is formed of at least two clocks, a GM and a slave, as in ITU-T G.8265.1. However, IEEE 1588 defines different kind of clock. The simplest ordinary clock (OC) has only one PTP port in a PTP domain. As an OC runs a BMCA, its port can be either in master or slave state. If after the BMCA process the port is in master state, this means that the OC is GM of the PTP domain. Otherwise the OC will be a slave, a leaf, an end node of the PTP hierarchy. A high-quality reference source is assumed feeding a GM, for example a GNSS receiver, as shown in Figure 2.19.

Synchronization Technologies 59

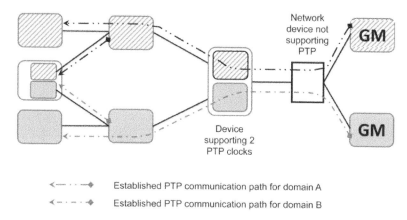

Figure 2.18. *PTP domain – one network, two domains with nodes in both domains*

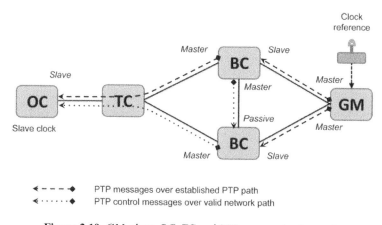

Figure 2.19. *GM, slave, BC, TC and PTP communication path*

Between the GM and the end slaves, the network nodes can be transparent to PTP, that is forwarding but not processing the PTP messages, or may assist PTP, providing on-path support. There are two types of such assisting nodes.

The boundary clock (BC) introduced in the first IEEE 1588 release [IEEE 1588-2002] has multiple PTP ports and runs a BMCA. BC thus participates in establishing the PTP hierarchy with at least one PTP slave port and one PTP master. Because a BC hosts multiple PTP ports, depending on the network communication topology, some PTP ports can be stated as passive. Moreover, a BC can become a GM (with multiple PTP ports).

The transparent clock (TC) introduced in the second release of IEEE 1588 [1588-2008] does not run any BMCA. Hence, TC ports have no specific state and do not participate in establishing the PTP hierarchy.

To distribute the timing reference from the GM, IEEE 1588 follows the same principle than any TWTT protocol as introduced in section 2.3.1. Some PTP messages allow a PTP slave port gathering the four time stamps to estimate the time offset from the master port. Figure 2.20 describes the PTP message exchange when clocks run the "delay mechanism". The peer delay mechanism, introduced in the second release of the IEEE 1588, will be described in section 2.4.3.

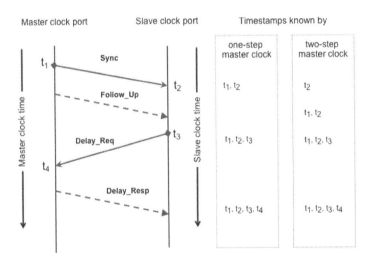

Figure 2.20. *PTP flow with delay messages*

The important messages for the synchronization process, called event messages (and shown as plain arrows in Figure 2.20), trigger the time stamps. Other messages, called general messages (and shown as dotted arrows), can also convey time stamps but do not trigger time stamping functions.

The first time stamp t_1 is recorded by the master clock port with master clock time when transmitting a Sync event message. The t_2 time stamp is recorded by the slave clock port with slave clock time at reception of the Sync event message. There are two ways for the slave clock port to receive this t_1 time stamp with smallest time error. Indeed, fluctuation in delays between the various time stamps generates errors.

Note that a fixed delay is acceptable on the condition the exact same delay exists in the reverse direction on the same interface (i.e. master clock and/or slave clock port). However, a difference in transmission and reception delays at one clock port would generate a time error due to the asymmetry.

For precise time (and frequency) synchronization, precise time stamping is recommended. IEEE 1588 states two principles clearly pictured in Figure 14 of [1588-2008].

The first principle highlights the importance of the generation of the time stamp. To reduce time error, time stamp should be recorded at the very last moment before the message is sent over the media, that is in optimum case, at the physical layer. This would greatly reduce any variation that might happen (e.g. when queuing packet or in link buffer) between time stamp generation and effective transmission over the link media. Any difference, as can exist in IEEE 802.3 10GE copper PHY, introduces an asymmetry. A similar requirement applies in the case of incoming event messages. The arrival time of the packet will be recorded as soon as it arrives.

The second principle is the specification of a time stamping point that will be identical for any transmission and reception of the protocol event messages. This latter principle can be a simple standard decision. IEEE 1588 specifies the time stamp point as the first bit after any transmission preamble. For Ethernet, this would be the Start of Frame (SOF).

The first principle however calls for best performance, but not mandates, to add some hardware assistance at the network interface of devices supporting IEEE 1588 clock. It particularly demands for some event messages, such as the Sync message, to record a time stamp (t_1 for Sync in Figure 2.20) and then to write this time stamp into the message payload at the same moment (from layer standpoint). Clocks implementing this "on-the-fly" method, called one-step clock, have been introduced in IEEE 1588-2008.

However, it is not always technically feasible or cost effective to implement a one-step clock. IEEE 1588-2002 has been developed with a two-step clock model only. A two-step clock will generate, for every Sync event message, a Follow_Up general message to transmit the t_1 time stamp. This eventually doubles the communication bandwidth between master and slave ports. The utilization of two-step clocks introduces other caveats particularly when using IEEE 1588-2008 TCs, as explained in section 2.4.3.

To complete the TWTT model, the slave clock port generates and records locally the t_3 time stamp when transmitting Delay_Req event message. Being useless to master, clock t_3 time stamp is not transmitted: one-step and two-step clocks behave the same. At the reception of the Delay_Req event message, the master clock port generates a t_4 time stamps. To transmit this t_4 back to the slave clock port, the master clock port generates a Delay_Resp general message.

The slave clock now has the four time stamps to estimate time via offset and delay computations. The formulas introduced in section 2.3.1 also express the basic assumption taken by TTWT calculations.

As described by Figure 2.21, the offset includes an error derived from the simple but necessary approximation that the actual delay from master to slave ($Delay_{MS}$) and actual delay from slave to master ($Delay_{SM}$) are identical, that is, symmetric.

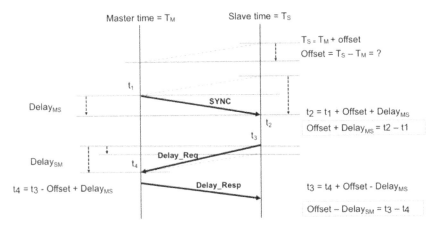

Figure 2.21. *Offset and delay calculations with PTP time stamps*

Indeed, without this approximation the number of unknowns would not allow us to solve the offset equation:

$$\text{Offset} = t_2 - t_1 - \text{Delay}_{MS} = t_3 - t_4 + \text{Delay}_{SM} \qquad [2.1]$$

Slave knows four time stamps (two from master clock and two from local clock) but cannot calculate the two distinct one-way delays without being time synchronized. Thus, TWTT offset and delay calculations will assume $Delay_{MS} = Delay_{SM}$:

$$\text{Round trip delay} = \text{Delay}_{MS} + \text{Delay}_{SM} = (t_2 - t_1) + (t_4 - t_3) \qquad [2.2]$$

$$\text{Mean (one-way) delay} = ((t_2 - t_1) + (t_4 - t_3))/2 \qquad [2.3]$$

From [2.1] and [2.3], the offset can be calculated (estimated) by:

$$\text{Offset} = t_2 - t_1 - \text{Delay} = t_3 - t_4 + \text{Delay} \qquad [2.4]$$

$$\text{Offset} = t_2 - t_1 - ((t_2 - t_1) + (t_4 - t_3))/2 \qquad [2.5]$$

$$\text{Offset} = ((t_2 - t_1) - (t_4 - t_3))/2 \qquad [2.6]$$

In practice, these delays, Delay_{MS} and Delay_{SM} are almost never equal. PTP using equation [2.3] can only detect variations in the differential delay. Any constant asymmetry will produce a time offset equal to half of the differential delay. Constant or slowly varying asymmetries can occur for a number of reasons, such as the two paths being physically different, or due to modulating over different lambdas on the same fiber.

Because master and slave ports, respectively, trigger Sync flow and Delay flow, two characteristics of PTP messaging have retained attention for telecom operations. First, PTP message intervals for both flows can be distinctly set, allowing a higher rate for Sync messages. Second, PTP communication might be established with Sync flow only. Those two characteristics were, in 2005, considered of high potential for frequency transfer in a telecom environment as an alternative ACR method (see sections 2.3.1 and 2.6).

To summarize at this point, two event messages, the Sync and Delay_Request, and two general messages have been introduced. Another general message, the PTP Management message, also exists in both versions of PTP. The management message is utilized between clocks and "administrative nodes" (a clock node can host an administrative node). As section 2.4.4 will detail, IEEE 1588-2008 introduces new event and general messages.

Last but not least, the fact that PTP defines "messages" is another important aspect of IEEE 1588. Indeed, this allows PTP to use various transmission modes and transport network options.

Now that some fundamentals (common aspects) of IEEE 1588 have been expounded, the following sections will present the two currently known iterations of IEEE Standard 1588, and will particularly focus on the limitations raised by the first version for telecom operation and on the extensions of the second release developed between 2005 and 2008.

2.4.2. IEEE 1588-2002: origin and limitations

For a detailed understanding of IEEE Standard 1588-2002, the reader should refer to work by John Eidson [EID 06]. Nonetheless, for the sake of this book, there are some points the authors would like to highlight.

The origins of IEEE 1588 development come from test and measurement (instrumentation and data acquisition), industrial automation or power generation, and received strong support from the automation industry leading to the adoption of the IEEE 1588-2002 standard by the Industrial Electro-technical Commission (IEC) as IEEE/IEC 61588-2004 (rev. 1).

Hence, specific application domains and targets have driven first release of IEEE 1588 work. Applications (e.g. data acquisition, military instrumentation, industrial robots and high-speed printers) and their targets, detailed in [EID 06], had sufficient shared objective to define common criteria:

– Achievable sub-microsecond accuracy (if required).

– Operation over control localized areas, such as local area network (LAN), with no intent to cover Internet or wide area networks, but supporting a PTP system.

– Administration-free operation allowing "plug-and-play" deployment of IEEE 1588 clocks (role of the BMCA explained in section 2.4.1).

– Designed for minimal resource requirement in devices supporting PTP (clock filtering or averaging might not be necessary).

Those criteria partly led to some technical choices for IEEE 1588-2002. Later, when the telecom community started considering (initially) this standard for frequency distribution, these criteria were not perceived as telecom friendly, triggering some new requirements for the new IEEE 1588 release, detailed in section 2.4.3. It is thus worth providing with few more details on this first release then considering its drawbacks from a telecom standpoint.

IEEE 1588-2002 introduced the OC, the BC and the administrative node. Collectively, the clocks form a distributed PTP system with PTP communication paths connecting those clocks.

Each clock and PTP port maintains groups (data sets) of clock or port attributes (members). Members can be static, dynamic or configurable. One data set, the default data set, represents the clock (i.e. own node clock attributes). To set and manage the hierarchy and its best master (clock source), each clock exchanges its default data set, that is its own clock characteristics or attributes. The result of the PTP Sync message exchange and BMCA will populate the dynamic members of the other data sets. This includes a parent data set that provides attributes of the GM clock.

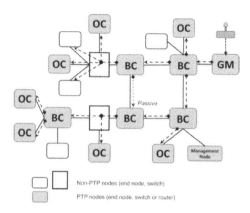

Figure 2.22. *IEEE 1588-2002 network*

In PTP version 1, the Sync message is in charge of transmitting the clock default data set members. As such, PTPv1 Sync message transmits both control plane (data set) and triggers generation of t_1 and t_2 time stamps. The BMCA runs on each PTP port at reception of the Sync message, and the Follow_Up message is mandatorily generated. Hence, in this release, default Sync interval is of 2 seconds (one Sync every 2 seconds and no faster than 1 second) for control plane and master to slave event message time stamping.

Beyond Sync message, PTPv1 messages include Follow_Up, Delay_Req, Delay_Resp and Management messages, all mandatory.

PTPv1 transport is based uniquely on User Datagram Protocol (UDP)/IP and principally on IP version 4 (IPv4) multicast transmission. The Internet Assigned Numbers Authority (IANA) assigned to IEEE 1588:

– two UDP port numbers to distinguish the PTP event (319) and the PTP general (320) messages;

– four IPv4 multicast group addresses (224.0.1.129 to 224.0.1.132) in order to manage four PTP domains, the default (PTP primary) and three alternate subdomains.

In IEEE 1588-2002, PTP management messages can utilize either multicast or unicast communication but any other message will use multicast. If this release of the standard does not prevent (and mention) IP version 6 (IPv6), no IPv6 multicast destination group address has been assigned.

IEEE 1588-2002 protocol had enough interesting characteristics to be considered in telecom operation. Originally, the interest was aimed at supporting frequency transfer for CES and mobile base station radio interface (see Chapter 1) through non-PTP communication network. It is important to understand the pros and cons of IEEE 1558-2002 in the context of such telecom operation as they introduce some of the enhancements of IEEE 1558-2008.

If utilization of IP as a transport mechanism is valuable for existing telecommunication networks with various transport modes, PTPv1 multicast transmission had one flaw in particular, one drawback and one caveat.

The flaw comes from the requirement for IPv4 time to live (TTL) and IPv6 hop limit (HL) to be set to 0, preventing IP equipment to accept PTP messages. Note that, because PTP uses IP transport and due to TTL/HL = 0, the IP device hosting the clock would discard any received PTP packet. Any implementation of IEEE 1588-2002 would thus have to disregard the TTL/HL value specification. More appropriately, an IEEE 1588-2002 clock shall, when sending a PTP packet, apply an appropriate TTL/HL value with a minimum of 1.

The drawback is specific to the "delay mechanism". Indeed, because the transmission of the PTP messages is based on multicast with one group address per domain, each slave port would have to communicate to master port broadcasting its message over the communication network. In other words, all Delay_Req messages would be received by all the slaves (but also by other nodes that would be in the same network, multicast domain and/or VLAN, not represented in Figure 2.23). From an IP standpoint, all clocks transmitting with same multicast group address, the PTP system would behave as a multipoint to multipoint distribution model, difficult to setup and maintain, wasting a lot of network resources and making the timing delivery inefficient in large networks, such as telecommunication. Figure 2.23 describes the effect in a short and small-scale network example.

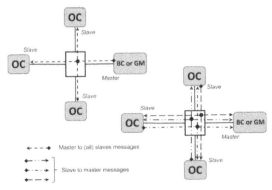

Figure 2.23. *PTPv1 multicast drawback*

Finally, the main caveat was the risk of having multicast forwarding engine in network node generating more PDV (multicasting single PTP flow to multiple clocks) than unicast forwarding could add to multiple individual (unicast) PTP flows. In other words, the risk for having PDV might be larger with multicast than with unicast communication depending on multicast forwarding implementation.

The transport and transmission limitations described above tend to confirm that IEEE 1588-2002 development was particularly aimed at Ethernet networks and/or full PTP support with single hop, link-based multicast transmission between clocks (and not over communication nodes), despite the fact that node-by-node PTP is not mandatory.

As introduced earlier, to remove as much uncertainty as possible from the network between master and slave (e.g. OCs), IEEE 1588-2002 assumes that each NE supports a BC function.

As shown in Figure 2.22 to bind its internal PDV, a node hosting the BC function should recover the time from its neighbor clock upwards to the GM and retransmit it downwards to slave one or multiple clocks. Addition of the support of BC capability would have led to a huge forklift upgrade in telecommunication networks.

To compensate for the lack of assistance (i.e. no BC) in wide area networks, mitigation of network noise, particularly the PDV but also any network rearrangement (e.g. due to network failure), will be handled by network engineering and end slaves.

From slave standpoint, the lack of network assistance leads to better filtering (i.e. more complex recovery servo). Only this topic led to huge parallel works at ITU-T and Alliance for Telecommunications Industry Solutions (ATIS) (refer to Appendix I of [G.8260] or [0900003]) and timing system manufacturers as described in section 2.3.1 (see also [SHE 09]). In those early days, ACR algorithms (see sections 2.3 and 2.6) were expecting a large amount of significant instants to recover accurate and stable frequency. When applying ACR behavior with a PTP system, this translates into using high-rate Sync message flow. IEEE 1588-2002 could theoretically permit high message rate. However, as mentioned earlier, Sync message also triggers BMCA driving for more intensive work.

From a network engineering standpoint, the lack of network assistance should lead to work on generating lowest PDV, defining proper quality of service, that is at the minimum, to give strict priority to PTP event messages. Combining higher rate and unicast communication leads to a new concern: the total amount of bandwidth that multiple high priority PTP flows would create. Reducing the bandwidth would lead to a reduction in the size of the Sync message.

Finally, as explained in Chapter 1, telecom synchronization network design arises from SONET/SDH network use. Strict specifications and rules were developed and acquired from experience to allow guaranteeing the timing service for both frequency and time delivery based on NTP (see section 2.6). In both cases, configuration is necessary. Contrariwise, IEEE 1588 was designed to be "plug-and-play", avoiding any timing configuration when a network is setup. BMCA was developed to allow hierarchy and GM selection to be performed with minimal to no preliminary work.

In summary, from a telecom perspective, IEEE 1588-2002 had some serious caveats and could not have been adopted as such. Modifications were necessary. In parallel, adoption of this first release in original markets showed some limitations. The IEEE Project Authorization Request (PAR) for IEEE 1588 version 2 was approved in March 2005.

2.4.3. *IEEE 1588-2008 and PTPv2*

The revision work on IEEE 1588-2002 started in February 2005. IEEE 1558-2008 was approved at the end of March 2008 and published the next July (and later as IEEE/IEC 61588 edition 2.0 2009-02). Among the participants of the working group, some were coming from the telecom industry. The initial intent was to enhance the protocol to allow a flexible, standard, packet-based frequency transmission to support circuit emulation (TDM pseudowires) and base stations (2G and frequency division duplex (FDD) 3G radio needs). Experience was based on adaptive clocking with one-way (master–slave) high-rate traffic. Those led to some changes of IEEE 1588.

During the development of this second release of IEEE 1588, two evolution trends appeared: one from work done at ITU-T SG15 Q13 in parallel to IEEE 1588v2 WG and one from Third-Generation Partnership Project (3GPP) evolution toward long-term evolution (LTE). These two trends reinstalled the interest in two-way transmission: the first because the reverse path (i.e. PTP Delay_Req message) could be used to recover frequency when uplink path presents lower PDV (see section 2.3.1) and the second because new mobile specifications were calling for phase synchronization between base stations. However, transfer of frequency and transfer of phase/time of day had distinct requirements and priorities.

Also, considering the enhancements expected from other markets and industries, IEEE 1588 version 2 added the following list of features:

– Improved accuracy of the time description with correction field (CF), achieving sub nanosecond ranges.

– Faster synchronization: smaller message interval.

– Shorter messages to reduce network bandwidth.

– New clock functions:

   - TC with end-to-end (E2E) and peer-to-peer (P2P) modes;

   - one-step clock mode (no need for Follow_Up).

– New delay measurement mechanism with new addressing.

– New transmission and transport options: unicast messaging and, for instance, direct mapping to IEEE802.3/Ethernet.

– Redundancy (e.g. Master Cluster and Alternate Master).

– More flexibility:

   - TLVs for standard message extensions: allow new features and options such as for management capabilities (Path Trace) or configuration (alternate timescale or master, unicast discovery, etc.);

   - new messages for control plane: Announce for hierarchy and master clock selection, Signaling (e.g. for unicast negotiation);

   - profiles (see section 2.5): with default profiles for conformance and interoperability;

   - alternate timescales (Arbitrary, in addition to PTP timescale).

– Security features (experimental solution only in IEEE 1588-2008).

From IEEE 1588-2002 implementation and PTP system deployment, it appears that a chain of BCs tends to decrease the performance of the synchronization distribution. Because each BC recovers the clock from the master clock then transmits its own clock down the chain, nonlinear noise was accumulated between GM and slaves. Moreover, because all BC run the BMCA, a BC tends to increase the resynchronization time during network reconfiguration.

TCs were introduced to eliminate these two effects by:

– not participating in hierarchy, that is not running BMCA;

– not terminating the clock from a master clock.

The ports of a TC have no state but assist any PTP communication paths established between any master and slave ports. A TC assists synchronization by correcting the delay variation of PTP event messages passing through the TC node.

A TC calculates the delay, namely the residence time (RT), between the ingress and the egress interfaces (TC ports) of the node. The RT is accumulated in a new CF.

As described in Figure 2.24, there are two types of TCs: the E2E TC and the P2P TC.

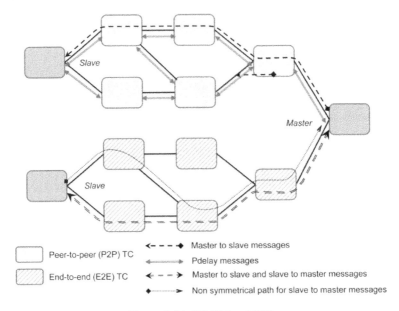

Figure 2.24. *TC E2E and P2P*

An E2E TC assists ports using a delay mechanism. It measures the RT of the event messages, that is Sync and Delay_Req. A slave clock or port will then be able to compute the correction. In the case of one-step clock, the TC will update the CF on the fly. In the case of two-step TC, for Sync messages, the RT will be accumulated into the CF of the Follow_Up message. For Delay_Req, the E2E two-step TC will have to maintain the RT value until the corresponding Delay_Resp message is received for adding the RT value. From telecom standpoint, two-step E2E TC raises two caveats:

– it requires the TC to be stateful, keeping in memory the CF value for every Delay_Req message passing through the TC;

– it mandates the master–slave and slave–master paths to be symmetrical (or congruent) in order to update the Delay_Resp.

The P2P TC does not assist the delay mechanism. Instead, P2P TC measures the propagation delay on its ports with a peer delay (pdelay) mechanism. It then adds the measured link or path delay to the transit time of the ingress Sync message. As shown in Figure 2.25, this PTPv2 pdelay mechanism is a TWTT mechanism and introduces a new set of messages as well as new addressing. The exchange of the Pdelay_Req and Pdelay_Resp event messages (and, if two-step clock, the Pdelay_Resp_Follow_Up) is established between a requestor port and a responder port.

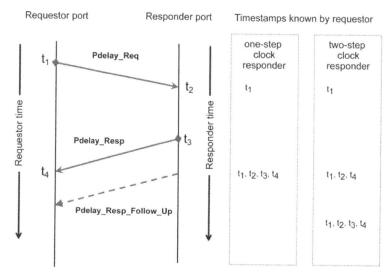

Figure 2.25. *Pdelay message flow*

Hence, the classical calculation of the mean link/path delay:

Round trip delay = (delay requestor to responder + delay responder to requestor) = $(t_2 - t_1) + (t_4 - t_3)$

Mean delay = $((t_2 - t_1) + (t_4 - t_3))/2$

The mean delays are appended to the CF value and thus accumulated with the RT values over the PTP path. OC and BC port, connected to P2P TC, whatever their state, have to support the pdelay mechanism, as shown in Figure 2.24.

Because the communication network or PTP hierarchy might change, and because pdelay is independent of the PTP path communication, every P2P TC port should perform the pdelay mechanism both as requestor and responder. This allows transmitting the link propagation delay correction whatever the direction of the Sync message. This would be necessary in ring or meshed topologies but might be useless in a linear unidirectional topology. This gives an advantage to P2P TC as, in the case of network reconfiguration, the path delay would already be known allowing fast PTP convergence.

As introduced earlier in this chapter, any asymmetry on the link introduces an error. With the new CF, any interface of a clock can be statically configured with a delayAsymmetry value that can be added to the CF. However this value (that can include asymmetric delay of a link but also of a time stamping mechanism such as with 10GE copper PHY) will be obtained by some means and is out of IEEE 1588-2008 scope.

To summarize the offset calculations that a slave clock can perform with PTPv2:

– offset (with delay mechanism, with or without E2E TC) = $t_2 - t_1$ – corrected mean path delay + CF (accumulated (RT + delayAsymmetry)) of Sync (or Follow_Up) message

– with corrected mean path delay = $[(t_4 - t_1) + (t_3 - t_2)$ – CFs in Sync (or Follow_Up) and Delay_Req messages]/2

– offset (with pdelay mechanism) = $t_2 - t_1$ – correction (accumulated (RT + mean delay + delayAsymmetry))

Before discussing the new addressing options, there are three points to know about the TCs:

– The larger the RT and/or lower the stability of the TC oscillator, the larger the time error generated by a TC is. To reduce the error in the correction, IEEE 1588-2008 suggests to either syntonize the TC to the GM (the TC might have to syntonize for each PTP domain) or to estimate its offset relative to the GM. With low pass through delay and/or stable oscillator, the TC can run in free-running mode.

– E2E and P2P TCs cannot be mixed in a PTP path because they utilize distinct mechanisms. This might lead to clouds of E2E or P2P TC in the communication network.

– For IEEE 1588, a "transparent" clock does not generate PTP message (except for management messages and for P2P TC, the pdelay messages). However, the TC updates the CF of PTP Sync and Delay (if E2E TC) messages. Hence, they will modify the communication packet or frame. There has been some discussion about a layer violation of TC. In its very first response, the IEEE 1588-2008 Interpretation Committee [1588-2008 IC] answered: "PTP does not attempt to change the behavior of the transport protocol". Some of the optional features (such as acceptable master table) might require that the source address of master port be preserved. The committee concludes by stating that writers of profiles be encouraged to highlight the expected behavior of TCs. Moreover, in November 2012, in response to an ITU-T Q13/15 liaison, the IEEE 802.1 working group stated that "Rigorous adherence to layering principles is fundamental to the continued growth of networking" confirming that a (layer 2) switch, running a TC function, will transmit frame with the Media Access Control (MAC) source address of the transmitting port. The IEEE 802.1 WG further suggests the utilization of higher layer entities in case the PTP master port source address has to be made available to PTP slave ports.

IEEE 1588-2008 supports new transport methods in addition to IP. This includes direct transport over Ethernet. Also, in order to prevent the pdelay messages being forwarded a single link further, new addresses were also required both for IPv4 and Ethernet. For Ethernet mapping, the reserved address for pdelay also allows us to transmit pdelay messages over ports blocked by (Rapid/Multiple) Spanning Tree Protocols. Table 2.3 summarizes the destination multicast addresses assigned IANA.

Ethernet and IP PTPv2 addressing (destination address)		IANA assignment	Comments
PTP primary for all except pdelay messages	MAC (Ethernet)	01-1B-19-00-00-00	From OUI 00-1B-19 assigned to IEEE I&M Society TC9.
	IPv4	224.0.1.129	Corresponds to PTPv1 default domain number.
	IPv6	FF0X:0:0:0:0:0:0:181	Value of X defines in section 2.7 of [RFC4291].
PTP pdelay for pdelay messages Note: might be used for all PTP messages in the scope of the address	MAC (Ethernet)	01-80-C2-00-00-0E	Allows transmission over Ethernet port blocked by any type of Spanning Tree Protocol.
	IPv4	224.0.0.107	TTL must be set to 1 and cannot be routed.
	IPv6	FF02:0:0:0:0:0:0:6B	HL must be set to 1 and cannot be routed.

Table 2.3. *Destination addresses assigned by IANA for PTPv2*

For IP transport, the UDP destination port numbers assigned for IEEE 155-2002 still apply to event and general messages, now including to the pdelay messages.

IEEE 1588-2008 introduces PTP domain field in the PTPv2 message header to allow the definition of up to 255 domains. Default domain uses number 0 and domain numbers 1–3 are reserved for backward compatibility to PTPv1. Users (or standard organization profile) can assign values from 4 to 127 when 127–255 are reserved. This increases flexibility in assigning and managing domains.

As an example, it is possible for management purposes to define different PTP domains from the same dedicated GM device, fed with a unique time source (i.e. with a unique timescale). The device would then have distinct PTP master ports and would behave logically as multiple GMs.

Another specific example comes from the adoption by ITU-T of IEEE 1588-2008 for packet-based transfer of frequency. In ITU-T Recommendation G.8265.1 [G.8265.1], detailed in section 2.5, the PTP domain is established uniquely between two clocks, a GM and a slave-only clock, by IP unicast communication. Any pair of source and destination IP addresses creates one unique PTP communication path and thus one PTP domain. Because communication paths are distinct and distinguishable, we can use the same PTP domain number but, as written before, different domain numbers could be assigned for management.

Besides the messages for the pdelay mechanism, IEEE 1588-2008 introduces two new general messages, Announce and Signaling, as well as TLVs to greatly improve flexibility and, optionally, the richness of the PTP system. The PTPv1 Sync message has been divided into a shorter Sync event message (74 octets vs. 164 including UDP/IP and Ethernet headers) and the new Announce message. Breaking up PTPv1 Announce into PTPv2 Announce and Sync messages allows:

– configuring distinct message interval values for Sync event message and Announce general message,

– increasing the Sync event message rate with control of its network bandwidth

– and having all the event messages with the same size, reducing the transmission delay asymmetry through non-PTP nodes.

The Announce message is now dedicated to establishing the PTP hierarchy. Obviously, new data sets and data set attributes such as for TC, mean path delay mechanisms, one-step clock support or timescale. As all PTPv2 messages, the Management message supports TLV that are now used to read and/or set data set attributes. What is not performed by Announce or Management messages can be performed by Signaling message. At the difference of Announce message that has a defined interval value, Signaling can be used at any time. As a noteworthy example, Signaling message is utilized for unicast negotiation.

Multiple TLVs are specified, enhancing the mandatory transmission in PTP header and each PTP message payload. Experimental, vendor and standard organization specific TLVs can be defined. Any PTP message can append any TLV. If not imposed by standardization, a receiver can ignore unrecognized TLV.

PTP message name	Message type (event or general)	IEEE 1588-2002 (PTPv1)	IEEE 1588-2008 (PTPv2)
Correction field (header)	NC (all messages)	No	Yes
Announce	General	No	Yes (control plane)
Sync	Event	Yes (control plane and t_1 time stamp)	Yes (t1 time stamp)
Follow_Up	General	Yes	Yes
Delay_Req	Event	Yes	Yes
Delay_Resp	General	Yes	Yes
Signaling	General	No	Yes
Pdelay_Req	Event	No	Yes
Pdelay_Resp	Event	No	Yes
Pdelay_Resp_Follow_Up	General	No	Yes
Management	General	Yes	Yes

Table 2.4. *PTPv2 messages and comparison with PTPv1*

Due to the interest of several industries, IEEE 1588-2008 introduces multiple new capabilities in the system. New clocks, PTP messages, data sets and data set attributes, various TLVs and options give a lot of flexibility. One obvious original objective was to maintain backward compatibility but the larger scope of applications forced the working group to reconsider the standard architecture at the end of 2006. Not to overweight the standard but to give each industry a way to optimize and tune PTP and PTP systems according to their needs and network environment, enabling interoperability and compliancy, the concept of "PTP profiles" was born.

2.5. The concept of "profiles"

IEEE Std 1588™-2008 [1588-2008] (referred to simply as IEEE 1588 in this section) introduced the concept of profiles. Profiles were introduced due to the wide use of IEEE 1588 by different industries with different requirements.

Profile as defined in IEEE 1588 is:

> "Profile: The set of allowed Precision Time Protocol (PTP) features applicable to a device."

The main objective of a profile is to allow applications to select IEEE 1588 attribute values and optional features to meet their specific requirements.

As stated in IEEE 1588 standard document, profile should define:

– the use of the default or an alternate BMCA;

– management options;

– path delay mechanism to be used (either delay request–response or peer delay);

– range and default values for attributes;

– transport mechanisms;

– PTP options.

Several organizations have been working on profiles for IEEE 1588. IEEE 802.1 defined a profile [802.1AS] to be used for bridged LANs (e.g. home network). IEEE C37 defined a profile [C37.238] to be used in power applications. ITU-T is working on profiles for telecom.

ITU-T developed the first profile, ITU-T Recommendation G.8265.1, for frequency synchronization. This profile was developed for use in networks without any PTP support (no BCs and no TCs). This profile was developed to support mobile backhaul applications that only need to support frequency synchronization.

By the time this book's publication, ITU-T was developing a second profile, ITU-T Recommendation G.8275.1, for time and phase synchronization. This profile was developed for use in networks with PTP support, BCs or TCs being supported in every single node over the planned PTP path. The first version of the profile supports BCs only, and TCs may be introduced in later revision of the profile (or in another profile). This primary time profile aimed supporting mobile backhaul applications that need time and phase synchronization.

In September 2012, several North American operators brought a proposal for ITU-T to work on a third profile that has partial support from the network. It was agreed to work on ITU-T G.8275.2 for time and phase synchronization with partial support from the network. A hypothetical reference model will need to be created and technical studies will have to be conducted to understand the performance limitations. The architecture will need to be defined to guarantee performance, as asymmetry in the network will translate into phase error.

2.5.1. *Frequency profile*

G.8265.1 defines the PTP parameters to be used to guarantee interoperability and it does not contain performance requirements. Performance requirements are defined in other ITU-T Recommendations.

Because the frequency profile does not have any PTP support from the network, then performance is dependent on the network, as the PDV will impact the performance of the slave clock. G.8265.1 is based on unicast mode; multicast mode is for further study.

IEEE Standard 1588™-2008 defines domains as "A logical grouping of clocks that synchronize to each other using the PTP protocol, but that are not necessarily synchronized to clocks in another domain".

G.8265.1 establishes a PTP domain by using unicast messages. In this profile, there is only a single master clock per PTP domain and the master clocks from different domains do not communicate with each other. The domain number from 4 to 23 is used.

The following types of messages are used in the frequency profile: Sync, Delay_Req, Announce, Follow_Up, Delay_Resp and Signaling.

As described in previous section, PTP defines several types of clocks and G.8265.1 only uses OCs (master and slave clocks).

PTP is a protocol designed to allow time synchronization to be achieved. For the frequency profile, only frequency synchronization is needed and therefore there is only a need for one-way operation. There are some advantages of using a two-way operation, and some current PTP implementation does make use of that. G.8265.1 allows for both one-way and two-way operation.

Regarding the use of follow-up messages for two-step clocks, G.8265.1 allows the use of both one-step and two-step clocks.

G.8265.1 uses the PTP mapping defined in IEEE 1588™-2008 annex D, transport of PTP over UDP over IPv4. An amendment was approved in September 2012 to include IEEE 1588™-2008 annex E, transport of PTP over UDP over IPv6 as an option for the frequency profile. A well-controlled packet network to minimize PDV is highly recommended for the implementation of this profile.

The use of unicast message negotiation is mandatory on the frequency profile. Slave clocks will make a request for unicast service from the master by sending a PTP signaling message in unicast.

Another important consideration on the frequency profile is the BMCA. Because of telecom requirements based on ITU-T G.781 for clock selection, the default BMCA from IEEE 1588 could not be used as it would select a grand master automatically. As mentioned in previous chapters, operators plan their synchronization networks and therefore an automatic selection of the grand master would be of concern.

To include key features for telecom such as timing protection and clock traceability, an alternate BMCA was developed. The alternate BMCA allows operators to manage and control their synchronization network as needed. Masters are always active masters and slaves are slave-only clocks.

As IEEE 1588 only allows one master to be active in one domain, the unicast message is used to separate PTP domains to allow each master to be isolated in its own domain. Also, the PTP masters do not exchange Announce messages.

Figure 2.26 shows Telecom Grand Masters (N x T-GM) connected to several Telecom Slave clocks (M x T-SC) in a packet network. Each one of these masters is on its own IEEE 1588 domain. Each one of the slaves may have access to several masters, and therefore they need to be able to monitor Telecom Masters in different IEEE 1588 domains.

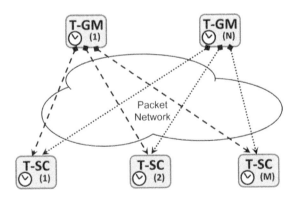

Figure 2.26. *Several masters connected to several slaves*

G.8265.1 defines Telecom Slave as shown in Figure 2.27, an adaptation of Figure 3/G.8265.1. The Telecom slaves may include several PTP slave-only clocks. The Telecom slaves participate in multiple PTP domains, but the slave-only clock that is part of the Telecom Slave only participates in one PTP domain. For the OCs, the alternate BMCA is static and provides a recommended state = BMC_MASTER and a state decision code = M1 for the master clock, and state = BMC_SLAVE and a state decision code = S1 for the slave-only clock. For example, if Telecom Slave 1 is locked to T-GM 1, and T-GM 1 fails, then the Telecom Slave 1 will switch to either T-GM 2 or T-GM 3 depending on their QL and local priority as defined in the master selection process.

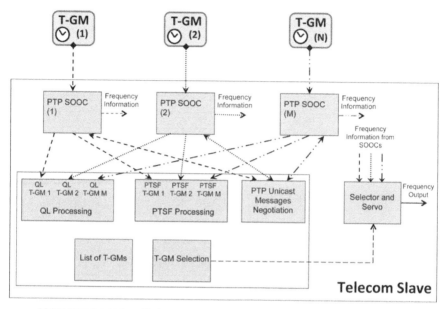

SOOC : Slave-Only Ordinary Clock

Figure 2.27. *Telecom Slave model*

The Telecom Slave model consists of several PTP slave-only instances, QL and Packet Timing Signal Fail (PTSF) processing blocks, PTP unicast messages enable, a list of T-GM with a local priority value associated with each T-GM, T-GM selection and a selector block to select the primary T-GM. It is important to note that the Telecom Slave model does not imply a specific implementation.

G.8265.1 also defines the master selection process, even though it is outside the scope of PTP profile as defined in IEEE Std 1588™-2008. However, for telecom applications, it is very important to define the process for Master selection. The Master selection is based on QL values as defined in [G.781]. The G.781 QL values have been mapped onto the clockClass attribute of IEEE 1588 as shown in Table 2.5, which is based on Table 1 of G.8265.1. It uses the PTP clock class values reserved for PTP profiles.

PTP clockClass	SSM QL	G.781		
		Option I	Option II	Option III
80	0001		QL-PRS	
82	0000		QL-STU	QL-UNK
84	0010	QL-PRC		
86	0111		QL-ST2	
88	0011			
90	0100	QL-SSU-A	QL-TNC	
92	0101			
94	0110			
96	1000	QL-SSU-B		
98	1001			
100	1101		QL-ST3E	
102	1010		QL-ST3/QL-EEC2	
104	1011	QL-SEC/QL-EEC1		QL-SEC
106	1100		QL-SMC	
108	1110		QL-PROV	
110	1111	QL-DNU	QL-DUS	

Table 2.5. *Mapping of quality levels to PTP clockClass values*

The Master selection is based on:

– QL;

– PTSF;

  - PTSF-lossSync;

  - PTSF-lossAnnounce;

  - PTSF-unusable;

– local priority;

The master clock with the highest QL is selected if does not have a Packet Timing Signal Fail (PTSF). If multiple master clocks have the same QL and they are not experiencing a PTSF-lossSync or PTSF-lossAnnounce or PTSF-unusable, then the master with the highest local priority is selected. If all the masters have the same highest local priority and QL, then an arbitrary master is selected.

2.5.2. Phase and time profile (ITU-T G.8275.1)

As mentioned in section 2.5, ITU-T is still working on the phase and time profiles. The application that this profile is targeting is LTE time division duplex (TDD). The phase accuracy requirement is given in 3GPP TS36.133 (2010) for LTE TDD: this is 3 µs for small cells (< 3 km radius) that is tighter than for large (> 3 km radius) cells (< 10 µs). Taking into consideration that this specification is on the radio interface between adjacent base stations, and if we consider the tighter specification of 3 µs, then the phase requirement is 1.5 µs with respect to an ideal reference.

With such stringent phase requirements, ITU-T Recommendation G.8275.1 [G.8275.1] will be based on a network that supports BC in every single node. The network with BC is still being studied in order to decide the number of BCs that can be cascaded and still meet the 1.5 µs of phase synchronization. ITU-T G.803 network architecture is being used as the starting point. Simulations are being done with 20 BCs in tandem, with a total of 60 clocks in the network. As the requirement of 1.5 µs is very tight, some proposals for a smaller network have also been made. At the time of this book's publication, 10 BCs in tandem have been provisionally agreed.

Another important point is that physical layer synchronization (e.g. SyncE) can be used in conjunction with BC. SyncE offers a reliable holdover in case of lost of IEEE 1588 packets.

G.8275.1 is based on multicast mode. The use of the default multicast address 00-1B-19-00-00-00 and the non-forwardable multicast address 01-80-C2-00-00-0E for all PTP messages is still under discussion. There have been proposals to use the IPv4 and IPv6 mapping, but these are still for further study.

The following types of messages are used in the time and phase profile: Sync, Delay_Req, Announce, Follow_Up and Delay_Resp.

One important difference between the frequency profile and the phase profile is the message rate. For the frequency profile, a minimum and a maximum message rate were defined. For the time and phase profile, the message rate is fixed to 16 packets per second for Sync, Delay_Req and Delay_Resp messages. This message rate was agreed for the case where physical layer synchronization (e.g. SyncE support) is used.

It has been agreed that G.8275.1 only uses OCs (master and slave clocks) and BCs. ITU-T G.8273.2 will contain the requirements for the telecom BC to be used in this profile. The use of both one-step and two-step clocks are allowed in this profile.

There are a series of recommendations, ITU-T G.827x, that are being developed at ITU-T as a result of the work being done to transport time and phase synchronization in telecommunication networks. For more details on these recommendations, see Appendix 1 of this book.

2.6. Other packet-based protocols

As discussed in previous chapters, the importance of packet-based timing distribution was introduced by:

– requirement for certain applications to obtain frequency or time synchronization;

– evolution from synchronous to asynchronous transmission that was not allowing physical frequency transmission to be an alternative without equipment upgrade;

– transparent timing transfer over third party (e.g. wholesale operator) networks;

– distinction between time synchronization need that requires a protocol with time stamp transmission and frequency that can use alternate physical solutions.

However, asynchronous packet networks have been principally designed for bandwidth efficiency with properties, particularly statistical multiplexing, that are not tailored to timing delivery. Solutions to support the synchronization needs of certain applications had to evolve to provide the expected services that include:

– availability to sustain planned network evolution;

– performance to support the application;

– confidence and dependencies in fulfilling the service demand;

– adaptability to network and application constraints (e.g. scalability, dissemination, redundancy and resiliency);

– operational manageability (e.g. deployment easiness and interoperability).

This section provides additional details on the packet protocols used in the context of packet timing in addition to IEEE 1588 PTPv2 as well as some historical perspective. It then compares those methods considering above points.

2.6.1. *Packet-based timing: starting with CES*

As discussed in Chapter 1, the first application for timing delivery in telecom packet networks was circuit emulation particularly for mobile wireless second-generation (GSM) base stations. Chapter 1 states that CES has a timing requirement and section 2.3 presents the methods (synchronous, differential and adaptive), inherited from ATM CES, available to support this timing requirement.

From the packet-based perspective, only differential and adaptive methods are considered because the synchronous method is fully independent of packet flow.

As shown in Table 2.6, different standard organizations have specified CES, also known as TDM pseudo wire (PW) in Internet Engineering Task Force (IETF). It will be noted that those standards define the emulation services, its encapsulation and the possible methods to achieve timing of the circuit.

Standard development organization	Specifications	Comments
IETF	RFC 4553 [RFC 4553]	SAToP, CESoPSN and TDMoIP support IP and MPLS encapsulations, adaptive and differential clocking.
	RFC 5086 [RFC 5086]	CESoPSN and TDMoIP define structured-aware service; SAToP defines structure-agnostic service.
	RFC 5087 [RFC 5087]	CEP is limited to MPLS encapsulation and differential clocking. It defines emulation of SONET/SDH circuits and services.
	RFC 4842 [RFC 4842] (obsoletes RFC 5143 [RFC 5143])	SAToP and CEP are in standard tracks when CESoPSN and TDMoIP are both informational (no consensus for unique solution supporting structured service).
ITU-T	Y.1413 [Y.1413]	ITU-T specifications distinguish MPLS and IP transport methods in distinct recommendations.
	Y.1453 [Y.1453]	The MPLS format does not support the differential clock method.
Metro Ethernet Forum (MEF)	[MEF3] CES definitions and framework	MEF defines a CES encapsulation directly over Ethernet (i.e. no IP or MPLS encapsulation).
	[MEF8] CES implementation	MEF specifications support structure-aware and structure-agnostic services, adaptive and differential clocking.

Table 2.6. *CES specifications*

As mentioned in section 2.3, adaptive and differential clocking methods aim to support the clocking of the circuit they emulate. This is referred as "service clock" in ITU-T Recommendation [G.8261] contrasting with the "network clock" usually provided by the carrier operator through the physical layer method (e.g. synchronous mode). In the case of CES, the timing transmission aims to follow the master–slave schema, that is one IWF node (tail end of the emulated circuit or PW) provides its service clock as a source to a peer IWF node (head end of the emulated circuit). The way the timing information is conveyed and recovered is distinct.

The differential clock method is about transmitting the difference between the service clock and the carrier reference clock. At the client interface of the source IWF, a counter disciplined by the reference carrier (network) clock reads the CDR value of service (client) clock frequency. This frequency is driven by the client (service) traffic. The value of the counter is attached to each CES packet as a time stamp. Standard Real-Time Protocol (RTP) header [RFC 3550] is used for that matter. Figure 2.28, which is an adaptation of Figure 2, Figure 1 and Figures 3 and 4 of, respectively, [RFC 4553], [RFC 5086] and [RFC 5087], shows a simplified example of TDM PW packet formats. Note that the location of the control word or field differs between the referenced IETF RFCs.

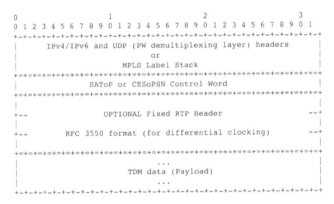

Figure 2.28. *Example of IETF PDH PWS encapsulation*

There should be one counter per client (service) interface. Because the time stamp value represents a delta from a clock reference, the remote IWF will have the same or similar reference available (see section 2.3). On the remote end of the emulated circuit, the client interface clock is disciplined by the servo recovering the clock, a DPLL adjusted to client frequency from the common clock.

The adaptive clock method also provides a means to transmit client (service) frequency. However, in CES adaptive mode there is no implicit information being transmitted. It will be considered that there is a direct relationship between the generation of the CES packet and the frequency of the client interface. Indeed, whatever the structure-agnostic or structure-aware service, the bits are filling the CES packet payload at the rate they arrived. As soon as the configured payload is full, the CES packet will be generated. The packet rate depends on the frequency of the circuit and on the payload size. In simple terms, for most common implementations (i.e. SAToP and CESoPSN), the flow rate would be:

– Unstructured E1 (SAToP): 2,048,000/8 bits/packet_size

– Unstructured T1 (SAToP): 1,544,000/8 bits/packet_size

– Structured E1/T1 (CESoPSN): N × 64,000/8 bits/packet_size

The packet_size value is fixed for any circuit and depends on the configured payload size: number of bytes for SAToP (default payload sizes: 256 bytes for E1, 192 bytes for T1 and1024 bytes for E3 or T3) and number of DS0 (N in above formula) for CESoPSN.

As an example, 256 bytes SAToP payload generates a stream of 1,000 packet/second.

When the remote (head end) IWF receives one or more CES flows, it will attempt to recover the original (service) frequency of each CES flow. A first method derived from ATM AAL1 CES relies on the de-jitter buffer necessary to absorb the jitter from the network (and to deliver TDM bit stream). In a nutshell, the method consists of maintaining the configured de-jitter buffer size up around a threshold, usually half the size. The output frequency will be adapted to the de-jitter buffer drain speed: if the buffer fills up and exceeds the threshold (overrun), the output frequency will be increased; it will be slowed down if emptying below the threshold (underrun).

A second method was required because the first one did not provide good enough results. This method consists of looking at the inter-packet rate that looks at the arrival of each packet of the emulated circuit. The frequency of origin could theoretically be derived from the flow rate that should be constant in perfect conditions and related to the client clock. Each packet arrival corresponds to a significant instant as described in [G.8260]. This instant should be marked (time stamped) precisely in order for the clock recovery algorithm to average (filter) the output frequency.

However, the CES packet transmission will be impaired by packet network jitter that generates PDV to the packet-based timing signal. The delay fluctuation can also append at the CES tail end. Indeed, the CES flow might leave the source IWF with some PDV caused by circuit encapsulation processing, by the forwarding engine within the equipment toward the network egress interface and by queuing and transmission delays at the network interface. Despite the fact those packets should be in the priority queue, if multiple client interfaces have the same (or very close) frequency, packets from the multiple emulated circuits can arrive at the same time at a common egress interface. This generates a battle or head of line blocking (HOL) at the network interface and thus can also create PDV.

Each head end of a CES will perform adaptive or differential clock recovery for its circuit. The objective is thus to support the CES service that, in ITU-T terms, target G.823 or G.824 traffic mask (see section 1.3.1). From a deployment perspective and particularly in mobile backhaul, few other points will be underlined. First, it was assumed that the timing service used for CES would not only fulfill the CES application but also support the base station radio frequency requirement. Second, the evolution to packet-based backhaul was driven by bandwidth requirement to support third-generation (WCDMA) deployment. In the early days of WCDMA, 3G base stations (NodeB) were using ATM transport, leveraging SDH/SONET transmission. In addition to TDM PW services (PWS) (CES), there was a need to support ATM PWS. But ATM PWS, unlike TDM PWS, is not capable of providing differential or ACR methods. Finally, cell sites aim to support both 2G and 3G base stations and often multiple E1/T1 and E1/T1 ATM (with IMA) were planned.

Those trends led to consideration of alternative packet-based timing solutions that would:

– be more flexible, providing timing to an entire cell site, for instance to a cell site gateway, and not to individual interfaces;

– permit better performance guarantee;

– offer better redundancy option.

Packet-based timing solutions independent of the PWS were thus considered.

The following sections discuss and compare three options: dedicated CES with no data, NTP and PTP.

2.6.2. *Dedicated timing TDM PW*

A first and simple approach for gaining flexibility was to dedicate a PW for frequency synchronizing a single piece of equipment. It suffices to utilize a dummy CES that is a TDM PW not attached to a TDM device, not transmitting data, but to generate a DS0 CES, for instance, with a frequency source common to the other TDM or ATM attached circuits, as in the case of a mobile base station controller

(BSC) and a radio network controller (RNC), or to a PRC traceable signal. The end equipment would have to recover the frequency from this circuit and distribute the signal to other interfaces of the end IWF.

If this solution allows leveraging IWF and limits the number of packet-based timing signals to monitor, it also presents some caveats. First, it is another PW and thus generates another CES flow for each end node. Second, it could, but usually does not, improve the flexibility of the location of the packet master (role performed by CES IWF) because there is no incentive to deploy CES-based master. Third, CES-based ACR, when they were initially deployed in their first generation, tended to target traffic interface performance and not synchronization interface performance. Hence, clock recovery for CES implementation often provided lower performance than other adaptive dedicated methods such as PTP or NTP. As such, dedicated CES does not always improve the performance guarantee except by, theoretically^2, allowing dedication of a CES IWF as timing master and positioning it closer to the CES IWF recovering the clock, hence reducing the path length.

However, the solution could provide a redundancy option common to all the PW services at the end node. Nonetheless, if its frequency transmission borrows a CES protocol, the solution is not a standard timing protocol. Moreover, utilization of CES traffic allows only frequency synchronization using only one-way traffic when TWTT protocols, such as PTP and NTP, allow, for frequency recovery, the slave node to consider downstream and upstream directions and allow time synchronization.

2.6.3. *NTP*

This section does not intend to describe in detail NTP as this is addressed by many documents and Internet sites and pages, starting with [MIL] or [NTP Org]. This section underlines some key elements of NTP and highlights some points of comparison with IEEE 1588 PTPv2.

NTP already has a long history (see the FAQ of [NTP Org] or [MIL 00] for further details) at IETF since the appearance of NTP [RFC958] in 1985. Since then, NTP has been improved multiple times and is largely deployed on the Internet, within operators and enterprises for many applications.

After NTPv1 [RFC1059] in 1988, NTPv2 [RFC1119] in 1989 and NTPv3 [RFC1305] in 1992, the latest NTPv4 [RFC5905] in 2010 made the previous

---

2 From author's knowledge, such model has not been used but remains technically valid.

versions obsolete but maintains backward compatibility. Beyond improvement in algorithms and protocol operation to provide better redundancy, security, stability and precision, other RFCs accompany the evolution of NTP:

- [RFC 1128] Measured performance of the NTP in the Internet System (1989)
- [RFC 1129] Internet time synchronization: the NTP (1989)
- [RFC 1165] NTP over the OSI Remote Operations Service (1990)
- [RFC 1589] A kernel model for precision timekeeping (1994)
- [RFC 1708] NTP PICS PROFORMA for the NTP version 3 (1994)
- [RFC 2783] Pulse-per-second API for UNIX-like Operating Systems, version 1.0 (1994)
- [RFC5906] NTPv4 autokey specification (2010)
- [RFC5907] NTPv4 management information base (2010)
- [RFC5908] NTP server option for DHCPv6 (2010)

Note also that RFC5905 (NTPv4) obsoletes previous simple NTP (SNTP) RFCs, that is [RFC4330] SNTPv4 (2006) that obsoleted [RFC1769] SNTP v3 (1995) and corrects [RFC2030] SNTPv4 (1996). SNTP is a "light" version of NTP as discussed later.

NTP, whose primary goal was to deliver time to many computers over the Internet and not, as PTP, to precisely synchronize devices in a limited scope, is based on long experience and has been proven to work in various use cases.

NTP goes beyond a protocol by defining modes of operation that set the relationship between NTP servers and clients and by defining algorithms, including clock recovery, an important difference to IEEE 1588.

NTP on-wire protocol utilizes the typical TWTT packet exchange (as shown by Figure 2.14), allowing us to obtain the four time stamps necessary to calculate the offset and the delay from the server. Without discussing the packet format, there is a unique format of the NTP packet used for all the operations of NTP. Supporting TLV extensions, NTP packet size is not fixed as the PTP event messages are.

Hence, the NTP packet includes all the information necessary to set the modes of operation, to exchange time stamps and to maintain a reliable synchronization structure. Beyond the protocol, NTP servers and clients maintain data structures (equivalent to PTP data sets).

Although NTPv4 does not define time stamp points, hardware time stamping can be used. If there is no description in the standard, NTP does not prevent an implementation to update the NTP time stamp fields "on-the-fly" for NTP packet sent. Many NTP servers perform this way and slave might also do. However, for equipment not able to do the same but able to time stamp accurately, an interleaved mechanism has been proposed. Too late to be discussed and added to [RFC5905], but described in [MIL], this mechanism works similarly to an IEEE 1588 two-step clock, allowing the node to capture time stamps in hardware and improving the accuracy of NTP.

NTP, which runs exclusively over UDP with IPv4 and, introduced in version 4, IPv6, can utilize unicast, multicast or broadcast (IPv4 only) transmission modes allowing, for instance, server announcement or server discovery by client.

Contrasting to default IEEE 1588 BMCA that establishes an automatic hierarchy, NTP allows various types of associations. These modes permit building redundant systems and, with NTP algorithms, improving resiliency and accuracy of the recovery. Indeed, servers can establish hierarchical but also same level relationships as shown in Figure 2.29.

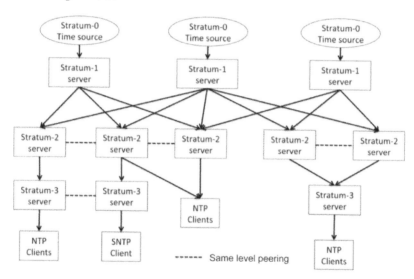

Figure 2.29. *NTP stratum hierarchy and same level peering*

Note that NTP stratum-1 server can be considered as GM when lower NTP servers can be considered as BCs, except the active associations between NTP servers of same level. NTP Stratum-0 would correspond to an ITU-T [G.8272] primary reference time clock (PRTC).

NTP describes algorithms to achieve distinct purposes. As pointed out in the introduction, NTP is particularly a set of algorithms aiming to improve robustness and accuracy. In particular, an NTP server or client can communicate with multiple servers and combine them. If IEEE 1588 does not exclude such ensembling, this is not a model which is considered. In contrast, IEEE 1588 properties assume a clock to be synchronized to one unique master clock at a time.

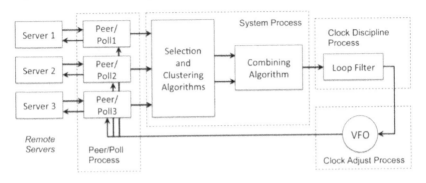

Figure 2.30. *NTP algorithms*

Full compliancy to NTPv4 implies supporting these algorithms, depicted in Figure 2.30 [NTP Org] (source: [MIL]), which aim to:

– maintain multiple peers (polling algorithm thru delay requests) with a Huff Puff algorithm correcting the asymmetrical delays for each peer;

– filter the values to further reduce some errors;

– select best *truechimers* and rejecting *falsetickers*, with detection of abnormal behavior such as byzantine type of failure (clock selection and clustering algorithms);

– combine (weighted average value);

– finally, discipline the local clock with loop filters to update the system clock.

SNTP uses the same networking options as NTP but does not feature all the algorithms of full NTP implementation. As such, SNTP clients do not aim for the same accuracy and reliability (single upstream server) but for lighter implementation, thus for simpler applications. SNTP can only apply to clients connected to SNTP primary server or to the leaves of NTP tree.

It is worth noting then that some implementations can also use NTP protocol aspects but with different algorithms and parameters. They however cannot claim full compliancy to NTPv4 as outlined in section 2 of [NTP v4].

In this respect, while [NTP v4] defines both a protocol and an algorithm to distribute time synchronization, the NTP on-wire protocol has sometimes been used to distribute a frequency reference. In this case, a specific algorithm to recover frequency was developed similar to the case of PTP. For these specific implementations, it would be more correct to refer to SNTP rather than NTP though. In fact, according to [NTP v4], an SNTP client is not required to implement the NTP algorithms specified in [NTP v4].

Because NTP typical implementations primarily targeted lower accuracy (above the millisecond range) and deemed as a complex implementation for a simpler environment, such as controlled LANs, with non-computers devices, some industries developed IEEE 1588-2002.

In summary, NTP is similar to PTP in many points. However, due to its original objective and evolutionary process, NTP was not able to propose at the required time the necessary options for high accuracy (e.g. time synchronization in the sub-microsecond range). Evolution of NTP toward tighter synchronization is possible but not yet endorsed by standard bodies. Suggestions have been proposed at IETF TICTOC [O'DO 08], but compared to the enthusiasm around IEEE 1588 from multiple markets, NTP may lack resources for its evolution today. NTP has, nevertheless, been adopted largely over the last few years and its algorithms have been built on strong experience. Moreover, NTP offers security mechanisms that do not exist with PTP today.

2.6.4. *Summary and comparison*

Table 2.7 summarizes the key distinction between the main protocol conveying timing events.

Packet-based method	Typical packet rate	Inter-packet range (CBR-like traffic)	Time stamp	Support of ToD (TWTT)
CES (and dedicated CES)	Some 100's to few 1,000's pps (related to CES payload)	Adaptive method	Yes, with RTP header for differential method	No

Table 2.7. *Key distinction between the main protocol conveying timing events*

(S)NTP	1 packet every 64 second (NTPv4: 16–1,024 second recommended range)	No	Adaptive*	Yes (non-NTP-compliant algorithm can be specific to frequency recovery)
PTP	1–128 pps	No	Adaptive*	Yes (profile dependent)

*Definition from [G.8260]:

"*3.1.1 adaptive clock recovery*: Clock recovery technique that does not require the support of a network-wide synchronization signal to regenerate the timing. In this case, the timing recovery process is based on the (inter-)arrival time of the packets (e.g. time stamps or circuit emulation service (CES) packets). The information carried by the packets could be used to support this operation. Two-way or one-way protocols can be used."

Table 2.7. *(Continued) Key distinction between the main protocol conveying timing events*

The beginning of this section 2.6 introduces some criteria. Based on these criteria, the Table 2.8 proposes a comparison of the packet-based mechanisms covered in this chapter.

Packet-based method	Availability to sustain planned network evolution	Performance to support the application	Confidence and dependencies in fulfilling the service demand	Adaptability to network and application constraints (e.g. scalability, dissemination, redundancy and resiliency)	Operational manageability (e.g. deployment easiness and interoperability)
CES adaptive	"in-band" solution when CES is required	Might be good	Dependency on: network PDV, beating effect, recovery algorithm	Limited by CES deployment	Integrated to CES application

Table 2.8. *A comparison of the packet-based synchronization mechanisms*

CES differential	Requires common reference clock	Very good for service clock	Very good confidence	Limited by CES deployment	
				Require synchronous network support	
Dedicated CES (adaptive only)	"out of band" solution possible when CES is required	As good as CES adaptive but master location can be adapted	Dependency on: network PDV, beating effect, recovery algorithm	Higher than default CES adaptive	Low flexibility
				May require adding CES IWF with reference clock	
(S)NTP	May require specific NTP server	Can be adapted to application	Dependency on: network PDV, recovery algorithm, intermediate NTP servers (if used)	May require addition of NTP servers with reference clocks	
PTP	Require PTP master	Can be adapted to application	Dependency on: network PDV, recovery algorithm, intermediate clocks (if used)	Require addition of PTP masters	

Table 2.8. *(Continued) A comparison of the packet-based synchronization mechanisms*

In summary, there are various packet-based solutions for transmitting frequency, time (including phase and time of day) or both. A solution might fit only a partial requirement (e.g. CES) or might not fit well into the network Total Cost of Ownership (TCO) that includes capital expense, operational expense and opportunity (i.e. ability to use same solution for other purposes). It is recommended, before adopting a solution, to proactively performed an analysis of:

– the applications (current and future) and their requirements;

– the current network capability;

– the evolution of network;

– and the impact (present and future) of the implementation of such solution in the network.

2.7. GNSS and other radio clock sources

Various solutions exist to provide the time of day and synchronize many points simultaneously. They can be global, providing (nearly) worldwide service or can be geographically limited, providing regional services.

Global distribution systems using satellite constellations are called GNSS. Regional distribution systems can use satellites or terrestrial infrastructure with broadcast transmitters. Systems based on satellites (space-based system) generally provide Position, Navigation and Timing (PNT). Frequency accuracy depends on the receiver and local oscillator, but the best systems can be better than 10^{-13} after 1 day of averaging. As a source of UTC time, accuracy from the best receivers with local cesium clocks can approach nanoseconds offset from the UTC source.

For all of those systems, this chapter will focus on the timing distribution aspect.

2.7.1. *Global and regional space-based timing system*

The term GNSS refers to different systems all of which provide PNT globally using similar underlying technologies. If the North American Global Positioning System (GPS) is the most known system, it is not the only GNSS currently operational. The Russian global orbiting navigation satellite system (GLONASS) is also fully operational. The Chinese COMPASS (BeiDou-2) is partially operational (today available as a regional system) and the European Galileo will be operational in coming years.

Galileo, though not commercially operational today, has four fully functional satellites in orbit to complete its initial operational validation (IOV). Galileo will provide some new services such as public integrity information that is not currently available from GPS and GLONASS. However, those two GNSS have plans to rejuvenate their services with new satellites and control segments and GPS modernization plans include integrity information as a public service.

China has launched its navigation system named BeiDou. In its current state, with few satellites, BeiDou-1 provides an experimental regional service. More satellites will be launched to complete the compass navigation satellite system (CNSS) or BeiDou-2, currently providing a regional service, which would then become a global service. India has planned to launch its Indian regional navigational satellite system (IRNSS).

GPS and GLONASS can be considered as the first generation of GNSS. To those first-generation GNSS, we can add different augmentation systems either space-based augmentation system (SBAS) or ground-based augmentation system (GBAS) that can be global, regional or local. There are now many such systems. The reader can see a number of publicly available GPS augmentation systems the US government has fielded from [USGOV AS].

Those systems were originally developed to improve navigation with GPS before the US government decided, in 2000, to end the use of the selective availability (SA) feature that was purposely degrading the civilian signal. Despite the fact the US government assured the SA will not be reintroduced, many countries or organizations preferred considering alternative solutions (e.g. IRNSS, BeiDou and Galileo). Today the augmentation systems keep improving GPS. The Russian system of differential correction and monitoring (SDCM) will support GPS and GLONASS.

Augmentation systems use ground stations checking the GPS signals and geostationary satellites or terrestrial radio transmitters to broadcast information such as corrections to monitored satellites and signals.

Galileo, BeiDou-2 and the modernization of GPS and GLONASS would become the second generation of GNSS. We can include IRNSS and Quasi-Zenith Satellite System (QZSS). QZSS is a Japanese regional GPS augmentation system, covering from Japan to Australia and compatible with signals from the modernized GPS. QZSS also includes a time keeping system (TKS) and the ability to use a two-way satellite time and frequency transfer (TWSTFT) scheme.

All PNT spaced-based systems have similar architectures but have technical differences.

2.7.1.1. *Overview of space-based systems*

A space-based navigation and timing system is composed of three main parts (also named segments), as shown in Figure 2.31: the satellite constellation or space part; the control part with multiple ground stations used for tracking, monitoring and commanding the satellites and the user part, which encompasses all the receivers and applications using one or more GNSS and optional augmentation systems.

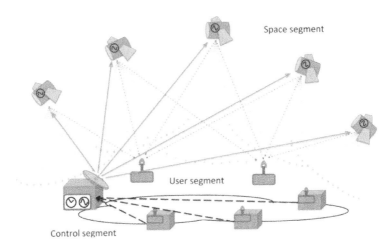

Figure 2.31. *Generic GNSS-based timing distribution*

The purpose of each GNSS is to provide PNT to users globally. This is achieved by broadcasting synchronized L-band timing signals from known locations. Users measure the time of arrival of at least four satellites and solve for their four unknowns: their position in space and the offset of their clock in time. For network synchronization, systems are stationary and require only time or frequency; antenna position is required only for setting up the system. Once antenna coordinates are known, time can be determined from any one GNSS satellite using the same coordinate system. Typically, however, time synchronization is relative to UTC. Each GNSS has its own timescale formed as a weighted average of clocks on satellites and in the control stations, which provides synchronization, but is not necessarily UTC. The satellites provide the offset of system time from a specific real-time UTC in the transmitted data. Note that the final UTC is a post-processed timescale. Any real-time UTC signal is a prediction of the final value, which is only defined after the fact. UTC from GNSS is relative to a particular national timescale. The most accurate of these differ from the final UTC by tens of nanoseconds.

Indeed, UTC is a processed value calculated from multiple clocks hosted by various laboratories (hence, the "C" of UTC) disseminated over the planet. The laboratories regularly exchange their own clock differences and UTC values to the international standards laboratory, the *Bureau International des Poids et Mesures* (BIPM), which generates the UTC. UTC is defined no sooner than about 1 month after the last measurement. No real-time signal can be an exact realization of UTC. However, clock laboratories (such as national labs) generate a real-time prediction, which is what must be used for any accurate requirement of time synchronization. Each GNSS refers to a specific laboratory. The GPS time refers to, and is steered by, the US Naval Observatory (USNO). Note there are other US clock laboratories (e.g. National Institute of Standards and Technology (NIST)). Hence, there is no unique steerable UTC for all the GNSS.

Moreover, UTC currently maintains a relationship with the Earth's rotation about its axis. This process requires adding or removing a second (the leap second) as necessary, though there is currently a discussion to discontinue this[3]. Except GLONASS that inserts leap seconds, the GNSS do not correct their time to match the rotation of the Earth. Instead they maintain a continuous timescale for synchronization in the specific GNSS with a fixed offset with Temps Atomic International (TAI; Atomic International Time in English) but transmit the offsets from UTC, including leap second information.

However, to provide traceability to UTC (maintaining a known difference between UTC and the GNSS time), a master control station will be connected (or participate) to one of the laboratories forming the UTC. Nevertheless, to the extent that the reference UTC timescale for a particular GNSS is steered by accurate cesium atomic clocks, a GNSS provides UTC (or TAI) frequency traceability. Note that the accuracy of cesium atomic clocks varies widely.

Satellites use a set of cesium, rubidium, and sometimes hydrogen maser, atomic clocks. The number and type of atomic clocks in each satellite depends on the system and generation of the satellite. Clocks on satellites generally run individually, one active and another replacing it when it is taken out of service. The onboard atomic clocks are used only for stability. GNSS require only that satellite clocks be predictable. GNSS control segments upload clock models that are used to give users the predicted offset of the actual clock signal. Users receive from satellites: the real-time clock signal, the offset of that clock from the GNSS system time, the offset of the particular GNSS system time from a national UTC and the position of the satellite at the time of transmission. The measurement (called the pseudo-range) is the time of arrival of the transmitted clock signal on the user's clock. This measurement is adjusted for the propagation time using the satellite's position at the moment of transmission and the user's position at the time of reception. Adjusting this value for the transmitted satellite clock offset allows the user to synchronize his/her clock to the GNSS system time. This in turn can be adjusted using the transmitted offset of that GNSS time from the relevant UTC.

The control segment is composed of multiple stations with various roles. Tracking and monitoring stations are dispersed over the coverage area, that is it will

---

[3] There is a discussion ongoing at ITU-R about maintaining UTC related to Earth's rotation.

be worldwide for GNSS. Their primary role is to transfer pseudo-range measurements of all visible satellites continuously to the master control station.

The master control station processes the measurements from ground stations around the world and produces predictions of the satellites' clocks and of their positions (usually called the ephemeris of a satellite). The master station then prepares uploads to each satellite with some regularity. For example, GPS currently uploads each satellite once-per-day, though contingency uploads are done for satellites that have clock or ephemeris behavior that is less predictable. There are generally several upload stations positioned around the globe to enable communicating faster with the satellites. Some systems are enabled with cross-link communications that can allow data transfer among satellites.

If the master control station determines that a satellite has an error condition, then there are various options. A contingency upload can be created, if the problem is simply that the current prediction being transmitted is no longer valid. Each satellite also gives an estimate of the expected accuracy of its signal. This value can be adjusted from the ground. Newer satellites also have an option for an integrity flag. A flag can be set that tells users that the integrity of the signal is unknown or compromised. In extreme cases, the satellite signal can be altered into a non-standard code that users cannot receive, but still allows the monitor stations to track the signal.

Detecting and correcting an error condition from the ground entails some delay. Inevitably, a failure in a satellite will cause transmission of an error signal to users until the control system can correct it. Some error conditions are monitored on the satellite. However, a system that requires an atomic clock can only be fully monitored with another atomic clock. No GNSS satellite currently does this onboard, hence only certain types of failures can be detected and mitigated onboard within a few seconds of occurrence. There is some on-going discussion around providing an onboard ensemble of atomic clocks.

GNSS master control stations do not maintain a version of UTC. Rather, national laboratories (e.g. USNO for GPS or National Time Service Center (NTSC) for Compass) or, for Galileo, precise timing facilities (PTFs) maintain a real-time prediction of UTC and monitor the transmitted GNSS timescale. The laboratory (or PTFs) then gives a prediction of the GNSS time versus the UTC(lab), which the GNSS broadcasts as data to satellites.

For the user segment, GPS was the first system freely available for international uses with a large number of receivers and public applications, hence its ubiquity. GLONASS is also for free international use but operational (and adequate quality) receivers have been lacking. Since 2011 GLONASS receivers have been more largely available (ex: iPhone4S). Multi-GNSS receivers (GPS/GLONASS as well as GPS/GLONASS/Galileo-ready) are now available. GNSS cooperation allows better availability and can improve the resiliency for the end user. Nations that own GNSS have been negotiating to make different systems that are at least compatible and (in some ways) interoperable.

The accuracy and stability of the timing signal from a GNSS receiver depends on various factors principally from the receiver, the antenna and the antenna cabling. The cable from the antenna will be carefully calibrated. To maintain the stability of this calibration, the impedance matching between the antenna cable and both the antenna and the receiver must be very good. Another option is to put the entire receiver at the antenna. In this case, since the entire system is exposed to open air, temperature-induced delay variations must be controlled. The receiver will correctly utilize the signal from satellites to recover the accurate frequency and time. This depends on various factors, such as the number of satellites, signals and carriers tracked by the receiver, and the detection and management of errors. The best systems can provide frequency accuracy approaching 10^{-14} after 1 day of averaging (i.e. short-term accuracy depends on the local reference of the receiver). For time, accuracy is measured relative to the GNSS UTC. Delay calibration of the receiver system is essential in this case.

2.7.1.2. *GNSS distinctions*

Each GNSS has its own characteristics with regard to their spaced-based, ground and user segments. For instance, the constellation, the frequencies and carriers are of particular importance in the context of telecom distribution and timing sources.

Each GNSS constellation has its own characteristics: average altitude of the satellite orbits, declination, period of revolution (when not geostationary or geosynchronous), number of satellites in orbit and in service, generation of the satellite, etc. Every receiver obtains necessary information (e.g. clock offset and ephemeris for GPS) either from the GNSS satellites or by an augmentation system.

As mentioned earlier, UTC is a post-processed timescale based on multiple clocks of various laboratories disseminated over the planet. Hence there is no single UTC available from all GNSS. The laboratories participating to the generation of UTC have offsets between each others. These offsets are measured and known but the "true" UTC cannot be known in real time.

GNSS	Owner	GNSS time	Number of satellites* in constellation (planned)/in operation (planned)
Global positioning system (GPS)	United States	GPS time has no leap seconds, thus a constant offset from TAI. Traceable to UTC (USNO) using transmitted parameters.	32/24
Global navigation satellite system (GLONASS)	Russia	GLONASS system time Traceable to UTC (RU) Use leap seconds	31/24
Compass navigation satellite system (CNSS; BeiDou 2)	China	Compass system time (Bei Dou Time (BDT)) Traceable to UTC (National Time Service Center (NTSC))	9 (30 + 5)/6
Galileo	European Community	Galileo system time (GST) has no leap seconds. Traceable to UTC using two precise timing facilities (PTFs) Provides offset with GPS time	4 (30)/(27; 2015: 18)
*At the date of the publication			

Table 2.9. *Summary of GNSS*

As discussed earlier, each real-time UTC GNSS time reference is a prediction of the true post-processed UTC. This offset cannot be known in real time as it is only defined at least a month later.

Hence, in terms of the traceability chain, time accuracy error accumulates:

– between any real-time source of UTC (typically a national lab) and the final post-processed UTC;

– between a national lab and the estimated offset of UTC from a particular GNSS timescale;

– between the true offset of UTC from the GNSS timescale and the broadcast value from a particular satellite based on the prediction of that satellite clock;

– between the broadcast value of UTC and the measurement in the receiver;

– between a receiver using one GNSS and a receiver using a different GNSS signal.

From a possible sub-10 ns accuracy between a receiver and its traceable UTC reference, the accuracy budget is set at a few hundred nanoseconds for a distributed time synchronization system. ITU-T G.8272 (PRTC) specifies an accuracy of 100 ns to UTC.

2.7.1.3. *Receiver operations for timing*

In this section, a GNSS receiver (which can support multi-GNSS frequencies and carriers) performing timing recovery has a fixed location (operator timing sources are not supposed to be mobile except units for field test and measurement).

The receiver will first detect satellites, obtaining, in case of GPS, the almanac then the more complete ephemeris for each visible satellite. Once locked, the receiver measures pseudo-range from each satellite: the received time according to the local clock, minus the transmitted time according to the satellite clock. Information such as error and ionospheric corrections can also be acquired. They can be received faster and more frequently, by other ways, such as from augmentation systems or from Assisted GPS[4].

With this information, the first step toward accurate synchronization for a receiver is to define its 3D position (latitude, longitude, altitude). To obtain such a position, at least four satellites (from the same constellation if multi-GNSS) are required, in order to solve the four unknowns: three for position and one for the user's time offset. If altitude was known, a usual trilateration would be sufficient. A receiver may be able to see seven or eight satellites, but for optimizing position there is a trade-off between using higher elevation satellites, which minimize noise, and maximize the orthogonality of the satellites to minimize geometric dilution of position. Using two frequencies (e.g. GPS L1 and L2) allows measurement of the ionosphere delay. Once the fixed coordinates of the antenna are known, accurate time can be received from a single satellite. In the case of multi-GNSS, a receiver's ability to track satellites from multiple constellations depends on how interoperable they are. Generally, the reference coordinate systems are different for different constellations, so the receiver needs a different set of coordinates for each GNSS, having obtained them by tracking at least four satellites from a given GNSS. In addition, the UTC source is different for different GNSS, thus it will be important

---

4 With Assisted GPS, the satellite ephemeris and almanac data are retrieved over the network (e.g. wireless with handset) and can be received at much lower S/N.

for the user to have a bound on the difference. Such information may be available from satellites for certain constellations.

Each satellite transmits time as the essential signal for operation. Short-term noise in the code requires filtering against the local oscillator for either time or frequency synchronization. Using the carrier to filter the code after performing carrier phase measurements ([PET 96], [LAR 98]) can decrease this noise. This is very similar to using SyncE to filter PTP in network time transfer [see sections 2.2 and 2.3].

However, while results can be very good, there is no limit how bad they can be either.

GNSSs have high availability and are reliable as systems. But, GNSSs evolve. New GNSSs such as COMPASS and Galileo are emerging, and GPS and GLONASS are being modernized. This will provide more and new satellites in the sky, new signals available, and allow better coordination between GNSS. Along with the evolution of the satellite and control segments, with multi-GNSS antennas and receivers and better filtering, such modernization will hopefully provide better resiliency, better accuracy and stability between redundant time sources in packet-based synchronization networks.

However, GPS (and other GNSS) vulnerability is now an accepted fact. New services in upcoming GNSS, such as integrity information that will be proposed as a public service on both Galileo and GPS, as well as different receiver system designs may be able to mitigate some of the vulnerabilities as discussed in Chapter 6. Nonetheless, GPS is currently the most reliable and ubiquitous source of precise time.

2.7.2. Regional terrestrial systems

Frequency and/or time are available from other ground-based systems. LOng-Range Aid to Navigation (LORAN) and other radio time broadcasting solutions are available in certain regions.

LORAN systems existed for decades with the first LORAN-A stations deployed in the 1940s. LORAN is a terrestrial navigation system operating in the low-frequency band, transmitting from ground antennas, hence covering a region. One primary and at least two other secondary fixed stations, all synchronized, using the same group repetition interval (GRI) and with known locations are necessary for a LORAN receiver getting service from this array.

The stations repeatedly exchange identifiable pulse signals, defined in microseconds. Delays between sets of pulse is specific for each GRI identifying an array. A fixed receiver, such as providing the source for a synchronization network, could get service from a limited number of arrays if any signal is reachable in the region.

To generate an accurate and stable pulse signal, LORAN stations are equipped with atomic clocks (usually a minimum of three cesium clocks) and referenced to a source of UTC. The LORAN signal is thus able to generate a very accurate timing service available in each array.

There are two current generations of LORAN in operation today: LORAN-C and eLORAN, latest evolution of LORAN. LORAN-C provides an accurate timing (frequency) service that has been largely used by US operators for backing up GPS in the context of SONET. LORAN-C provides a Stratum-1/G.811 frequency standard reference. eLORAN transmits additional information including traceability to UTC and leap second support, which allows listening to all receivable transmitter signals improving the accuracy. With a precision better than 50 ns to UTC, eLORAN is considered a powerful and independent alternative or backup to GNSS for time synchronization.

As for GNSS, LORAN chains (arrays) are steered from a specific UTC lab, usually related to the government in charge of the array. But the LORAN system clock will remain independent of any GNSS time system and thus is really an independent and alternative source.

The LORAN technical strength in providing a much stronger signal makes it a suitable alternative or backup to GNSS. Combining a GNSS/eLORAN antenna and receiver is possible, though the frequencies and signal structures are very different. Because it is regional, LORAN needs to be funded, maintained and operated by governments. The number of LORAN arrays is limited worldwide. With the emergence of new or modernized GNSS and the current economy, the budget for LORAN is under risk. As an example, the US government has decommissioned LORAN-C in 2007 with some possibility to modernize LORAN-C and to move to eLORAN. In short, LORAN cannot be considered as a timing solution within the US until the government revises its decision. However, LORAN remains a critical service in some regions, such as for maritime navigation in Europe.

CDMA, the 3GPP2 mobile wireless technology, when available, can also be a source of time. Every CDMA base station is synchronized by a GPS receiver that gives traceability to UTC (USNO), but also the same level of uncertainty that we get from GPS and the base station's receiver. Moreover, if the base station frequency is steered by the GPS receiver and so might provide Stratum1 long-term accuracy, the 3GPP2 base station radio interface specifications (3GPP2 C.S0010-B and C.S 0002-C – see Chapter 1) do not require such quality. Hence, the CDMA signal might not have the same quality from a frequency standpoint. Finally, because it is disciplined by GPS, CDMA cannot be a backup to GPS clock source.

There are finally various broadcast time sources available in low-frequency bands (approximately from 40 kHz to 20 MHz) from single transmitters deployed in many countries that can transmit over long distances (e.g. 2,000 km). Those radio time signals (ex: DFC77, WWVB, TDF or BPM) provide usually ToD to devices such as radio clocks or watches. They could be used as a source for time servers, but have limitations such as interference sensitivity, unknown delays from transmitters and such low absolute accuracy. Moreover, clock references are distinct. In case of relative time synchronization between two receivers, the accuracy will degrade fast with the distance.

2.7.3. Comparison

	Type of infrastructure for distribution	Signal characteristics	Service	Pros/cons
GNSS	Satellite constellation Terrestrial support infrastructure	Microwave L-band (in 1–2 GHz band) Low power Line-of-sight (LOS) propagation (obstacles block the signal)	ToD; receiver can derive frequency	Pros: high accuracy and availability, ubiquitous (global) Cons: can be jammed and might be spoofed; hosted by country or federation of countries; unavailable inside buildings or behind obstacles
LORAN	Terrestrial broadcast	Low frequency (in 100 kHz band) High power Ground wave propagation (diffracts around obstacle)	Frequency and ToD Note: eLORAN has better time accuracy than LORAN-C	Pros: high accuracy, difficult to jam, signal available indoors Cons: regional service (limited coverage worldwide); very few eLORAN arrays (only one operational today, in UK; trials in France [URS eLORAN]), less accurate than GNSS
Others	Terrestrial broadcast	From 40 kHz to 20 MHz Various modulation types	ToD; receiver can derive frequency	Pros: available Cons: poor accuracy and propagation uncertainties

Table 2.10. *Brief comparison of some sources for time reference*

Besides atomic clocks from frequency, there are multiple time sources being broadcast. They all can be steered to UTC frequency with varying accuracy (i.e. based on the cesium definition of the second) but cannot be steered by the same, unique UTC. Instead all time system transmitters, ground stations or satellites, will have to connect to one of the UTC real-time sources to steer their time reference.

The time source is a critical element when engineering precise time synchronization, in particular, when distributed sources could backup each other.

2.8. Summary

This chapter gave a general overview of technologies available to transport frequency, phase and time synchronization. The deployment of each technology depends on the target application and target accuracy.

Physical layer synchronization, such as SyncE, can be used for frequency synchronization with a very high level of accuracy and stability, and it is a proven technology. However, it cannot be used for time and phase synchronization, as it can only transport frequency. Another aspect is that it may not allow a transparent transport of timing (SyncE timing cannot be carried transparently across a packet network, as SyncE carries timing node-by-node, so the timing is terminated at each router/switch).

Packet-based technologies, such as IEEE 1588™-2008, are becoming very popular to transport frequency, time and phase; however, new architectures need to be deployed and understood to be able to deliver the proper synchronization performance. Section 2.6.6 summarizes and compares the different packet-based protocols (CES, NTP and PTP).

GNSS type technologies, such as GPS in North America, have been used for frequency, phase and time synchronization with very high levels of accuracy, and is a proven technology. Section 2.7.3 compares GNSS with Loran.

Table 2.11 gives a high-level comparison the different technologies addressed in this chapter.

Technologies	Advantages	Disadvantages
Physical layer synchronization (e.g. Synchronous Ethernet, SDH, SONET)	– Proven technology – Reliable and low cost – High-frequency accuracy	– Does not support time and phase synchronization – Needs to be deployed in every node of the network – May not allow timing transparency (e.g. SyncE in packet networks)
Packet-based synchronization (e.g. PTP, NTP)	– Supports time and phase synchronization – Allows for timing transparency	– Performance is dependent on the network – For reliable performance, it needs to be deployed in every node of the network
GNSS (e.g. GPS in North America)	– Proven technology – Reliable – Frequency, time and phase, high accuracy synchronization – Widely available (ubiquitous)	– Expensive (Capex and Opex) – Requires line of sight of satellites, not always possible in urban canyons – Very low signal strength making it vulnerable to unintentional and intentional interferences, such as radio frequency interference (RFI) and spoofing
LORAN	– Proven technology – Reliable – Frequency, time and phase, high accuracy synchronization	– Not available everywhere – Need eLORAN for time traceability with an even lower availability

Table 2.11. *Technologies comparison*

2.9. Bibliography

[0900003] ATIS-0900003, Metrics Characterizing Packet-Based Network Synchronization, 2010.

[802.1AS] IEEE Std 802.1AS™-2011, Timing and Synchronization for Time-Sensitive Applications in Bridged Local Area Networks, IEEE Computer Society, March 2011.

[802.3] IEEE Std 802.3™-2008, Carrier Sense Multiple Access with Collision Detection (CSMA/CD) access method and Physical Layer specifications, IEEE Computer Society, December 2008.

[1588-2002] IEEE Std 1588™-2002, IEEE Standard for a Precision Clock Synchronization Protocol for Networked Measurement and Control Systems, IEEE Instrumentation and Measurement Society, November 2002.

[1588-2008] IEEE Std 1588™-2008, IEEE Standard for a Precision Clock Synchronization Protocol for Networked Measurement and Control Systems, IEEE Instrumentation and Measurement Society, July 2008.

[1588-2008 IC] IEEE 1588-2008 Interpretation Committee page. Available at http://standards.ieee.org/findstds/interps/1588-2008.html.

[C37.238] IEEE Std C37.238™-2011, IEEE Standard Profile for Use of IEEE 1588™ Precision Time Protocol in Power System Applications, IEEE Power & Energy Society, July 2011.

[G.781] ITU-T Recc. G.781, Synchronization layer functions, 2008.

[G.803] ITU-T Recc. G.803, Architecture of transport networks based on the synchronous digital hierarchy (SDH), 2000.

[G.810] ITU-T Recc. G.810, Definitions and terminology for synchronization networks, 1996.

[G.811] ITU-T Recc. G.811, Timing characteristics of primary reference clocks, 1997.

[G.812] ITU-T Recc. G.812, Timing requirements of slave clocks suitable for use as node clocks in synchronization networks, 2004.

[G.813] ITU-T Recc. G.813, Timing characteristics of SDH equipment slave clocks (SEC), 2003.

[G.823] ITU-T Recc. G.823, The control of jitter and wander within digital networks which are based on the 2048 kbit/s hierarchy, 2000.

[G.824] ITU-T Recc. G.824, The control of jitter and wander within digital networks which are based on the 1544 kbit/s hierarchy, 2000.

[G.825] ITU-T Recc. G.825, The control of jitter and wander within digital networks which are based on the synchronous digital hierarchy (SDH), 2000.

[G.8260] ITU-T Recc. G.8260, Definitions and terminology for synchronization in packet networks, 2012.

[G.8261] ITU-T Recc. G.8261, Timing and synchronization aspects in packet networks, 2008; see also Amd.1 (2010).

[G.8261.1] ITU-T Recc. G.8261.1, Packet delay variation network limits applicable to packet based methods (Frequency Synchronization), 2012.

[G.8262] ITU-T Recc. G.8262, Timing characteristics of a synchronous Ethernet equipment slave clock (EEC), 2010; see also Amd.1 (2012) and Amd.2 (2012).

[G.8263] ITU-T Recc. G.8263, Timing characteristics of packet based equipment clocks (PEC) and packet based service clocks (PSC), 2012.

[G.8264] ITU-T Recc. G.8264, Timing distribution through packet networks, 2008; see also Amd.1 (2010) and Amd.2 (2012).

[G.8265] ITU-T Recc. G.8265, Architecture and requirements for packet based frequency delivery, 2010.

[G.8265.1] ITU-T Recc. G.8265.1, Precision time protocol telecom profile for frequency synchronization, 2010; see also Amd.1 (2011) and Amd.2 (2012).

[G.8275.1] ITU-T Draft Recommendation G.8275.1, Precision time protocol telecom profile for phase/time synchronization, 2012.

[I.363.1] ITU-T Recc. I.363.1, B-ISDN ATM Adaptation LayerSpecification: Type 1 AAL, 1996.

[RFC 4553] IETF RFC 4553, Structure-Agnostic Time Division Multiplexing (TDM) over Packet (SAToP), June 2006.

[RFC 4842] IETF RFC 4842, Synchronous Optical Network/Synchronous Digital Hierarchy (SONET/SDH) Circuit Emulation over Packet (CEP), April 2007.

[RFC 5086] IETF RFC 5086, Structure-Aware Time Division Multiplexed (TDM) Circuit Emulation Service over Packet Switched Network (CESoPSN), December 2007.

[RFC 5087] IETF RFC 5087, Time Division Multiplexing over IP (TDMoIP), December 2007.

[RFC 5143] IETF RFC 5143, Synchronous Optical Network/Synchronous Digital Hierarchy (SONET/SDH) Circuit Emulation Service over MPLS (CEM) Encapsulation, (obsoleted by [RFC 4842]), February 2008.

[RFC 5905] IETF RFC 5905, Network Time Protocol version 4, June 2010.

[TR 101 685] ETSI TR 101 685, Transmission and Multiplexing (TM), Timing and synchronization aspects of asynchronous transfer mode (ATM) networks, 1999.

[Y.1411] ITU-T Recc. Y.1411, ATM-MPLS network interworking – cell mode user plane interworking, 2003.

[Y.1413] ITU-T Recc. Y.1413, TDM-MPLS network interworking – user plane interworking, 2004; see also Corrigendum 1 (2005).

[Y.1453] ITU-T Recc. Y.1453, TDM-IP interworking – user plane interworking, 2006.

[BRE 02] BREGNI S., *Synchronization of Digital Telecommunications Networks*, John Wiley & Sons, June 2002.

[EID 06] EIDSON J., *Measurement, Control, and Communication Using IEEE 1588 (Advances in Industrial Control)*, Springer, April 2006.

[FER 08] FERRANT J.-L. et al., "Synchronous Ethernet: a method to transport synchronization", *IEEE Communications Magazine*, vol. 46, no. 9, pp. 126–134, 2008.

[FER 10] FERRANT J.-L. et al., "Development of the first IEEE 1588 telecom profile to address mobile backhaul needs", *IEEE Communications Magazine*, vol. 48, no. 10, pp. 118–126, 2010.

[GAR 05] GARDNERF M., *Phaselock Techniques*, John Wiley & Sons, July 2005.

[GOL 12] GOLDIN L., MONTINI L.,"Impact of network equipment on packet delay variation in the context of packet-based timing transmission", *IEEE Communications Magazine*, vol. 50, no. 10, pp. 125–158, 2012.

[HUS 12] HUSTON G. (APNIC), "Protocol basics: the network time protocol", *Internet Protocol Journal*, vol. 15, no. 4, pp. 2–11, 2012. Available at http://www.cisco.com/web/about/ ac123/ac147/archived_issues/ipj_15-4/ipj_15-4.pdf.

[JOB 12] JOBERT S., HANN K., RODRIGUES S., "Synchronous Ethernet to transport frequency and phase/time", *IEEE Communications Magazine*, vol. 50, no. 8, pp. 152–160, 2012.

[LAR 98] LARSON K., LEVINE J., "Time transfer using the phase of the GPS Carrier", *IEEE Trans. On Ultrasonics, Ferroelectrics and Frequency Control*, vol. 45, pp. 539–540, 1998. Available at http://www.boulder.nist.gov/timefreq/general/pdf/1220.pdf

[LOM 05] LOMBARDI M., CELANO P., POWERS E., "The potential role of enhanced LORAN-C in the national time and frequency infrastructure", *Proceedings of the 2005 International Loran Association Meeting*, 2005. Available at http://tf.nist.gov/timefreq/general/pdf/2105.pdf

[MEF3] Metro Ethernet Forum Technical Specification #3, Circuit Emulation Service Definitions, Framework and Requirements in Metro Ethernet Networks, April 2004.

[MEF8] Metro Ethernet Forum Technical Specification #8, Implementation Agreement for the Emulation of PDH Circuits over Metro Ethernet Networks, October 2004.

[MIL 00] MILLS D., "A brief history of NTP time: confessions of an Internet timekeeper", August 2000; available at: http://www.eecis.udel.edu/~mills/database/ papers/history.pdf

[MIL 06] MILLS D., "Computer network time synchronization: the network time protocol", CRC Press, March 2006.

[O'DO 08] O'DONOGHUE K., "NTP/1588 protocol review (gap analysis)", TICTOC WG, IETF#71, Philadelphia, 2008. Available at http://www.ietf.org/proceedings/71/slides/tictoc-8/tictoc-8.htm.

[PET 96] PETIT G., THOMAS C., "GPS frequency transfer using carrier phase measurements", *Proceedings of the 1996 IEEE International Frequency Control Symposium*, 1996.

[SHE 09] SHENOI K., *Synchronization and Timing in Telecommunications*, BookSurge Publishing, September 2009.

[BIP] Bureau International des PoidsetMesures (BIPM) portal: http://www.bipm.org/en/home/

[ESA] EuropeanSpace Agency (Galileo): http://www.esa.int/esaNA/galileo.html.

[FAA] US Federal Aviation Administration (FAA) about GPS, http://www.faa.gov/about/office_org/headquarters_offices/ato/service_units/techops/navservices/gnss/gps/.

[GLA] U.K. General Lighthouse Authorities; eLORAN News, http://www.gla-rrnav.org/radionavigation/eloran/international_work.html

[MIL] David Mills blog site, http://www.eecis.udel.edu/~mills/; suggested bookmarks: http://www.eecis.udel.edu/~mills/ntp.html; http://www.eecis.udel.edu/~mills/exec.html (how NTP works? Exec summary); http://www.eecis.udel.edu/~mills/ntp/html/warp.html (how NTP works? Details).

[NTP Org] NTP Project portal: http://www.ntp.org/; FAQs: http://www.ntp.org/ntpfaq/.

[NIS] U.S. National Institute of Standards and Technology (NIST): Time and Frequency Divisionportal: http://www.time.gov/.

[PNT] National Executive Committee for Space-Based Positioning, Navigation, and Timing (PNT Committee): http://www.pnt.gov/101/.

[RFSA] Russian Federal Space Agency (GLONASS): http://www. glonass-center.ru/en/.

[USG] U.S. Government information about the Global Positioning System (GPS) and related topics: http://www.gps.gov/systems/gps/; http://www.gps.gov/systems/augmentations/.

[USN] US Naval Oceanography; time portal: http://www.usno.navy.mil/USNO/time.

Chapter 3

Synchronization Network Architectures in Packet Networks

3.1. The network synchronization layer

The network synchronization (NS) layer refers to the multiple timing paths within a network that form the synchronization network. However, before details of the synchronization layer can be described, it is necessary to provide some background to the general concepts of network modeling and, in particular, the concept of a network layer.

3.1.1. *Network layers and abstraction*

The basic function of a telecommunication network is to provide the ability for users to exchange information. While seemingly simple in concept, in reality this represents a complex distributed system. The individual components that form the network are in themselves complex and may have evolved over time. While the task of transferring user traffic may seem simple, the level of coordination required of the various systems is significant.

In a system, a precise behavioral specification is needed in order to allow the realization of the system. In the telecommunication network environment, where the components from different sources may be present in the network, equipment and network standards are developed. One single standard cannot define the network. Indeed, the network is continually evolving with the introduction of new technologies and functionality, resulting in the network being defined through multiple standards.

As the complexity of the technology used within the network increased, it was recognized that formal methods should be used in order to help bound the scope of standards development. To achieve the desired network behavior, the functions provided by the network are described in terms of an abstract model, with the interactions of the various components precisely specified. Abstract network models have been used for many years to define computer networks. For example, the Open Systems Interconnection (OSI) was first described around the 1980 time frame.

The OSI reference model developed in the early 1980s defined an abstract model that can be used to describe communications between computer systems. The basic process needed to achieve communications between an end user and a computer system was described in terms of seven distinct layers, each performing a distinct function. While the model described seven layers, not all layers needed to be implemented. For example, an Internet Protocol (IP) router typically only needs to implement OSI layer 3 and an Ethernet switch, only layer 2. An end user would, in most cases, implement the topmost layers. Although this provides an abstracted model, it only describes the behavior of a system and does not define the implementation. Figure 3.1, illustrates the model representing the connection of a user to a server while connected over an intermediate router. In this specific case, note that the router implements only the protocols up to and including layer 3 (e.g. IPv4, IPv6), while the end user and data storage application would implement the higher protocols.

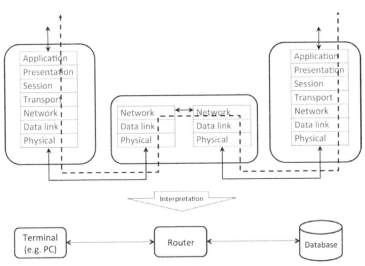

Figure 3.1. *Terminal to database example*

Readers are referred to the OSI model itself (the OSI model is contained in International Telecommunication Union Telecommunication Standardization Sector (ITU-T) Recommendation X.200).

A major construct in the OSI model is the concept of a layer. As discussed above, individual functions are grouped within a layer. However, within the context of a computer communication model, where the functions are generally implemented via protocols, communication is between peer instances of the protocol. A "layer network" is a term that is often used in telecommunications to describe the layer at which connectivity is provided. In Figure 3.1, a network resulted at layer 3 (the IP layer) since the router is providing connectivity between two end points at the network layer.

Within the OSI model, each layer would encapsulate the information presented from the layer immediately above it, process as necessary and then pass the information to the next lower layer. A communication path through a network could be between functions at the same layer (peer layer).

Certain limitations exist when applying the OSI model to general telecommunication network applications. For example, in the model of a computer communication network, the model can only define a physical layer between adjacent NEs. However, in a telecommunication network application, this is overly simplistic, in that the simple link may actually consist of a complex network capable of supporting a variety of different client signal types. Since the need to define a model in abstract terms was seen as invaluable for the development and understanding of network behavior, the ITU-T developed a similar layer based model in G.805, which provided the capability to define network layers as needed in a fully recursive manner. Layers are frequently described as client or server layers to reflect the relative positions of the layers. Additional differences may be seen between the ITU-T model and the OSI model to reflect the generalization. Most notably, transitions between layers are performed using an adaptation function, and overhead added using a trail termination function. This simply reflects the fact that in some cases, information associated with a specific layer function may be carried in a separate channel.

The adaptation and trail termination functions are fundamental components of the ITU-T Recommendation G.805 model and, together with an access point, define the boundaries of a layer network. Specific details of these functions are typically found in technology-specific equipment recommendations, such as ITU-T Recommendations G.798 for optical transport network (OTN) elements, and G.783 for synchronous digital hierarchy (SDH) network elements. For a specific technology, specific aspects related to the network architecture are also defined. For example, SDH architecture is defined in ITU-T Recommendation G.803. These general concepts are also applied to synchronization in ITU-T Recommendation G.781 as discussed later in this chapter.

Within the G.805 model, the concept of a layer is defined as the collection of access points. A network connection exists between connection points. Graphically, these components are represented by special symbols as shown in Figure 3.2. Often, however, diagrams omit the access and connection points for simplicity. The information that is carried over the G.805 model is referred to as "characteristic information" (CI). The specific separation of adaptation and trail termination results in a network that is capable of being defined to support multiple client signals, independent of the technology used within the core of the network.

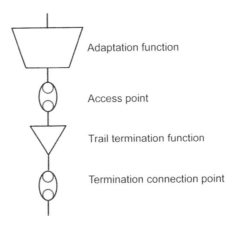

Figure 3.2. *Basic G.805 component representations*

A significant function required in the telecommunication network is the need to cross-connect signals to affect a switching function. This is needed for both packet and TDM networks. Within G.805, this cross-connection function is called a subnetwork or matrix connection as shown in Figure 3.3. This function allows switching of the characteristic information (CI) associated with the layer in which the subnetwork connection is defined.

The general concepts between the OSI and ITU-T models are similar. However, due to the use of specific graphical symbols, the resulting models appear to be different. The G.805 model is defined in a manner that allows the model to be recursive. If necessary, details of lower networks can be hidden if they are not needed. For example, an underlying OTN network need not be shown on a network model if the intent is to understand the requirements necessary to multiplexing at a packet level.

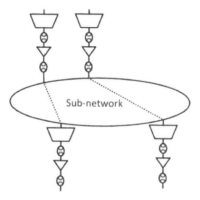

Figure 3.3. *G.805 subnetwork or matrix connection*

Although the G.805 model was initially developed to model TDM systems such as SDH as defined in G.803, the general principles have also been applied to data networks. For example, the ITU-T describes Ethernet as a two-layer model. The Ethernet MAC layer functionality (OSI data link layer) is represented the ETH layer, while the Ethernet physical layer is represented as the ETY layer. An example of a simple Ethernet model is shown in Figure 3.4, which is an adaptation of Figure 4/G.8010 (see [G.8010]). It illustrates some of the graphical constructs used in a G.805 representation. In this specific case, the concepts of trails and connections are extended to "flows" (e.g. a network flow (NF)) to reflect the nature of a packet network. Note that the model does not show the ETY (physical) layer connection. Further details of the ITU-T Ethernet layer architecture can be found in ITU-T Recommendation G.8010.

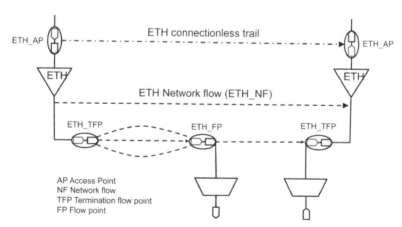

Figure 3.4. *ETH layer network example (unicast flow)*

3.1.2. *The synchronization layer*

Both the OSI and ITU-T models were developed to describe an abstract model to define the carriage of traffic. The layer model concepts have also been used to describe the synchronization network. This allows a formalized description of the synchronization network and helps draw parallels to underlying transport networks, which have also been described in similar terms.

The general architecture for frequency distribution is based on a hierarchy of clocks operating in a master/slave configuration and is described in G.803. The overall objective is to distribute the frequency provided from a primary reference clock (PRC), defined by ITU-T Recommendation G.811, along a chain based on a combination of clocks defined by ITU-T Recommendations G.812 and G.813. In general, timing distribution at the network level requires multiple timing signals to be provided to minimize the impact of network failures.

The G.805 model has also been used to describe the synchronization network. Two distinct layers have been described: the NS layer and the synchronization distribution (SD) layer. All clocks are located in the SD layer. The chain of clocks essentially provides a timing trail through the network. The network synchronization (NS) layer represents matrix connection functions that are used to provide protection. The control of the matrix connections is defined in terms of the sync selection function defined in G.781. The PRC at the head of the timing trail is providing a frequency within 10^{-11} of Universal Time Coordinated (UTC) frequency. The objective of the NS layer is to provide the necessary protection in order to maintain an estimate of the UTC frequency at each NE.

An example of the synchronization network layers is shown in Figure 3.5, which is an adaptation of Figure 5/G.781, where the timing trail from a PRC traverses three NEs. In this case, only a single trail is shown, but in practice the NS matrix connection may be used to provide interconnection of multiple trails to minimize any degradation due to network failures. The G.781 selection algorithm that is implemented in each NE will select the input reference based on a preprovisioned set of criteria. In the event of a failure, including the change in input quality, the synchronization selection algorithm will be run, resulting in the election of a new timing source. In the event of failure of all inputs, the selection algorithm will select the holdover.

Note, while the transport layers are shown, not all transport networks are suitable for supporting the clock trail. For example, clock trails may not be transported over an underlying network layer if it introduces excessive degradation. For example, pointers will preclude carriage of the trail over a PDH signal such as an E1 carried over and SDH lower order path. Acceptable transport layers include SDH (line) and Ethernet (ETY).

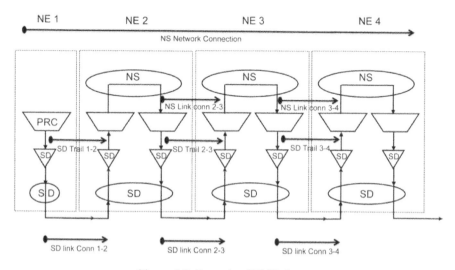

Figure 3.5. *Example of NS/SD layers*

3.2. Functional modeling

One of the aspects of standards development, particularly with regard to equipment specification, is the need to define functional models for NEs in a manner that is coordinated with the overall network architecture and requirements. It must be noted that the description defined by the functional models is only descriptive and does not mandate any particular way to implement the functions.

The layer models define the specific functions required in order to achieve the goal of transport synchronization. The specific functions must be defined in sufficient detail in order to define any necessary functionality that needs to be included in an implementation. In some cases, the functional models appear to be quite simple. For example, in the case of a two-port Ethernet media access control (MAC) relay, the model becomes almost trivial as shown in Figure 3.6.

In the previous work in Q13/15, notably the work on circuit emulation service (CES) and Synchronous Ethernet (SyncE), initial functional models were developed and included in ITU-T Recommendation G.8264. These covered the case of frequency distribution only. For some technologies, especially SyncE, the models were essentially those found in SDH with minor changes (e.g. the synchronization status message (SSM) mechanism). The target architecture, in this case, is defined in G.803, with some aspects defined in G.781.

118 Synchronous Ethernet and IEEE 1588 in Telecommunications

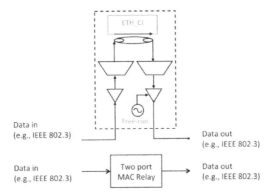

Figure 3.6. *Simple model of a two-port MAC relay*

In the case of packet-based synchronization mechanisms, since the timing flows in the network were no longer constrained to a single link, a supplementary architecture was needed. For packet-based frequency, architectural aspects were initially described in ITU-T Recommendation G.8265. The precision time protocol (PTP) telecom profile in ITU-T Recommendation G.8265.1 is based on this architecture. In progressing the development of the architecture, wider aspects such as protection methods required development of more advanced clock models than what was needed for the physical layer clocks. Figure 3.7, which is an adaptation of Figure I.3, G.8264, provides an example of the network architectural models where the processes to extract time and phase are added (via a specific adaptation function) to a network capable of providing physical layer timing.

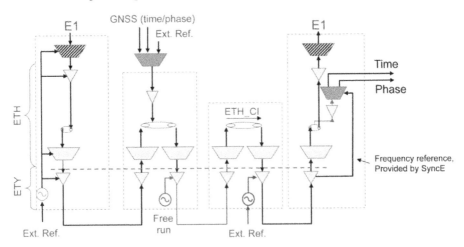

Figure 3.7. *Extensions of the models to address packet mechanisms*

The development of models to support physical layer synchronization was a fundamentally straightforward adaptation of existing models used in SDH. However, in the case of time/phase distribution, new clock types are being developed that process time at the packet layer and frequency at either the physical layer or at the packet layer. The equipment models become slightly more complex in that they need to accommodate both the packet and physical networks. However, we can see from Figure 3.8 that the functional model for a boundary clock begins to clearly define the functions that need to be implemented and how they interact.

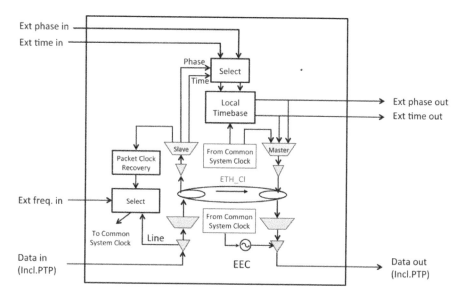

Figure 3.8. *Example of boundary clock modeled using G.805 constructs*

3.3. Frequency synchronization topologies and redundancy schemes using SyncE

3.3.1. *Introduction*

This section deals with the transport of frequency in SyncE networks; section 3.6 of this book provides an introduction to topologies and redundancies for the transport of phase and time.

As explained earlier in this book, SyncE has been specified to interwork with SDH; as a result, topologies and redundancy schemes for SyncE will be very similar to what has been done in SDH. Network topologies for frequency distribution will also include a mix of SDH and SyncE as the network evolves.

The key aspect in synchronization topologies is the protection of the timing transport during network rearrangement; the main risk is the generation of a timing loop.

A timing loop is a condition where the output signal of a clock is connected, generally through network rearrangement, so that it becomes the input reference to this clock. This condition must always be avoided, as all equipment that is part of this loop will eventually drift away in frequency, thus becoming isolated from the rest of the network. Further, individual clocks may not be able to track the input signal causing them to enter a free-run state, further exacerbating the effect of the failure.

Timing loops are extremely dangerous for synchronization networks because they are difficult to detect. Indeed, the equipment involved in these loops does not generate alarms as long as they are locked to an input and their timing signal remains within the acceptable range. When an alarm is generated by one or several pieces of equipment of the loop, it is generally not easy to identify the source of the problem.

The synchronization network topologies have been investigated in the 1990s and described in the European Telecommunications Standards Institute (ETSI) and ITU-T standard documents. ETSI published a document on synchronization network engineering [EG 201 793] which explains how to control the protection of the reference frequency transport in rings; several examples for SyncE given in this book are based on the synchronization network engineering ETSI document written for SDH.

ITU-T G.803 Appendix I provides guidelines on synchronization network engineering.

3.3.2. *Network topologies*

The purpose of a synchronization network is to provide a reference timing signal to all equipment in the network, from the reference source to the end user equipment.

Two different topologies can be implemented. The first is a combination of SDH and SyncE equipment, where SyncE equipment is added to existing SDH networks already carrying synchronization.

The second is a pure SyncE network where the main application is mobile backhaul. In this application, most of the chains are expected to be shorter than in SDH networks. Long chains are naturally less reliable than shorter chains. It is important to remember, when designing a synchronization network, that Global Navigation Satellite System (GNSS) sources might also be unavailable for a period of time, either due to an extremely rare failure in the system or simply due the use of jammers (see section 6.2); during such period of unavailability, long chains may occur when the synchronization network has no other source than a central PRC.

This section will first concentrate on the SDH equipment clock-Synchronous Ethernet equipment clock (SEC-EEC) level, as many new SyncE networks may not include synchronization supply units (SSUs); an extended topology with SSUs will be described in a second step.

Timing references for SDH equipment can be Synchronous Transport Module (level N) (STM-N) line signals, or 2.048 MHz, 2.048 Mbit/s or 1.544 Mbit/s via specific external synchronization inputs. Not all SyncE equipment will have such specific synchronization inputs to receive a direct timing reference from an SSU, but this might be useful in some cases.

The network topologies can be linear chains, rings or mesh networks. Each section between two NEs can be configured to provide either unidirectional or bidirectional transport of timing. The choice depends on the ability to implement an automatic protection of the timing transport.

As SyncE is based on SDH for the transport of timing, the protection of timing is based on the use of the SSM for traceability as defined in section 2.2.1.

3.3.3. *Redundancy and source traceability*

Redundancy is mandatory in telecommunication networks since any failure can affect a huge amount of interconnected equipment. SDH equipment should be able to receive timing from at least two independent trails; as SyncE is intended to also fit within the same network and be fully compatible with SDH, SyncE also follows this rule.

In addition to hard failure conditions, protection allows automatic reconfiguration of the synchronization network based on changes in the traceability of the reference. In SDH, the SSM transports the quality level (QL) of the clock that is the source of timing. The SSM has been described in section 2.2.1 of this book and is specified in G.781. Although the SSM is very useful in some networks for automatic restoration, it is important to note that the QL provides an indication of the type of clock that is timing the chain. It indicates whether the clock will operate within a specific frequency range, for example ± 4.6 ppm for a QL-SEC or $\pm 1.10^{-11}$ for a QL-PRC. QL gives no information on the level's noise of the input signal noise. The QL received at the output of a chain of clocks is the same as that of the input.

Traceability information is of great help to allow automatic protection in the synchronization network as described in the next section. However, one aspect that should also be considered is the risk of creation of timing loops.

Prevention of timing loops is one of the main benefits of using SSM; the SSM algorithm together with the generation of the quality-level Do not Use/Don't Use for Sync (QL-DNU/DUS) SSM codes in bidirectional lines is key to preventing timing loops and is described in the next section.

In SDH, the QL is the only information transported by the SSM. During the development of SDH, allocation of overhead was scarce and only 4 bits were allocated to the SSM. However, with Ethernet and the new Ethernet synchronization messaging channel (ESMC) used to transport SSM in the SyncE frames, there is potential for new methods that, if standardized, could further aid in the prevention of timing loops. Some examples include adding a source identifier or recording the traversal of all equipment in a chain between the source and the end equipment (similar to IP trace route). This type of functionality has been discussed during the SDH period, but could not be implemented due to lack of available bits in the SDH SSM field.

3.3.4. *Use of SSM in real networks*

This section gives practical information on how to configure rings and linear chains so that the SSM algorithm might manage protection without risk of timing loops.

3.3.4.1. *DNU, DUS*

The most important SSM code to prevent timing loops is the code 1111 representing QL-DNU for ETSI and QL-DUS for the American National Standards Institute (ANSI). This code is generated on the transmit port of a bidirectional interface when its receive port has been selected to be the source of synchronization for the equipment clock, either the SEC or EEC.

Figure 3.9 shows an example with two NEs either add-drop multiplexers (ADMs) or a SyncE switch. In this case, they are connected in tandem via a SyncE link through their bidirectional synchronization interfaces, East and West. The two arrows within each NE show how the selected timing is regenerated on the output interfaces of the NE. Within the NE, the plain arrows show the timing active and the dotted line arrows show the timings that are not selected at the moment but that have been configured to be selectable during a protection operation.

Synchronization Network Architectures 123

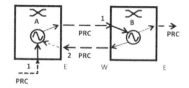

Note: all figures of section 3.3 use the following symbols

- - - - -▸ Ref. timing   1,2 Priorities
— —▸ SyncE   E,W East/ West
↘ Selected synchronization interface
↘ Non selected synchronization interface

Figure 3.9. *Bidirectional link without DNU/DUS*

To have the ability to propagate the timing from west to east, or vice versa in case of protection, the NEs are able to select the reference synchronization from either the West or the East interface. All output interfaces are physically locked to the internal clock synchronized on the selected input.

If the NE A of Figure 3.9 loses its PRC input signal, it will consider the signal present on its East interface: in the absence of a QL-DNU/DUS SSM code on this interface, a timing loop would be generated between NEs A and B (Figure 3.10).

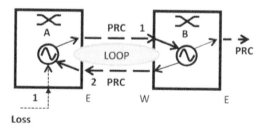

Figure 3.10. *Example of timing loop*

The generation of a QL-DNU/DUS SSM code on the transmit part of the interface selected by NE B will allow NE A to reject this signal as a timing source (see Figure 3.11).

Note: in Figure 3.11 and the figures that follow in this section 3.3, the equipment clock represents either an SEC or an EEC: the QL-SEC represents the timing generated by an SEC or an EEC in the case of a SyncE equipment.

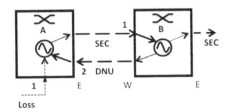

Figure 3.11. *Removal of a timing loop situation*

3.3.4.2. *Unidirectional linear chain*

In some parts of the network, it might be possible to configure the equipment so that the reference timing is transmitted only in one direction; this is the typical case in the access part of the network where the end user does not have a reference timing. In case of a failure in the chain, the end part of the chain will be locked to the equipment in holdover mode, located after the failure (NE B in Figure 3.12).

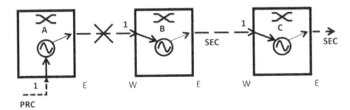

Figure 3.12. *Unidirectional chain*

This situation is not always a severe problem since, in these cases, data and timing follow the same links and the data that need synchronization are also lost due to the link failure.

3.3.4.3. *Bidirectional linear chain*

The first use of the SSM algorithm was to allow the transport of timing on a chain of equipment in one direction and then to reverse the direction in the case of a failure. In this case, restoration can be performed automatically without the risk of forming a timing loop.

Figure 3.13 shows a chain of NEs, again either SDH or SyncE equipment, connected via bidirectional links:

– Equipment A is configured as externally timed where the synchronization has QL corresponding to PRC. The external interface is given priority 1, and its East interface with SSM code and priority 2 (as shown in Figure 3.13).

– Equipment D is also configured to be externally timed, also with QL corresponding to PRC, but with priority 2. The West interface is given higher priority as shown in Figure 3.13.

– Equipment B and C are configured with two synchronization interfaces with SSM code.

The highest priority timing reference is provided on the left equipment, and this reference is propagated to the next equipment (Figure 3.13).

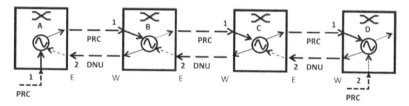

Figure 3.13. *Bidirectional linear chain*

In case of failure of this left reference (to NE A), it is expected that the second priority reference will provide its timing from the right to the left part of the chain, without causing any timing loop, even momentarily.

This is performed by the SSM algorithm, which generates the "1111" QL-DNU/DUS SSM code meaning that the signal must not be used for timing. The protection of the chain requires several steps since the SSM algorithm is performed independently on every equipment of the chain.

Step 1: The first priority reference disappears from equipment A (see Figure 3.14)

Equipment A has been configured with only two input ports carrying timing, its only choice is to lock to the signal coming from B when its first priority fails. This signal coming from B carries a QL-DNU/DUS SSM code, since B is locked to A.

Equipment A has no other choice than going to holdover; as a result, it will generate a QL-SEC/EEC SSM code on all its output synchronization interfaces.

Step 2: Equipment B propagates the new status of equipment A (see Figure 3.15)

Equipment B receives a QL-SEC/EEC SSM code on its left interface and a QL-DNU/DUS SSM code on its right interface; it selects the best code, that is SEC/EEC. It continues to be locked to equipment A and maintains the QL-DNU/US SSM code on its left interface that is selected as its source of timing. The only change is that now it generates a QL-SEC/EEC SSM code on its synchronization interfaces.

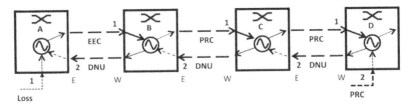

Figure 3.14. *Loss of the active reference*

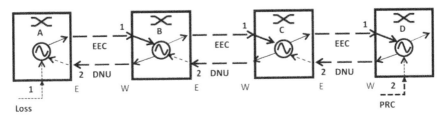

Figure 3.15. *Change of the SSM downward*

Step 3: Equipment C receives the new status of equipment D.

Equipment C gets a QL-SEC/EEC SSM code on its West interface; it receives a timing reference on its East interface. According to the SSM algorithm, specified in G.781, it selects the best quality input.

The important point is that it stops generating a QL-DNU/DUS SSM code on its West interface since it no longer selects this interface as a reference. It now generates a QL-PRC SSM code.

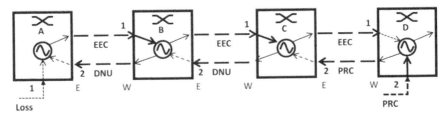

Figure 3.16. *Selection of the other PRC*

Step 4: Equipment A detects a new SSM code on its East interface

Equipment C (then B, then A) gets now a QL-SEC/EEC SSM code on its West interface and a PRC SSM code on its East interface. According to the SSM algorithm, equipment C selects the East interface and generates a QL-DNU/DUS SSM code on its East interface and a PRC SSM code on its West interface.

Equipment A gets now a QL-PRC SSM code on its East interface but still does not get any signal on its external interface. According to the SSM algorithm, it selects the East interface.

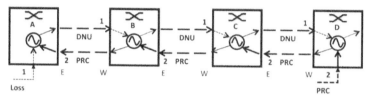

Figure 3.17. *Timing restored*

This final step ends the automatic restoration of this chain. Timing loop formation was prevented during the restoration. Note that if the East reference were of SSU quality, the restoration process would have been similar.

Note according to the G.803 there could be up to 18 NEs between equipment A and equipment D. All equipment would have a similar behavior as equipment B; each of them will perform a step identical to step 2 propagating the information to the right side and a step identical to step 4 propagating the information to the left side.

3.3.4.4. Rings with a single external synchronization reference

In some parts of the network, it is not possible to receive two independent synchronization references, for example in access networks. A ring with a single synchronization reference can, with proper engineering, restore the synchronization in the ring when a failure occurs in the ring but not in the case of a failure of this synchronization reference. Note that the loss of timing might have no consequence when the timing and traffic are transported on the same physical media.

The configuration of such ring is similar to that of a linear chain with external synchronization references at each end; the only difference is that the two ends of the chain are in the same equipment. This equipment will not be configured to recover timing from the chain since it is the only source of reference timing in the chain; this provides a very safe architecture that cannot generate any situation of timing loop.

Figure 3.18 shows a configuration of such a ring with priority 1 given to the West interface and priority 2 to the East interface.

The configuration of equipment with external reference timing (NE A in Figure 3.18) is very simple. In the figure, equipment A has only one source of synchronization, a dedicated external synchronization interface. It generates the SSM code on its two line interfaces part of the ring and is not configured to receive timing from the two interfaces of the ring.

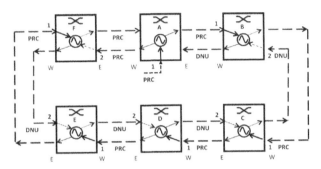

Figure 3.18. *Synchronization on an external timing*

In case of failure of a cable, the chain is automatically reorganized to provide a timing reference to all equipment (Figure 3.19).

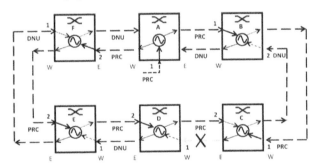

Figure 3.19. *Restoration after a failure*

3.3.4.5. *Rings with two, or more, external synchronization reference*

When a ring is part of the backhaul network, the loss of timing reference might be unacceptable due to the importance of the equipment affected by a single failure; in this case, it is recommended to provide two independent external synchronization references.

Such a ring needs to be carefully configured to prevent the generation of timing loop under any network failure condition. One solution is to configure one NE of the ring without a synchronization interface in the east direction and another NE without a synchronization interface in the west direction; this will prevent timing loops in the clockwise and counterclockwise directions, as explained in the next example.

Figure 3.20 provides an example of the propagation of timing being protected, clockwise and counterclockwise, from two simultaneous external references. Both equipment F and B receiving synchronization reference by an external synchronization interface, only one external interface is selected as reference for the ring at any time.

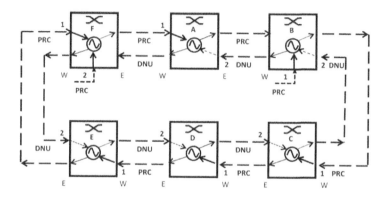

Figure 3.20. *Ring with two external synchronization signals*

When the external interface of equipment B fails, it still receives a QL-DNU/DUS on its East interface. QL-PRC SSM code is present on its West interface but the selection of this timing would immediately create a timing loop. This is the reason for not configuring this West interface as a synchronization interface. If the external interface on equipment F fails also, the same risk of timing loop exists. As for equipment B, the solution is to not configure the East interface as a synchronization interface as depicted by the figure. Network synchronization restoration will be delayed by the transmission of SSM node-by-node, but timing loops are prevented.

3.3.4.6. *Meshed networks*

Meshed networks, without SSU can be split into rings and linear chains, but careful design is needed to prevent timing loops.

Appendix III of G.803 provides guidance on managing meshed networks.

3.3.5. *Networks involving SSUs*

Since SyncE is based on SDH standards in order to allow interoperability between these two technologies, it is possible that some networks might implement a long synchronization chain as described in the G.803 synchronization network reference chain, with SSUs and SECs/EECs.

The SSM algorithm in general has not been specified to operate at both the SSU and the SEC/EEC levels, as the operation, especially in large networks, had not been fully studied. There was considerable risk of timing loop formation in these cases. Appendix III.7.1 of G.803 details the reasons why the SSM should not operate at the SSU level in complex networks. The main reason is that the SSM algorithm operates independently on each equipment and that the management of a network becomes more and more complex when the network size grows (the implementation of SSUs might imply a synchronization network of large dimensions).

However, in very well hierarchically organized networks, where there is no risk that timing generated by an SSU could be present at the input of the same SSU after any network reconfiguration, it might be possible to operate the SSM over the two SSU and SEC/EEC levels.

Note also that for option II SONET networks, there are some specific topologies containing both SEC and SSU clock equipment, for which the application of the SSM algorithm is described in G.781.

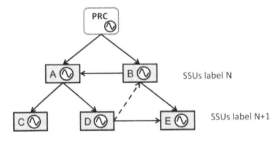

Figure 3.21. *Example of timing loops between SSUs*

It should finally be noted that G.803 Appendix III.6.1 provides guidance how to plan synchronization network with SSUs; SSMs which are not used by the SSUs, but the SSM algorithm might control the links between SSUs, composed with SECs or EECs. The basic principle is to assign to each SSU a label N representing level number N of SSUs in the synchronization chain (or hierarchy) from the PRC to this SSU and make sure that an SSU with a label N will never be locked to a signal issued from an SSU with a label higher than N. Figure 3.21, which is an adaptation of Figure III.3b/G.803, provides an example. The link from SSU D to SSU B is forbidden because it could create a timing loop between SSUs A, D and B.

It is recommended to carefully review G.803 Appendix III.6.1 before designing synchronization networks with SSUs.

3.3.6. Classical errors during SSM configuration

3.3.6.1. Bidirectional parallel lines

The configuration of two bidirectional linear chains in parallel might lead to timing loops, as is warned in ITU Recommendation G.781. The SSM algorithm is specified to return a QL-DNU, or a QL-DUS in SONET, on the interface that has been selected and only in this one. In Figure 3.22, the interface with priority 2 (P2) of equipment B sends a QL-PRC SSM code back to equipment A since its receive port is not selected as the source of timing.

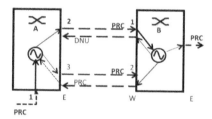

Figure 3.22. *Network with two parallel bidirectional lines*

If the signal at interface P1 of equipment A fails, equipment A immediately selects the signal present at interface P3, which generates a timing loop.

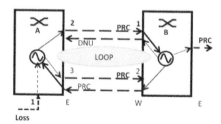

Figure 3.23. *Generation of a timing loop with parallel links*

To avoid the risk of timing loop, only one of the parallel lines has to be configured to transport timing in both directions.

Section 5.13.2 of G.781 states that it could be possible to define a new SSM algorithm, which specifies that all ports, known to have the same timing source, will return a QL-DNU/DUS SSM code as soon as one of these ports is selected for synchronization. But the related solution has not been fully described in the standards.

This condition can be quite common in the case of SyncE networks due to the Link Aggregation Groups (see [IEEE 802.1AX]) used in the case of Ethernet technology. ITU-T Recommendation G.8264 Amendment 2 provides some warning in this respect.

3.3.6.2. Interconnection between different networks

It is a common practice to generate a QL-DNU/DUS SSM code on synchronization interfaces between two networks operated independently by two different operators. Indeed, normally, timing is not distributed across different administrative domains because:

– neither of the two operators has a full knowledge of the entire synchronization network; and

– a protection operation in one network, unknown to the other operator, could result in a timing loop as described in Figure 3.24.

In this simple example, network 2 has a backup reference on NE B delivered by NE A of network 1 and network 1 has a backup reference on NE C delivered by NE B of network 2. In case NE A and B lose their PRC reference, a timing loop would be generated. This looks obvious but there might be several NEs between A and C, and B and C.

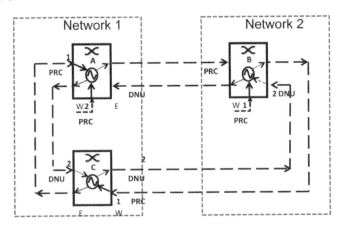

Figure 3.24. *Interconnection between two operators*

3.3.6.3. *Multiple failures*

In mesh networks, many failures may occur successively and it might be difficult to predict all situations and configure the network correctly; SSM must be used in reasonably sized networks.

3.3.7. *Conclusion on synchronization topologies*

The SSM algorithm has been primarily designed to prevent timing loops during protection on a linear chain of equipment where a timing reference is available at both ends of the chain; one of the main applications is the protection of timing in rings.

The SSM selection algorithm and messages play a major role in the synchronization topology since it allows automatic reconfigurations in case of equipment or fiber failure; but it is important to warn the reader that the SSM algorithm by itself will *not* prevent timing loops. It remains the responsibility of the synchronization network planner to correctly configure the SSM algorithm in all the network equipment in order to effectively prevent timing loops.

Many SDH networks use the SSM to provide automatic configuration at the SEC level, but this is not as simple as plug and play since it requires a careful synchronization network design. The same constraint applies to SyncE networks.

3.4. The IEEE 1588 standard and its applicability in telecommunication networks

As stated in Chapter 2, IEEE 1588 defines the PTP and related system properties, but it does not define performance requirements for specific applications. PTP is a protocol that can be used by many applications. Therefore, for specific architecture and performance requirements to be met, specific standards bodies must develop their own specifications and profiles. For telecommunication networks, ITU-T has been working on several recommendations to address network architecture, requirements and IEEE 1588 profiles. It is very important to understand specific applications and performance targets.

IEEE 1588 allows transport of synchronization over packet networks; however, other aspects, in addition to the timing protocol, are important and need to be taken into consideration. Network requirements, packet slave clock requirements and network limits for packet delay variation (PDV) are among the aspects that need to be studied. For frequency synchronization in telecommunication networks, ITU-T developed a series of recommendations that are intended to be used when deploying IEEE 1588 in telecommunication networks.

134 Synchronous Ethernet and IEEE 1588 in Telecommunications

The ITU-T Recommendations for frequency synchronization are as follows:

– G.8261 defines the network limits.

– G.8261.1 defines the PDV network limits.

– G.8263 defines the packet slave clock requirements.

– G.8265 defines the network architecture and protection mechanisms.

– G.8265.1 defines the IEEE 1588 profile.

– G.8260 contains the definitions and terminology for synchronization in packet networks.

ITU-T is also working on a series of Recommendations (the G.827x series) for phase and time synchronization in telecommunication networks. These Recommendations are intended to be used when deploying IEEE 1588 in telecommunication networks when precise phase and/or time of day synchronization are necessary.

As mentioned earlier, the need for protection mechanisms is very important in telecommunications networks. IEEE 1588 does specify a best master clock algorithm (BMCA) that allows the network to reconfigure in case of a master or network failure. The BMCA will allow the network to reconfigure to elect a clock as a master clock in a network, which differs from practices normally used in the operation of a telecommunication network. Additionally, in telecommunication networks, protection mechanisms are designed to allow a slave device to be synchronized to distinct masters that are geographically separated in the network.

3.5. Frequency synchronization topologies and redundancy schemes using IEEE 1588

The distribution of frequency synchronization using IEEE 1588 for telecommunication networks is based on architectures that employ "ordinary clocks" as defined in IEEE 1588. In this type of architecture, a slave cannot become a master, and a master cannot become a slave. Figure 3.25 shows general packet network architecture.

The architecture shown in Figure 3.25 assumes a generic packet network that does not have NEs with IEEE 1588 support. Hence, there are neither boundary clocks, nor transparent clocks deployed in this network. Therefore, this network is susceptible to PDV that may impact the clock recovered at the slave nodes. For this type of architecture, it may also be difficult to guarantee a symmetric network that would normally have an impact on the accuracy of time distribution. But, as the first profile defined by ITU-T only relates to frequency synchronization, the asymmetry will not affect the recovery of precise frequency.

Synchronization Network Architectures 135

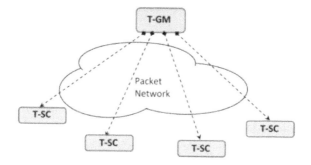

Figure 3.25. *General architecture for frequency distribution*

The packet master clock may be frequency locked to a PRC traceable clock or to a synchronous network (e.g. SyncE, SDH and SONET). The telecom grandmaster (T-GM) clock will use this clock (a physical layer signal) to generate timing packets to be sent to the telecom slave clocks (T-SC). The T-SC uses these timing packets to recover a clock that is traceable to PRC. Network operators should engineer the network to allow consistency of the slave clock performance. This engineering should be performed, if possible, along with the packet network and not *a posteriori*.

Figure 3.26 shows a packet network with timing provided by a traditional synchronization network.

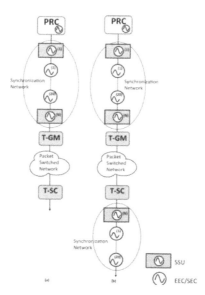

Figure 3.26. *Synchronization from a traditional SONET/SDH/SyncE network feeding a packet network*

In the network showed in Figure 3.26, extra care should be taken: as the clock used by the T-GM clock is being fed by a synchronization network, there is wander and jitter accumulation through the chain of EECs/SECs clocks.

Figure 3.26(b) shows a network where part of the traditional synchronization network has been replaced by a packet network. For this architecture, it is important to note that the output of the telecom slave clock needs to meet the EEC interface network limits, defined in section 9.2.1 of G.8261. For this type of network, special care must be taken as the PDV may affect the performance of the clock at the output of the telecom slave clock, making it very difficult to meet the specified network limits.

3.5.1. *Redundancy schemes using IEEE 1588*

As mentioned in section 3.3, synchronization redundancy is mandatory in telecommunication networks. Therefore, there is a need for the packet slave clock to be able to switch to a backup packet master clock in case of a loss of connectivity to the T-GM or clock failure. It is desirable that the Telecom Slave clock (T-SC) monitors two or more T-GM clocks in the network. Figure 3.27 shows primary and secondary T-GM clocks sending timing packets to a T-SC through a packet switch network.

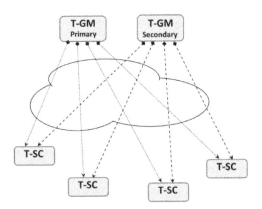

Figure 3.27. *Frequency protection*

In Figure 3.27, the T-SC is locked to the primary T-GM clock. The secondary T-GM clock is also active and may send timing information as requested by the T-SC. Timing information can include all the PTP messages or only the Announce messages. In such a mode, the T-SC would have all the expected necessary synchronization information from the secondary T-GM clock for switching to this secondary packet master in case of a failure of the primary packet master. It is important to note that the default BMCA described in IEEE 1588 was designed to allow only one active grandmaster clock within a PTP domain. As mentioned in section 2.5, ITU-T developed an alternate BMCA for the frequency profile to allow a redundancy scenario as depicted in Figure 3.27, and therefore the alternate BMCA allows for two or more active T-GM clocks.

As mentioned in section 3.3, reconfiguration of the synchronization network is based on the traceability information provided by the SSM QL of the source of timing. Therefore, there was also a need to support the transport of SSM QL when using the IEEE 1588 protocol to allow deployment of a hybrid network as depicted in Figure 3.28.

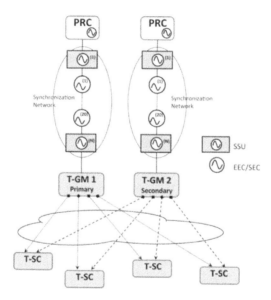

Figure 3.28. *Hybrid network with a traditional SONET/SDH/SyncE network feeding a packet network*

The SSM QL value that is being carried in the synchronization network can be passed to the packet network using the clockClass attribute in IEEE 1588 (see details in section 2.5). This allows the master selection algorithm to also use the SSM QL of the clock to select the active T-GM clock.

As described in G.8265.1, the master selection algorithm will select the T-GM clock with the highest QL that is not experiencing a Packet Timing Signal Fail (PTSF) condition. The PTSF indicates that there is a loss of Announce or Sync messages.

In the example depicted in Figure 3.28, the T-SC receives Announce messages from both the T-GM1 and T-GM2 clocks. The Announce messages carry the QL value in the clock class attribute. If the T-GM1 has the highest QL value and it is not experiencing a PSTF signal failure, from T-GM1, then the T-SC will lock to T-GM1. If both T-GM1 and T-GM2 timing messages do not have excessive PDV and they both have the same QL value, then the T-SC will make the decision based on G.781 priority. It is important to note that this priority is different from IEEE 1588 priority attribute. The T-GM priorities are set locally in the telecom slave clocks and this is done by configuration of the telecom slave clocks.

ITU-T G.8265 describes several additional functions that are needed for the T-GM clock selection, which are as follows:

– Temporary master exclusion – lock out function.

Slaves should be able to exclude temporarily a master from a list of acceptable masters.

– Slave wait-to-restore time function.

The slave will switch to a backup master if the primary master fails or if the slave stops receiving timing packets from the primary master. If the primary master becomes available again, then in order to make sure that the primary master is stable, the slave should wait for a period (the wait-to-restore time) before switching back to the primary master.

– Slave non-reversion function.

In the event of a failure, or loss of connectivity to the primary master, the slave will switch to the backup master. If non-revertive switching is active, then the slave will not switch back (revert) to the primary master once it becomes available again.

– Forced traceability of master function.

If the external input reference to the master clock has no QL value, then the operators may decide to provision a specific traceability value at the input of the masters. In this case, the master will "force" the GM to send the QL value to the slave clock that was provisioned at its input.

– Packet slave clock QL hold off function.

If the slave clock loses all its input references, then the operator may delay the transition of the QL value at the output of the slave clock. This needs to be carefully analyzed since it depends on the quality oscillator that provides the holdover function of the slave clock. If the slave clock has a very good holdover performance, then the delay of the transition of the QL value will allow operators to limit the downstream switching in the network.

– Slave output squelch function.

If the slave has external output synchronization interface (e.g. 2 MHz), a squelch function may be implemented.

This function could be used in cases where the slave clock loses all its input references and the slave is providing clock to an end device that implements very good holdover which is better than the slave holdover itself. By squelching the signal, the end equipment will no longer receive a clock from the slave device and it will go into holdover.

For example, usually a base station implements a very good holdover that may be better than the holdover at the slave clock. If the slave clock supplying synchronization to the base station loses its input reference then it will go into holdover, and it will continue to supply a clock (holdover quality) to the base station. On the other hand, if the slave squelches the output of the slave clock, then the BS will no longer receive a synchronization signal, and therefore will go into holdover that has a better quality compared to the slave clock holdover.

3.6. Time synchronization topologies and redundancy schemes

The distribution of time and phase synchronization over packet networks can be based on two main techniques as described in Chapter 2 and in [G.8271]:

– Deploying PRTCs where a time synchronization reference is required (typically by means of a GNSS receiver).

– Using packet-based methods to distribute the time synchronization reference from one or more centralized or distributed PRTCs.

Note: in many applications (see Chapter 1), only phase synchronization is required. A practical way to achieve phase synchronization is by means of distributing a time synchronization signal. Therefore, this chapter focuses on time synchronization solutions and the related architectures.

A combination of the two approaches mentioned above is also possible. Moreover, it is also possible to combine packet-based methods with physical layer based frequency synchronization transport (e.g. via SyncE). This option, as described later more in detail, is being considered in order to enhance the resiliency in the network.

In the case of packet-based methods, the support from intermediate nodes is considered as a key aspect in order to increase the performance. This is also the reason why the IEEE 1588 protocol is preferred over NTP where high accuracy is required. In fact, as described in Chapter 2, IEEE 1588 focuses on solutions able to remove the dependency from PDV by means of specific support in the intermediate nodes (using either boundary clocks, or transparent clocks, or a combination of both), which although not mandatory in every node, is highly recommended in order to achieve the highest performance, particularly for NEs susceptible to generating PDV and asymmetry (see [GOL 12]).

Details on the various time synchronization architectures and their redundancy schemes are provided in the following sections.

3.6.1. *Locally distributed PRTC*

An architecture to distribute time and phase based on locally distributed PRTCs is shown in Figure 3.29, which is an adaptation of Figure 1/G.8271.

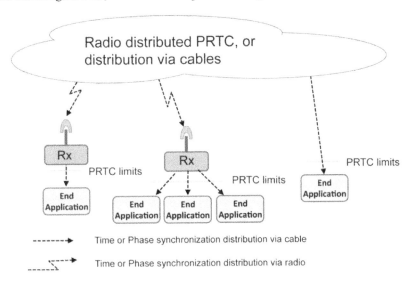

Figure 3.29. *Locally distributed PRTCs*

In this case, the synchronization network is very simple. Equipment directly generating a reference timing signal compliant with the PRTC specification (G.8272) is locally available in the site where this signal is required. The reference timing signal can be distributed from this equipment by means of suitable interfaces (e.g. 1PPS, see [G.8271]).

A typical implementation is based on a GNSS receiver (e.g. a GPS receiver). As an example, this has been the typical setup used for Code Division Multiple Access (CDMA) networks.

As indicated in Chapter 4, the use of satellite systems presents some drawbacks (e.g. issues related to the antenna installation), this is why this solution needs to be complemented, for example by means of the packet-based solution as described in the following section.

3.6.2. *Packet-based method*

The distribution of time synchronization via packets based methods implies a more complex architecture.

An example is shown in Figure 3.30, which is an adaptation of Figure 3/G.8271 (see also [G.8271]).

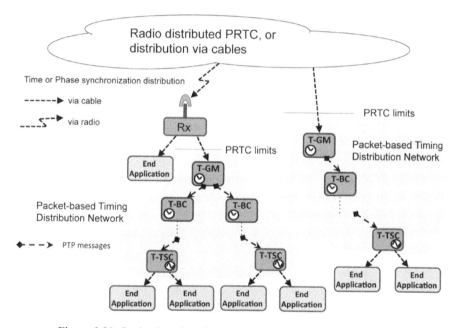

Figure 3.30. *Packet-based methods, with timing support from the network*

Timing is distributed via packets from a site where a signal compliant with the PRTC quality is present. Where high accuracy is required, this typically uses IEEE 1588 PTPv2. In this case, the PRTC reference is used by the grandmaster to generate the PTP timing messages as described in Chapters 2 and 4.

In the example shown in Figure 3.30, timing support is provided in every node by means of boundary clocks in order to increase the accuracy. The performance that can be achieved with this arrangement is currently being studied over a network reference model with 10 boundary clocks and the results of these studies will form the specification for the network limits to be included in G.8271.1, currently under preparation within ITU-T Q13/15.

Alternative arrangements where not all the nodes support IEEE 1588 (i.e. "partial timing support") are also being considered, but these will require careful analysis in order to verify if they can meet the stringent requirements required by some mobile applications (e.g. in the sub-microsecond range). Such models require careful consideration of non-PTP-capable nodes for their PDV and asymmetry generation (see [GOL 12]).

It should be noted that the PTP protocol can be used to distribute both frequency and time synchronization. Nevertheless, the combination with physical layer synchronization (typically SyncE) is also considered. In this case, the physical layer carries frequency synchronization and the packet layer (PTP) carries time synchronization.

The main advantage in combining PTP with SyncE is that it can enhance overall reliability as discussed in the following section.

3.6.3. *Resiliency and redundancy schemes*

Architectural and redundancy aspects related to time synchronization are being studied as part of the G.8275 specification. At the time of this book's, preparation only some high-level considerations can be provided.

In contrast to frequency synchronization, the requirements to maintain time offsets in the order of microseconds may place additional constraints on a system. Maintaining accuracy in terms of time offset, expressed in terms of absolute phase, when the timing reference is lost may require holdover performance substantially better than that required when considering frequency applications. This, of course, will ultimately translate into additional cost.

As an example, a local oscillator based on rubidium technology can maintain in holdover an accuracy within a few microseconds over a period of several hours. However, the cost associated with this type of solution might not be always acceptable.

As already mentioned, a combination of the locally distributed PRTC method with packet-based method is one possible way to increase the reliability of the solution. An example is shown in Figure 3.31.

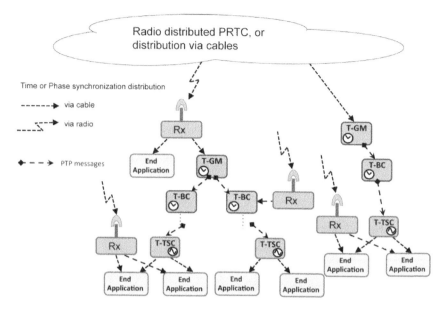

Figure 3.31. *Combined distributed PRTC and packet-based method*

By distributing the PRTCs closer to the access, two technical benefits should be underlined:

– first, it is possible to minimize the impact of failures in the PRTC chain (fewer users are impacted when a PRTC fails);

– second, the risk of creating uncontrolled asymmetries in the network is minimized because of the reduction in the number of hops between the PRTC and the end user.

Moreover, the combination of the two approaches in general also results in providing multiple references (direct GNSS-based reference and packet-based reference) in some important points of the network.

As in case of frequency synchronization network, the network should always be provided with redundant masters in order to enhance resiliency in the network (see Figure 3.32). As shown in the figure, the PTP clocks are normally synchronized by T-GM1. A secondary T-GM clock (T-GM2) would be available in case the primary master is failing.

ITU-T is currently studying the details on how this protection would actually work (e.g. in terms of BMCA).

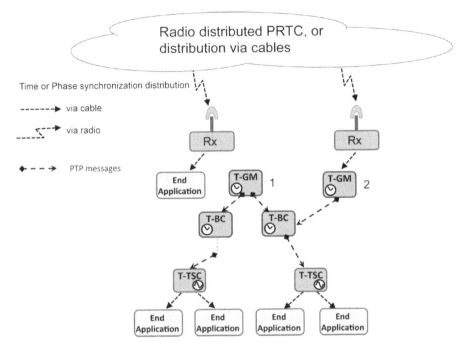

Figure 3.32. *Redundant PRTC and telecom grandmasters*

A PTP master with a stable oscillator (rubidium) may also be considered as an option to maintain acceptable accuracy for some period (several hours) when the PRTC reference (e.g. GPS signal) is lost.

In terms of GNSS, several options will ultimately be available such as GPS, GLONASS, Compass and Galileo, as described in section 2.7.

Combining various GNSS technologies can be considered as a way to enhance the reliability at the PRTC level, although this may not always help in cases where the issue is intentional or unintentional jamming (GPS and Galileo operate in the same range of frequencies). This is further discussed in Chapter 6.

SyncE could also be used as a complementing technology to either provide redundancy to the PRTC (e.g. when GPS signal is lost), or directly provide a reference timing signal for the end user so, that when PTP traceability is lost, a time synchronization signal can be maintained by means of the external stable frequency synchronization reference over relatively long periods; without the need of implementing an expensive oscillator, for example based on atomic technology (see Figure 3.33).

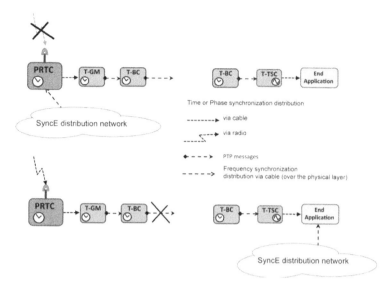

Figure 3.33. *SyncE enhancing time sync resiliency*

When SyncE is combined with PTP, the concepts of congruency and coherency become relevant. Congruency means that the SyncE timing signal and the PTP timing signal follow the same path from the PRC and PRTC, respectively. Coherent signals means that PTP and SyncE are generated by the same source.

These two concepts are described in Figure 3.34.

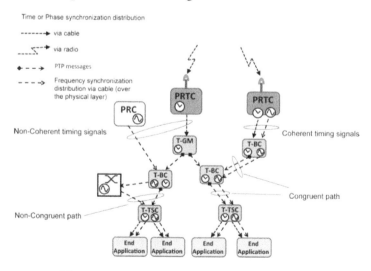

Figure 3.34. *Concepts of coherency and congruency*

The need to deploy a congruent or coherent network is currently under study. In general, it could be said that a coherent architecture could minimize the error during failure conditions (loss of PTP traceability) where the time sync is controlled only by the frequency synchronization signal. However, network and timing engineering might not allow configuring a coherent network. Moreover, congruency and coherency conditions in the network may introduce a possible undesired dependency between time and frequency distribution when the intention (with regard to the frequency synchronization signal) is also to enhance the resiliency. In this case, the same failure might impact both time and frequency distribution.

In practical terms, it is also uncertain if a network can always guarantee coherence or congruency under all conditions.

Further considerations on time synchronization deployments are provided in Chapter 4.

3.7. Bibliography

[802.1AX] IEEE Std 802.1AX™-2008, IEEE Standard for local and metropolitan area networks – Link Aggregation.

[1588-2008] IEEE Std 1588™-2008, IEEE Standard for a precision clock synchronization protocol for networked measurement and control systems, IEEE Instrumentation and Measurement Society, 2008.

[EG 201 793] ETSI EG 201 793 V1.1.1 (2000-10), Transmission and multiplexing (TM); Synchronization network engineering.

[G.781] ITU-T Recommendation. G.781, Synchronization layer functions, 2008.

[G.783] ITU-T Recommendation. G.783, Characteristics of synchronous digital hierarchy (SDH) equipment functional blocks, 2006.

[G.798] ITU-T Recommendation. G.798, Characteristics of optical transport network hierarchy equipment functional blocks, 2012.

[G.803] ITU-T Recommendation. G.803, Architecture of transport networks based on the synchronous digital hierarchy (SDH), 2000.

[G.805] ITU-T Recommendation. G.805, Generic functional architecture of transport networks, 2000.

[G.810] ITU-T Recommendation. G.810, Definitions and terminology for synchronization networks, 1996.

[G.812] ITU-T Recommendation G.812, Timing requirements of slave clocks suitable for use as node clocks in synchronization networks, 1998.

[G.813] ITU-T Recommendation G813, Timing characteristics of SDH equipment slave clocks (SEC), 1996.

[G.8010] ITU-T Recommendation. G.8010, Architecture of Ethernet layer networks, 2004.

[G.8260] ITU-T Recommendation. G.8260, Definitions and terminology for synchronization in packet networks, 2012.

[G.8261] ITU-T Recommendation G.8261, Timing and Synchronization aspects in Packet Networks, 2008.

[G.8261.1] ITU-T Recommendation. G.8261.1, Packet delay variation network limits applicable to packet based methods (frequency synchronization), 2012.

[G.8263] ITU-T Recommendation. G.8263, Timing characteristics of packet based equipment clocks (PEC) and packet based service clocks (PSC), 2012.

[G.8264] ITU-T Recommendation. G.8264, Timing distribution through packet networks, 2008.

[G.8265] ITU-T Recommendation. G.8265, Architecture and requirements for packet based frequency delivery, 2010.

[G.8265.1] ITU-T Recommendation. G.8265.1, Precision time protocol telecom profile for frequency synchronization, 2010.

[G.8271] ITU-T Recommendation G.8271, Time and phase synchronization aspects in packet networks, 2012.

[GOL 12] GOLDIN L., MONTINI L., "Impact of network equipment on Packet Delay variation in the context of packet-based timing transmission", *IEEE Communication Magazine*, 2012.

Chapter 4

Synchronization Design and Deployments

4.1. High-level principles

Massive architectural changes are occurring in telecommunication networks driven by customer needs and demands for bandwidth-hungry services. As packet-based switching and routing technology has been deployed and changed the nature of networks, a similar evolution is also impacting the resulting synchronization architectures. The obvious expectation in packet networks would be that any requirements for network timing would decline. However, this is certainly not the case. One of the key areas driving these changes is mobile phone communications and the drive toward real-time data services. For the next couple of decades, mobile networks will certainly become one of the key uses of the new synchronization architectures.

Most of the networks deployed today – and historically – have required frequency synchronization, more correctly called "syntonization". However, going forward, time and phase synchronization will be required. The distinction between frequency and phase/time synchronization has been covered in the introduction to this book. Certain applications have, in the past, required time which has commonly been served with protocols, such as Network Time Protocol (NTP) and stand-alone Global Navigation Satellite System (GNSS) receivers. However, it is again the challenge related to mobile networks that drives this take-up of time/phase. This chapter will describe some of the issues and challenges faced for both frequency and time/phase synchronization from a design perspective. A brief history of the evolutions of telecommunication networks and their impact on synchronization can be found in Chapter 1. This chapter starts off by describing some general principles and uses frequency synchronization examples. However, aspects of time/phase are introduced where pertinent. As the chapter progresses, more time and phase aspects and challenges are brought into the text.

Synchronization network design and the resulting deployment models have some simple guiding principles, but also considerable and subtle hidden complexity that needs to be understood. In essence, the objective is to derive a stable source of frequency or time, transport over a transmission medium or set of transmission mediums and then lock an application to that source of frequency or time, as described in Chapter 1. An application will essentially track the source but with the addition of any noise introduced by the medium. Synchronization design principles when applied to traditional networks based on Time Division Multiplexing (TDM) or increasingly more prevalent packet networks are the same, recovery of the stable; source of frequency or time.

A very simple model of this principle is shown in Figure 4.1. Frequency sources will be an atomic clock, which will generate the frequency. These may be owned, e.g. a cesium beam frequency standard, or may not be owned, e.g. when located in a satellite system such as the Global Positioning System (GPS). When GPS is used, a receiver provides access to a signal traceable to an atomic clock. The medium is a piece of network technology that distributes synchronization signals. At the lowest level in the network are transport mediums such as a fiber section, copper connection or a radio system. All these transport mediums, even a section of fiber, will have characteristics that will influence the transport of synchronization and introduce noise as well as clocks that are in general cascaded when transporting the synchronization reference. However, networks are more complex and are built from layers of technology, such as optical transport equipment, transmission equipment and packet-based equipment; all these influence the transport of synchronization differently. Finally, the application will recover the synchronization and effectively be the sink for the synchronization and use it in order to perform its designed function. Even in this final stage, noise will be introduced. Different applications often carried over the same network will have different requirements, all of which have to be considered and balanced as presented in [GIL 07].

Very few networks are designed to carry only synchronization signals. Essentially, the network is a complex system design that carries information; synchronization signals are just one form of information (synchronization networks can be considered from this perspective as overlay networks). Synchronization may be a requirement to allow both the network equipment and the end application to work efficiently (for instance, as explained for TDM networks in the next sections). However, it may only be required for the end application. Regardless of the type of technology and the information carried, equipment and their clocks will introduce noise that will disrupt the transport of synchronization information.

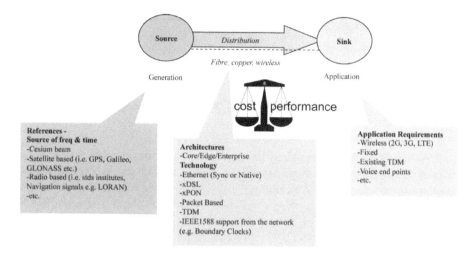

Figure 4.1. *Balance of reference source, architecture and technology and application requirements*

Synchronization design is often considered an afterthought, with the design of the network to carry revenue-earning traffic developed first. This often results in synchronization design being added afterwards. Such an approach can lead to either a less than optimal design or considerably higher cost. The synchronization design should be considered as a key component of any design and given the same level of consideration from the start.

Development of a synchronization design for a green field (i.e. a completely new deployment) may be totally different to what can practically be achieved when considering a brown field deployment (e.g. migration between technologies).

Many existing networks are capable of distributing or carrying frequency synchronization only; hence, most current changes are, in effect, brown field as long as only frequency synchronization is required. Technology migration may occur, but frequency synchronization requirements may remain the same and could be addressed by new solutions providing a similar performance as the previous ones, for example Synchronous Digital Hierarchy (SDH)-based transport provides similar synchronization quality as Synchronous Ethernet (SyncE)-based transport. However, this changes with the introduction of time/phase distribution because many existing networks do not have the capability to transport phase/time with high accuracy, as will be discussed further in section 4.4.

4.1.1. *Network evolution*

4.1.1.1. *Existing TDM-based technology*

Another issue to consider is the evolution of different networking technologies and how it impacts synchronization design. A key area to understand is the capability delivered by existing TDM-based networks and their synchronization design.

Existing TDM-based technology has provided a stable base on which to transport frequency synchronization. Not only did the applications or services served require synchronization, but also the technology itself for inter-working. A classic example of the synchronization network is the Public Switch Telephony Network (PSTN) built on 64 kbit/s TDM switching. All the switches require frequency synchronization to minimize data slipping through buffers. The switches are connected together using TDM-based transmission connections, typically E1 2.048 Mbit/s or DS1 1.544 Mbit/s hierarchies. These transmission connections based on Plesiochronous Digital Hierarchy (PDH) effectively provide a transparent pipe through which not only does the information flow from the switches but also the frequency used to provide synchronization to the switch. Ultimately, this synchronization frequency is also provided to the Digital-to-Analog (D/A) conversion at the edge of the PSTN in a remote concentration unit. Further details can be found in Chapter 1.

These PSTN networks were typically fully owned by an incumbent operator, who owned the complete network with core and access transmission and service switching. Connections between networks would be traceable to different reference clocks, and hence, there would be "slips" of the data between these networks on these international or even national links as historically national networks have been broken up. It is the reason why the use of reference clocks with high accuracy is important to minimize the occurrence of slips in these cases.

4.1.1.2. *Migration from TDM to Ethernet*

Migration from TDM-based networks to Ethernet-based networks has increased in recent years as Ethernet has become a more mature technology. Chapter 1 provides details on some of the technical aspects. This section will concentrate on setting the scene to illustrate some of the design and deployment challenges.

Ethernet started out as a technology designed to connect computers, terminals, workstations and peripherals together in the local environment, commonly called the Local Area Network (LAN). Typically, LANs are privately owned networks within a building limited by distance to a couple of kilometers. Over time, Ethernet has migrated out of the LAN environment and into the Wide Area Network (WAN) environment. This development has been driven by the lower cost of Ethernet interfaces. The commonly held view is that Ethernet interfaces have a much lower cost than comparable TDM-based interfaces such as SDH. This may be correct, but SDH has developed a very rich set of operations and maintenance capabilities over time which are only now starting to be established in Ethernet. Synchronization could be considered as part of these inherent capabilities and will be discussed later in this chapter. However once these considerations are factored in cost can be equivalent. Another area for comparison is the bandwidth capability on the interface and its flexibility.

Another key reason to consider Ethernet interfaces in WAN is the ability to carry multiple services and thus the resulting impact of a lower unit cost for all service types. Moreover, the ability to carry high bandwidths becomes available. Communications providers are not only under pressure to reduce costs, but also increase bandwidths for new service types. Typically, networks have been built to carry a certain type of service or traffic that can be achieved very efficiently to a wide scale, but if a new network has to be built for every service or customer segment as a whole, this becomes inefficient. Efficiency can be created by carrying different types of services over the same network – convergence of services and networks.

4.1.1.3. Synchronization evolution

Evolution of the synchronization architecture from TDM networks to packet-based networks could, on initial inspection, be seen as an important change that requires fundamental study. However, this needs to be caveated by the requirement to understand what is trying to be achieved, for what purposes and with what technology.

An initial high-level analysis of different technology types will clearly show that TDM technology can provide a stable line code at the physical layer from which to recover a synchronization reference signal. This signal can be used to synchronize an oscillator in each respective item of equipment through the synchronization chain. Thus, from a synchronization perspective, TDM networks can form a stable, deterministic distribution that is not impacted by traffic load or traffic variation.

Packet-based technology, in contrast, was designed to operate in an asynchronous mode. Typically, there may be no underlying line codes at the physical layer carrying a synchronization reference signal traceable to a primary reference clock. Indeed, such systems allow a design where each equipment in the traffic delivery path would have an oscillator that can be free running. In such equipment, the signals carrying the consecutive packets of information are, in general, transmitted at a different frequency than the one used by the previous equipment, because each equipment may rely on a local free running oscillator to generate output signals; the offset between these frequencies is compensated by adjusting the "holes" between packets.

Although such an approach allows the underlying transport infrastructure to operate and packets of information to be sent over this infrastructure, it would not allow, for example, frequency synchronization to be transported at the physical layer for use by other applications or equipment. Many applications exist that require frequency synchronization and stability. A couple of common application examples are mobile base stations and TDM emulation end points.

Mobile base stations require frequency synchronization to maintain the radio spectrum being used within the required limits. This is in order to avoid interference, ensure smooth handovers and finally to allow efficient use of the allocated spectrum and maintain good customer experience. Base stations are typically located on the edge of a providers' network.

TDM emulation end points provide an Inter-Working Function (IWF) between the packet network and the TDM network. These require frequency synchronization to meet TDM timing requirements. These end points will typically be at the edge of providers' network where they will connect to TDM-based equipment or provide a gateway into a TDM-based network.

In both these cases, the application is on the edge of the providers' network. Clearly, the requirement for frequency synchronization is moving away from an infrastructure and application requirement of the core network in the TDM world toward a requirement of the edge application only in the packet-based world. But in contrast, the ability to transport synchronization from the center to the edge declines as the infrastructure evolves toward a packet-based architecture. Assuming that synchronization can no longer be transported over the packet-based network, this evolution can be best visualized as a disk with an increasingly large hole in the center. An analogy can be drawn to a doughnut without a hole versus a doughnut with a hole when describing the need to have access to a synchronization reference [GIL 06]. The dough covers the area where synchronization is required and the jam is the primary reference (i.e. the PRCs). The hole describes both the declining requirement of the infrastructure to use synchronization and its inability to transport it toward the end applications located at the edge. This evolution is shown in the following figures.

The following sections provide a very simple illustration of this evolution to represent some of the issues in terms of synchronization transport. It quite clearly and deliberately does not cover many other issues that result from TDM to packet-based networks evolution and that are the subject of many other bodies of work. This illustration justifies why having the ability to carry synchronization over packet-based networks is important and why the techniques described in sections 4.3 and 4.4 have been developed.

4.1.1.3.1. Simple TDM network

Figure 4.2 shows a very simple network, which could represent a typical 64 kbit/s switched network based on TDM carrying voice. The five switches in the center of the network are highly meshed and represent the core network. These will typically be the trunk-switching layer connecting large geographic areas together, such as large cities and towns. Switches toward the edge of this core are the local switches; these will typically be connected to two or more core switches. Local switches provide the aggregation point for the telephones. The telephones provide the terminal for the voice service.

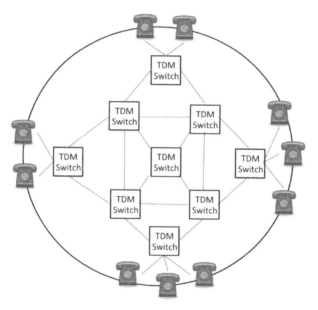

Figure 4.2. *Simple switched network carrying voice*

Each switch carrying or generating E1 signals as represented in Figure 4.2 generally requires frequency synchronization (note: it does not apply to high-level PDH containers). Figure 4.3 shows the underlying synchronization connectivity of this simple switched network. In this simple network, the switch in the core is chosen as the source of frequency synchronization. A suitable primary source (e.g. cesium clock) is connected to this node and corresponds to the Primary Reference Clock (PRC). Suitable links are then chosen to pass synchronization onward to the core nodes where the oscillators/clocks within the switches are synchronized. In turn, these core nodes will then pass synchronization on toward the local nodes and their internal oscillators via selected links. Finally, the local switches pass synchronization onward to the telephones where the frequency is used to lock the D/A converter. In this simple network, only the active transmission links are shown with the direction of synchronization flowing down the link from core to edge.

Within this network, almost every device and every switch requires a synchronized frequency to work correctly and pass the voice without any degradation. Clearly, the entire network is covered with this disk of synchronization. Using the doughnut analogy, this is a solid doughnut with the PRC being the jam in the center.

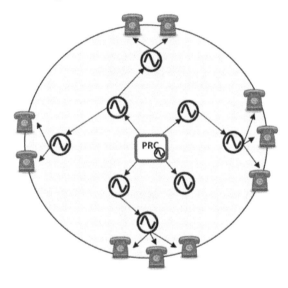

Sync Only Paths

Figure 4.3. *Main synchronization paths for simple switched network*

4.1.1.3.2. Simple hybrid TDM/packet network

To gain efficiencies within this simple network, the voice services can be treated as data and combined with other data sources. Typically, the core voice switches can be removed and replaced with a soft switch that will perform the function of determining where and how to connect voice calls. The existing voice services can remain connected to their parent local 64 kbit/s switch. However, the connections back from these switches into the core of the network are now delivered via routers, which also carry other data services, thus creating efficiencies.

Figure 4.4 illustrates this evolution. The five 64 kbit/s TDM switches in the core have been replaced with routers. To keep the model simple, they are all still connected together using the same connections.

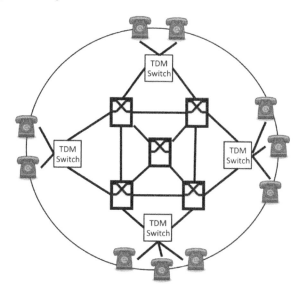

Figure 4.4. *Simple switched network carrying voice with core replaced by data routers*

Routers and packet switches do not, as such, need synchronization to work. However, the remaining voice switches and telephones carrying the voice service do. Figure 4.5 represents one model for how this has changed the synchronization network. The core routers now have free-running oscillators and no ability to pass synchronization from the previous centralized PRC down to the Core routers and onward to the local switches as shown in Figure 4.3.

A hole has now opened up in the center of the disk where synchronization is neither required nor can it be transported. Essentially, a hole has appeared in the doughnut.

However, the local switches do still require synchronization as described in Chapter 1. The only result is to push the source of the synchronization closer to the edge. This may for instance require more PRCs to be deployed and then essentially creates a cost.

It also creates other potential challenges as it may increase the uptake of less suitable sources such as wider scale deployment of GNSS-based technologies.

Clearly, deploying many PRC functions in the edge of the network is not cost effective in general; hence, new solutions that allow carrying frequency synchronization over packet-based networks were required. Some of them will be discussed in section 4.3.

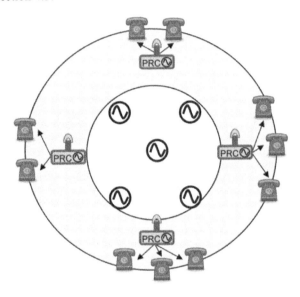

Figure 4.5. *Change in synchronization for a hybrid network with a mix of core routers and edge-located TDM switches*

Traditional voice as represented in the models just discussed is only one aspect of this evolution. It could be argued that this method of transporting voice is declining, and hence, the problem will not exist in a few years. Therefore, it is worth exploring other changes that are taking place in the architecture and their impact on the underlying synchronization architecture. One such change is mobile backhaul.

4.1.2. *Typical mobile networks requirements and evolutions*

Before looking at the high-level evolution that is happening driven by mobile networks, using the simple models presented in section 4.1.1, it is worthwhile discussing the evolution of the mobile base station connectivity, which follows the TDM to packet-based networks evolution, and what is driving these requirements and their impact on the Next-Generation Network (NGN) synchronization.

4.1.2.1. Evolution of base station connectivity

A major mobile requirement is cost reduction of the backhaul. Initial 2G/3G base station designs used native TDM interfaces only. These interfaces were originally served by DS1/E1 connectivity, but this connectivity has migrated to hybrid circuit emulation solutions relying on packet-based networks. Such an approach allows the existing interfaces to be supported using a lower cost base built on Ethernet connectivity. This approach also provides better flexibility to increased bandwidth.

As long as TDM interfaces are used at the base station, TDM synchronization performance requirements clearly need to be maintained and some operators have achieved this by continuing providing a DS1/E1-leased line served connection. In other cases, Adaptive Clock Recovery (ACR) principles as described in Chapter 2 have been used. Regardless of the solution, these will have a finite lifespan as they effectively require the use of TDM-based signals.

Recent growth in intelligent mobile devices has driven the demand for bandwidth and thus for direct Ethernet-based interfaces into existing base stations. As Long-Term Evolution (LTE)/4G base stations are deployed, Ethernet-only interfaces carrying Internet Protocol (IP) will dominate and at this point new synchronization solutions may be required.

4.1.2.2. Impacts of TDM to packet migration on mobile networks

Mobile base stations have always been key users of synchronization. There are many ways in which this can be achieved. Typically, either by recovering synchronization over TDM-based connectivity from an upstream source with physical layer solutions similar to those described in section 4.3.2 or by deploying some form of GNSS source at the base station. The choice would be down to a number of factors, such as who owns the connectivity (owned by the mobile operator or use of leased line), are there concerns over the use of GNSS (e.g. cost of GNSS), etc.

With the migration from TDM to packet, without the use of new timing techniques, synchronizing mobile base stations becomes problematic; it is therefore useful to describe the impact of such migration. Expanding on the simple model presented in the earlier sections, Figure 4.6 now shows a backhaul network entirely based on packet routers providing connectivity to mobile base stations.

As described earlier, the routers do not require synchronization to operate. Because these routers have free-running oscillators and no ability to pass synchronization from any centralized PRC, the only solution would be to deploy GNSS-based technology acting as PRCs at the base stations. This is illustrated in Figure 4.7.

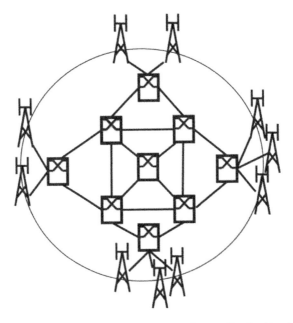

Figure 4.6. *Simple fixed/mobile network providing backhaul*

Similarly to the hybrid TDM/packet case presented in section 4.1.1.3 for traditional voice services, a hole has now opened up in the center of the disk where synchronization is neither required, nor can it be transported. The hole has, however, become much larger than the hybrid TDM/packet case shown in Figure 4.5. As synchronization can no longer be delivered from a centralized PRC located with a network switch in the center of the network, providing economies of scale, a greater number of PRCs may now be required. The jam (i.e. the PRCs) needs to be spread through the disk of the doughnut.

In the case of large mobile networks, this can be many tens of thousands or even hundreds of thousands of sites. Without suitable methods of transporting synchronization from a source to the point of consumption, corresponding to the base station in this case, considerable cost, complexity and issues of security of supply can exist.

With the cost and bandwidth drivers to migrate from TDM networks to packet networks based on Ethernet, clearly there was a requirement to also develop new techniques of transporting synchronization over packet networks to solve this synchronization hole. This was one driver to develop SyncE, whose concepts are introduced in [FER 08], and Precision Time Protocol (PTP) technologies, to give additional choice against the GNSS-only options as has been expanded in Chapter 3.

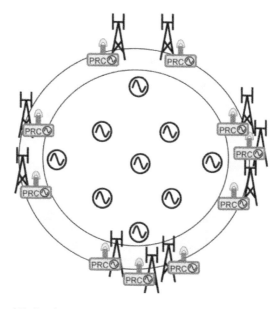

Figure 4.7. *Synchronization provided at edge for mobile base stations*

A number of use cases can be explored for mobile networks. Migration from TDM-based backhaul to Ethernet-based backhaul in a fully owned network versus migration in the case of a leased line example. These cases are discussed in the next sections.

4.2. MAKE or BUY network synchronization strategies

In many businesses, MAKE or BUY strategies may be applicable and can be part of the strategic discussions of how to build a network to carry services. Advantages and drawbacks can be found within both approaches, depending on the context, e.g. time to market, responsibility versus flexibility, cost and independency. It is also the case for synchronization networks.

Key decisions for a network carrier are the MAKE or BUY synchronization strategies. In the synchronization context, MAKE or BUY can apply to various aspects; it may be related to the synchronization reference (for example: buying a PRC and designing a synchronization network versus leasing a synchronization reference with a guaranteed performance and reliability), it may also be related to the primary reference source used (for example: using a cesium-based PRC versus a GNSS-based PRC) and it is finally closely related to MAKE or BUY connectivity strategies considered by the network operator. These models can exist for both frequency and phase/time synchronization, sometimes with different considerations, for instance, because of the different levels of synchronization quality expected.

Many factors form part of this decision but probably some of the key factors are size of the carrier and scale of its network versus the level of risk it wishes to carry. Risk or perceived risk tends to increase as the size of the operator increases. This may be based on business risk, for example, loss of supply impact service or risk of the technology being stretched beyond its technical limits due to the physical size of the network.

Another key factor is the type of network technology deployed and its ability to transport synchronization. This may partly be dependent on the ownership of the deployed technology or parts of that technology. It may be fully owned by one operator or a mix of providers. For example, the network may be based on a managed service, may be managed bandwidth, have leased lines or may be built using managed fiber or even dark fiber. Some of these key aspects will be covered later when discussing key packet network architectures.

4.2.1. *Relationships between MAKE or BUY strategies for network connectivity and synchronization*

Actually, it is worth differentiating two fields where MAKE or BUY discussions could apply and impact synchronization operations:

– *with regard to the connectivity*: network connectivity of an operator may fully rely on its own network or may be based partly on leased lines provided by another carrier operator. This scenario is very common in mobile networks, where some mobile operators buy connectivity from carrier operators. The choice is, in general, made independently of synchronization operations, but may have an important impact on them.

– *with regard to the synchronization reference*: the reference signals provided to the end equipment of an operator, such as base stations, may come from the same operator or from a different operator. Again, advantages and drawbacks apply with each approach, the key point being to provide a timing reference with the appropriate quality to the end equipment. Different considerations may well apply in the case of frequency synchronization versus phase/time synchronization.

Table 4.1 summarizes possible relationships between the two fields described above.

	MAKE connectivity	BUY connectivity
MAKE synchronization reference	Applicable scenario, examples are fully owned large networks (e.g. incumbent fixed operators)	Applicable scenario, examples are, for instance, mobile operators relying partially on leased lines provided by a carrier operator, but willing to provide their own timing reference to their base stations (see note 1)
BUY synchronization reference	Scenario not applicable in general in telecoms	Applicable scenario, examples are mobile operators relying partially on leased lines provided by a carrier operator, where a synchronization service belongs to the managed service of this carrier operator (see note 2)

Note 1: A typical example is when frequency synchronization is carried by 2.048 Mbit/s, transported transparently over leased lines.

Note 2: This was not necessarily always common practice in existing TDM networks (although it does exist), but may become more common in packet-based networks and increasingly when accurate phase/time synchronization is targeted.

Table 4.1. *Relationships between MAKE or BUY scenarios*

The three applicable scenarios depicted in the table are illustrated just after. Depending on the synchronization technology used, there are different means to build these different scenarios (these technical solutions will be discussed later in this chapter).

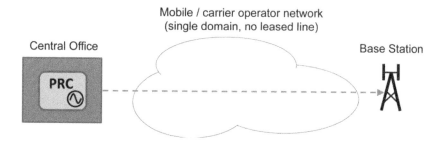

Figure 4.8. *Illustration of "MAKE connectivity/MAKE synchronization reference" scenario*

164   Synchronous Ethernet and IEEE 1588 in Telecommunications

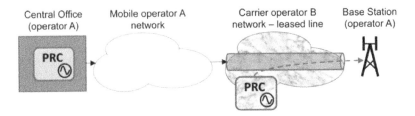

Figure 4.9. *Illustration of "BUY connectivity/BUY synchronization reference" scenario*

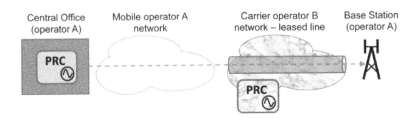

Figure 4.10. *Illustration of "BUY connectivity/MAKE synchronization reference" scenario*

One last example may also be applicable: it is the case where synchronization is not delivered by the network, but by means of solutions outside the transmission network, such as GPS or satellite systems more generally (e.g. future Galileo). In this case, the "BUY connectivity/MAKE synchronization reference" scenario may be applied as illustrated in Figure 4.11.

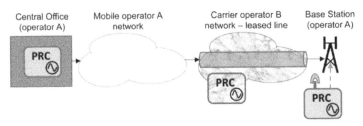

Figure 4.11. *Illustration of "BUY connectivity/MAKE synchronization reference" scenario – GPS/Galileo alternative*

The scenarios involving multiple operators can be analyzed both from the perspective of the carrier and of the mobile operators. For the mobile operator, the critical point is that their end equipment (e.g. mobile base stations) would be properly synchronized, whatever the scenario or technical solution. For the carrier operator, the objective is either to deliver a timing reference with the proper quality or to ensure that the timing reference provided by the mobile operator is transported "transparently" through his network without too much degradation. These two views are discussed hereafter.

Clearly, different considerations may apply in the case of frequency synchronization versus phase/time synchronization. The requirements related to frequency synchronization are indeed less stringent in general than those related to accurate phase/time synchronization. Therefore, there may be a lot more flexibility to fulfill these frequency requirements using various approaches, compared to accurate phase/time synchronization for which providing a synchronization source from a well-engineered network as a synchronization service may be the only technical alternative to GNSS-based solutions.

First, from the mobile operator perspective, it is important to stress that there is no technical issue in the case a where carrier operator provides a synchronization reference to a mobile operator as long as the reference provided has the proper expected quality. This case is mentioned in Table 4.1 ("BUY connectivity/BUY synchronization reference"). Where timing can be provided from carrier networks, this becomes more of a commercial rather than technical issue.

Indeed, even if the references provided to the base stations of the mobile operator do not come from the same source (assume, for instance, that some of the base stations receive a reference directly from the mobile operator and the others from a carrier operator as part of a managed service/leased line), as long as all the delivered references have the appropriate required quality, the network operations will not be impacted.

Typically, Frequency Division Duplex (FDD) mobile networks require that the base stations are synchronized with an accuracy within +/− 50 ppb from the Coordinated Universal Time (UTC) frequency. As long as all the references are traceable to a PRC and the distribution does not degrade the signals too much (i.e. does not generate too much wander), the base stations will be correctly synchronized, regardless of whoever is providing the reference and even if different PRCs generate the signals delivered to the base stations.

Theoretical exceptions might exist in the case where some of the signals delivered to the end application are not PRC-traceable, but are plesiochronous. In practice, these legacy cases (for instance, interconnection of Private Automatic Branch eXchange (PABX)) are no longe observed in telecommunication networks in principle, because even plesiochronous applications are synchronized normally with PRC-traceable signals or can be configured with this mode of operation.

Second, the notion of "timing transparency" is applicable to the "BUY connectivity/MAKE synchronization reference" scenario. Depending on the technical solution used, timing transparency is characterized differently:

– *in the case where physical layer synchronization methods are used (e.g. SyncE)*, then timing transparency implies that the timing information of the physical layer of the client signals carried over the leased line is not lost and requires in general particular functions to be supported by the network nodes of the carrier operator (this can be achieved, for instance, by justification process). This type of solution is possible in some cases when frequency synchronization is needed (e.g. SyncE over Optical Transport Networks (OTN) with appropriate timing transparent mapping, SyncE over dark fiber, PDH over SDH, etc.), but is not applicable in general to accurate phase/time synchronization delivery (except with dark fiber). However, timing transparency is more common with PDH client signals than with SyncE client signals.

– *in the case where packet-based methods are used (e.g. IEEE 1588 in end-to-end mode)*, then timing transparency does not require particular functions dedicated to synchronization to be supported by the network nodes of the carrier operator because synchronization is transmitted as traffic via packets. However, without functions dedicated to synchronization, this type of solution can address in general frequency synchronization needs when the requirements are not too stringent (e.g. +/– 50 ppb requirement of mobile base stations), but is much more challenging for accurate phase/time synchronization delivery. This is due to the fact that the timing reference is highly impacted by packet jitter, also called Packet Delay Variation (PDV), with these kinds of techniques.

Finally, as previously mentioned, it is important to stress that the "BUY connectivity/MAKE synchronization reference" scenario raises important challenges when stringent phase/time synchronization is needed at the end, application; indeed, the degradations of the timing reference makes it, in general, incompatible with the stringent phase/time synchronization requirements of mobile networks (e.g. 1 microsecond accuracy). Therefore, according to today's knowledge, the need for very accurate phase/time synchronization requires relying either on its own network, on a specific phase/time reference provided by the carrier operator or on GNSS systems. Studies have recently been initiated to understand and clarify the impact of utilizing IEEE 1588 over networks that do not provide any intermediate support in packet network. It is too early to discuss this approach whose specification was just starting when this book was written.

4.2.2. *MAKE or BUY network synchronization source strategies*

Another key decision that must be addressed, which is linked to the MAKE or BUY synchronization strategy in general, is the synchronization source to be used as the network reference. These sources can be broken down into two key groups: off-air based reference sources and network-located reference sources.

Off-air systems can be divided into two major categories. They can be satellite-based or terrestrial radio systems-based. A number of different types exist in both categories. These are discussed in the following sections.

The network-located reference sources, which are often based on cesium clocks owned by the network operator, are not detailed here. Their related architectures are presented and discussed in Chapter 3 and in sections 4.3 and 4.4.

4.2.2.1. *Off-air–based synchronization sources relying on satellite systems*

One source of reference for synchronization that should be considered is the use of what are typically called off-air systems. These systems allow the network operator to derive their frequency from a reference source that someone else may own. Typically, another party owns the atomic clock standards and transmits a frequency or time code based on these standards using radio signals over the air, hence the title "off-air." This has a number of advantages, but also some key disadvantages, all of which have to be considered when determining a suitable reference source for deployment.

4.2.2.1.1. Considerations when using satellite-based reference sources

Satellite-based reference sources have the atomic clocks located in satellites. Frequency or time code information is transmitted from these satellites on dedicated frequencies. These frequencies are transmitted at very low power and are recovered using dedicated receivers. The atomic clocks located in the satellites are constantly monitored and corrected from dedicated ground stations. One or a number of ground stations will be responsible for creating the frequency or timescale, which is transmitted to the satellite on a frequent basis. Within the ground station, many more atomic clocks, cesium beam frequency standards and hydrogen maser clocks will be used to create the frequency scale. These frequency sources will be traceable within known limits to national timescales.

Common satellite-based systems, typically called GNSS, are GPS, GALILEO and GLONASS. A detailed description of these typical satellite technologies is given in Chapter 2, section 2.7; therefore, the actual principles of such technology are not further discussed here. However, what is relevant is the applicability of such systems in the synchronization design and their value in deployment scenarios.

168 Synchronous Ethernet and IEEE 1588 in Telecommunications

4.2.2.1.2. Advantages and disadvantages of satellite-based synchronization sources

One of the key advantages in deploying such systems is that it allows the expensive source (i.e. the atomic clock) to be owned and maintained by someone else rather than deploying dedicated cesium beam frequency sources.

Other advantages may come from distributing the reference more widely across the network. It allows the distance from source to sink to be shortened. This has the advantage of flattening the network, thus reducing noise accumulation from lengthy distribution chains. Such an approach can also simplify the synchronization planning. Specifically, designed distribution to these injection points would no longer be required. However, there is the increased cost of this distributed reference to be considered.

Another reason may be from a planning perspective. It may be a design requirement to segregate the network into specific dedicated regions all with their own source.

Cost of delivering the reference should also be considered. Although a typical synchronization receiver qualified for telecommunication networks may be at a lower cost than a cesium beam standard, the cost of the antenna and its installation and connection to the receiver must be considered. This installation cost and connection in building should not be underestimated. In some extreme cases, with very large complex buildings or lengthy connections between antenna and receiver, the cost can be comparable to a cesium beam frequency standard. In some cases, the addition of a GPS antenna to an existing structure may also need to be added to the operational costs. Some buildings may not be owned by the operator, and new antennas may result in additional year-on-year license costs of leasing the space, which will add to the OPerational EXpense (OPEX) budget.

It should be noted that the use of off-air systems to inject a timescale may be the only method of delivering time or phase information. Indeed, a cesium beam frequency standard provides a frequency but has no time base or knowledge of phase. Off-air systems provide their own traceable source of time that can be referenced to other time bases so that their relative offsets can be determined.

Although off-air signals, such as GPS, are almost universally available and very reliable, this does create a weakness. Essentially, it is assumed they always work. With modern life increasingly using satellite navigation and GPS in smart phones and tablets and our personal experience suggesting they always work and are correct, it is difficult to perceive that they can go wrong or be interfered with. However, that is the case. The signal transmitted by the GPS is very weak and can easily be disrupted by accidental interference or deliberately jammed. In fact, this is relevant to all off-air signals based on GNSS. Security weaknesses of GNSS systems are discussed in Chapter 6.

4.2.2.1.3. Typical user cases of satellite-based reference sources

Two typical user cases of GNSS based systems exist. The first user case is for generating the reference within the core of the network. This can be at a central location providing the main PRC of the network. This source is then distributed down through the network in a hierarchal fashion to the eventual application.

The second user case may be at the edge of the network again for generation, but where no synchronization is carried over the network connection. In this case it provides a local source of PRC. A typical example would be providing the reference source to a mobile base station.

4.2.2.2. *Off-air–based synchronization sources relying on terrestrial radio systems*

Reference sources based on terrestrial radio systems are far less common. However, for completeness, some information is provided.

Similar to satellite solutions, these again have reference atomic clocks typically located with the transmitting site. Frequency, or time code, information is transmitted on dedicated frequencies. However, unlike satellite-based systems, these frequencies are transmitted often at very high power and low frequency. The frequency is recovered using dedicated receivers.

The atomic clocks will typically be constantly monitored and corrected from dedicated ground stations; these may be the transmitting stations or another reference site. Typically, these ground stations may have some form of connection to national timing labs or even be associated to these timing labs. So, the transmitting ground station can be kept traceable or at least within known limits to national timescales.

Common systems are LOng-RAnge Navigation (LORAN) and national time frequency standards such as DCF (Deutschland long wave signal emitted from Frankfurt) and MSF (UK low-frequency time signal and standard frequency radio station based on the NPL timescale UTC (NPL)). Further descriptions are given in Chapter 2, section 2.7.2.

Compared to GNSS systems, the use of low frequencies with these systems may allow better in-door penetration of the radio signals. However, these systems are not available worldwide compared to GNSS and hence not widely used today in telecommunication networks. Also, consideration must be given to the lack of redundancy/resilience whereby the loss of the transmitter site will result in a loss of reference. Hence these systems can never be considered as the sole source of reference in a telecommunication network.

4.2.3. *Fixed/mobile network scenarios*

One of the key applications driving connectivity on fixed networks is the demand for bandwidth from current and future mobile networks. It is useful to understand in a little more detail the various scenarios that can exist to connect the components in a mobile network so as to better understand the challenges faced for synchronization. Some aspects have already been presented in Chapter 1. Figure 4.12 provides a very simple model. In this model, a switch provides the connection point for all the base stations connected together within its domain of control. At each base station location, a circuit may connect directly onto the base station or may connect to some form of Customer Premises Equipment (CPE) providing an operational demarcation point.

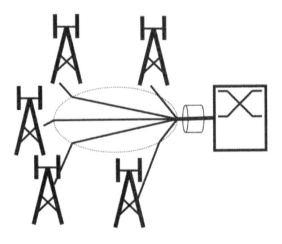

Provider A - MNO

Figure 4.12. *Simple mobile radio access network connectivity*

Most discussions on the synchronization for this Radio Access Network (RAN) connectivity will use this simple model, which shows the network as a simple "network bubble". On simple inspection, the conclusion could be reached that the mobile network operator owns all the backhaul and all the access. This is in fact correct; some operators are combined fixed and mobile operators. However, as explained earlier in section 4.2, the reality is much more complex. The Mobile Network Operator (MNO) may have some owned access (e.g. MicroWave (MW) point-to-point connections) and backhaul. However, to get reach, the MNO may also sub-contract to a fixed line provider to obtain some form of long-distance connectivity (Provider A). For the final access link (e.g. MW point-to-point, fiber or copper access type system) to the base stations, the MNO may have to sub-contract yet again to another connectivity provider (Provider B), see Figure 4.13.

4.2.3.1. *Mobile network operator provides synchronization*

The normal model of synchronization distribution would see the MNO make and inject their synchronization source into the RAN switch using options described in section 4.3 and transparently pass this down to the various base stations. In the TDM world where these connections are based on PDH clients within a transmission system, this is relatively easy as these links are transparent to physical layer timing. Figure 4.13 expands the simple RAN model to show this.

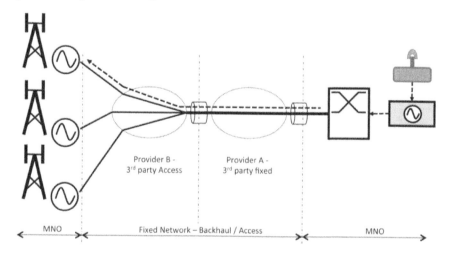

Figure 4.13. *Simple mobile RAN connectivity with additional fixed access to provide reach, MNO-owned synchronization*

Moving toward Ethernet and packet-based solutions, MNO-owned synchronization can start to create some challenges and some real deployment issues. As we have already discussed in Chapter 1, Ethernet and packet-based networks are no longer fully transparent to timing in many cases. In some cases, there may be no timing transparency between the MNO switch and the base station. These aspects will be further discussed in sections 4.3.2.3 and 4.3.3.1.4, together with synchronization techniques applicable over packet-based networks, which have been described in detail in Chapter 2.

Even if techniques such as IEEE 1588 end-to-end (discussed in section 4.3.3 and described in further detail in Chapter 2 and in [FER 10]) are used, dependent on how the fixed network is implemented, PDV will exist, and full transparency as accepted in the TDM world may no longer exist. At a protocol level, IEEE 1588 end-to-end will be transparent from a protocol perspective, but it is not fully timing transparent from a timing perspective. Some short links with few hops may exist, which have very good performance. However, other links may exist that have many hops and, by implication, much worse performance. However, this simple rule (i.e. counting the number of hops) is not strictly correct as it will be dependent on the actual technology of the equipment, its operation and the loading. This loading will be the actual traffic load, the traffic mix and the variation of these factors. Tolerance of the packet slave clock is another key factor, but is very dependent on the actual implementation; some packet slave clock designs may indeed tolerate high levels of PDV especially when the end application requirement is not stringent (e.g. when the packet slave clock provides frequency synchronization to a mobile base station). Minimum tolerance has recently been standardized in the case of frequency synchronization, see International Telecommunication Union - Telecom (ITU-T) Recommendation G.8263 [G.8263].

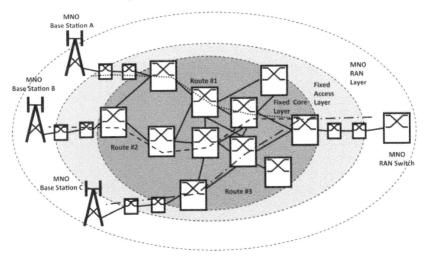

Figure 4.14. *Simple mobile RAN connectivity with mixed technology – MNO-owned synchronization*

Exploring the hop count issues further, a simple model showing the mobile RAN connectivity across one type of fixed network scenario is illustrated in Figure 4.14. This model is based very much on the mixed technology scenario as discussed in the ITU-T Recommendation G.8265 [G.8265], section 7.4, which talks about the use of different technologies in the core and access, and the impact of PDV on the ability of slave clocks to derive a suitable frequency reference. This mixed technology scenario is expanded further to illustrate some of these issues.

The MNO-RAN layer contains both the mobile switch controllers and the base stations; this is the outer layer of this network bubble.

The next inner layer of this network bubble contains some form of fixed access layer. In this example, it will be assumed that these are just point-to-point connections. However, these could have some form of active packet processing inside or may be formed from Digital Subscriber Line (xDSL) or Passive Optical Network (PON) technology; this could certainly be the case for connectivity to some of the base stations. The final inner layer of this network bubble contains Layer 2 packet switches and Layer 3 routers. So, this model is not complicated further (the actual Layer 1 connectivity is not shown), and it will be assumed that the access layer does not have any active packet processing.

Three routes are provided across this network, routes #1, #2 and #3.

1) *Route #1 – from MNO switch to base station*

Synchronization source – MNO switch – access – switch – router – router – switch – access – CPE – base station;

2) *Route #2 – from MNO switch to base station*

Synchronization source – MNO switch – access – switch – router – router – switch – switch – access – CPE – base station;

3) *Route #3 – from MNO switch to base station*

Synchronization source – MNO switch – access – switch – router – switch – access – CPE – base station.

It can clearly be seen that these routes through the network may consist of a concatenation of different technologies with different numbers of the specific technology types per route. The PDV may, therefore, be different based on the operation of these technologies. For example, routers may inject much more variability in the route than just switches. But this will be dependent on implementation of the technology, with potential differences between equipment vendors and based on the use for which the specific equipment is engineered. Clearly, a carrier scale device may have considerably different performance to an enterprise scale device as highlighted in [GOL 12].

Furthermore, the aggregate PDV may vary when different mixes of technologies are deployed. Route #1 has dual routers in the core compared to route #3, which only has a single router. Route #2 has dual switches after the routers, whereas route #1 only has a single switch.

4.2.3.2. *Fixed line operator provides synchronization*

Another option exists where the MNO can buy the synchronization as part of some form of managed service from a fixed provider. In this case, the MNO now buys or has a commercial agreement to use synchronization that is not under their direct control. This scenario is worth investigating to show some of the advantages that can be derived in how the performance may be improved and made more deterministic. Leaving aside any commercial issues, the question is: does this create a technical problem? Some simple models are used in section 4.2.3.3 to discuss this.

4.2.3.3. *Ownership of synchronization reference in multiple operator scenarios*

Supporting many different operators' synchronization across another providers' network requires timing transparency. On initial inspection, each operator requires their own reference clock. However, we know that transporting timing in the packet world may be more complex compared to the TDM world. Given these issues, it is useful to look at the requirements in a multiple operator scenario using mobile networks as a case example.

While relying on timing transparency has been quite common in the past with TDM networks when only frequency synchronization was required for mobile networks, there are cases implying a need for stringent phase/time synchronization, such as Time Division Duplex (TDD) mobile networks, for which timing transparency becomes more challenging. The TDD example will be further analyzed in section 4.2.3.3.1, before discussing generally the case where only frequency synchronization is required in section 4.2.3.3.2.

4.2.3.3.1. Example of TDD mobile networks with phase/time requirements

For the purposes of this illustration, a particular example will be chosen, which is also discussed in [GIL 11]; in this case a LTE TDD wide area base station with a frequency requirement of +/− 50 ppb accuracy at the air interface and a time/phase requirement of 1.5 µs accuracy in relation to UTC also at the air interface. In this simple example, there are three base stations that may belong to three different mobile operators, BS#1, BS#2 and BS#3.

Taking the frequency requirements of +/− 50 ppb at the air interface, in practical terms, this may correspond to a requirement of +/− 16 ppb at the input interface to the base station. Taking the 1.5 µs time/phase requirements at the air interface, in practical terms, this may correspond to a requirement in the order of 1 µs at the input interface to the base station. This is due to the noise budget and holdover budget allocated to the base station.

In practical terms, it could be assumed that the primary synchronization reference used by each operator is based on GPS to give traceability to a commonly used reference source, based on UTC timescale/frequency reference. This is shown in Figure 4.15 where each base station is illustrated with a different color to represent the different timing domains of the mobile operators.

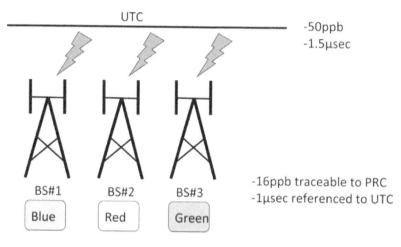

Figure 4.15. *Common traceability multiple base station/operator environment*

Keeping this in mind, three initial scenarios can be envisaged that primarily consider the aspects of the time/phase synchronization challenge and the possible impact on the network.

Scenarios

Three initial cases can be considered:

– Case 1: combined fixed and mobile operator;

– Case 2: separate fixed backhaul operator and mobile operator – MNO's timing reference is used (i.e. the fixed operator carries the mobile operators' timing domains);

– Case 3: separate fixed backhaul operator and mobile operator – fixed operator's timing reference is used (i.e. the fixed operator provides synchronization capability from his fixed domain to the MNOs).

In these three scenarios, the reference clock may be placed in various locations within the fixed/mobile network. It is assumed that packet clocks need to be added on the chain from reference to application on a hop-by-hop basis to deliver the required time/phase performance (this technical option is discussed in section 4.4.2 based on technologies introduced in Chapter 2).

These clocks will be colored differently for each operator, and the "clock" will be described as a packet clock to create a generic model for architecture discussion. Furthermore, the reference clock is just described as the reference clock but in essence is a PRC as well as a Primary Reference Time Clock (PRTC) locked to an easily accessible source of UTC, e.g. GPS-UTC. Figure 4.16 describes scenarios where the reference clock of the fixed network operator (PRTC on the right in the figure, owned by the fixed network operator) will be used to deliver synchronization to the base stations (cases 1 and 3) and a scenario where the reference clock of the various mobile operators (PRTC of the left in the figure, owned by the mobile network operators) will be passed through the fixed network hop-by-hop (case 2).

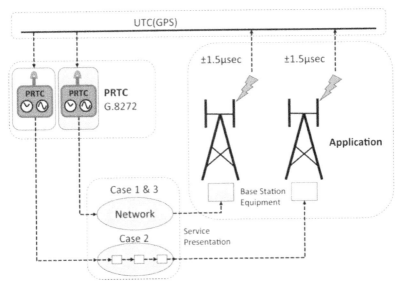

Figure 4.16. *High-level timing model*

Focusing on the network and the three high-level cases, some aspects of each are discussed. Note that in the following three cases, the usual IEEE 1588 clock types' terminology (such as Boundary Clocks (BCs) or Transparent Clocks (TCs)) is not used within the diagrams because these cases are written to be generic to discuss some of the issues of supporting multiple clocking/time domains within the fixed/mobile environment. The different timing domains are colored in *Blue*, *Red* and *Green*.

Case 1

In this case, the fixed and mobile operators are one and the same.

The reference clocks *Blue* can be located in the fixed portion of the network or in the mobile portion (i.e. switch #2). It is entirely probable that these physical locations may well be one and the same.

In this case, the synchronization reference *Blue* is carried over the fixed network to the base stations and suitable packet clocks *Blue* are deployed hop-by-hop (see Figure 4.17).

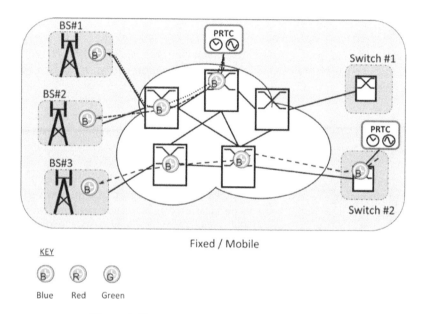

Figure 4.17. *Fixed and mobile same timing domain*

In this case the fixed/mobile infrastructure effectively becomes one timing domain, that is *Blue*.

All the base stations can be traceable to the same reference clock or separate clocks, that is, physical equipment. But in essence, they all need to meet the same requirements, that is, traceability to a common timescale, such as provided by the GNSS-based systems.

Advantages

– ease of implementation and flexibility;

– transport fixed and mobile possibly built from one vendor;

– mobile may be highly integrated into fixed.

Disadvantages

– model may not be easily transferable to separate fixed/mobile networks;

– lack of demarcation/Service-Level Agreement (SLA) monitoring between fixed and mobile infrastructures;

– not all fixed operators have a mobile network/not all mobile operators have a fixed transport network.

Case 2

In this case, the fixed and mobile operators are separate operators. The reference clocks *Red and Green* are located in the mobile operators' network (i.e. switch #1 & #2, respectively).

In this case, the synchronization references *Red and Green* are carried by the fixed backhaul provider over his network to the respective mobile operators' base stations.

In this case, the fixed provider has, in general, to support suitable packet clocks *Red* and *Green* hop-by-hop (see Figure 4.18). In some cases, multiple packet clocks *Red* and *Green* now might have to be supported in the same network equipment, depending on the type of support for synchronization implemented in the network equipment.

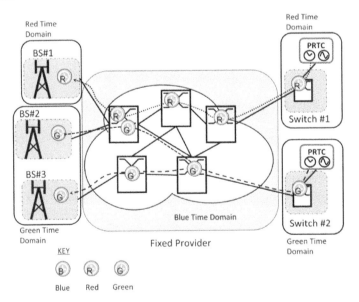

Figure 4.18. *Fixed supporting multiple/other operator timing domains*

In this case, the fixed infrastructure is one timing domain (i.e. *Blue*) that participates in the timing of other timing domains (i.e. *Red and Green*). Effectively, the fixed infrastructure has to support as many timing domains as required.

The base stations of different mobile operators are now traceable to different reference clocks, that is, physical PRTC equipment. But in essence, they all need to meet the same requirements, that is, traceability to UTC.

Advantages

– supports multiple sources/operators;

– provides the illusion that mobile/other operators are solely responsible for their own timing.

Disadvantages

– Fixed operator has to support multiple sources/multiple operator domains (complexity/how many?).

– May create undue architecture limitations, for example suitable packet clocks deployed end to end.

– Demarcation/monitoring (needs to happen on packet timing flow on line interfaces carrying also traffic rather than dedicated physical timing interfaces).

– The synchronization reference of other operators is processed by the fixed operator (e.g. the messages carrying synchronization are processed and possibly modified); the mobile operators are therefore not solely responsible for their synchronization in practice.

– Operation under failure conditions. Clearly, if the fixed provider's packet clock fails or misoperates, clarification is needed on the operation between the operators and some form of agreement may be required in an SLA.

– If timing is part of an encrypted packet flow, identification and access to the flow by the fixed operator is needed to process the synchronization messages.

– Problems within the fixed network will impact other operator timing. This becomes a challenge if the fixed provider's packet clock fails or misoperates as inherently the transport medium is modifying another operator's timing data. This is in contrast to the PDH world where bits were passed transparently. Of course, this aspect would need to be covered in SLAs.

Case 3

In this case, the fixed and mobile operators are separate operators, but the fixed operator is providing a managed service (including timing) to the mobile operators. The reference clock *Blue* is located in the fixed operator's network.

In this case, the synchronization reference *Blue* is carried by the fixed backhaul provider over its network to the respective mobile operators' base stations.

In this case, the fixed provider has to support suitable packet clocks *Blue* and these are deployed hop-by-hop (see Figure 4.19).

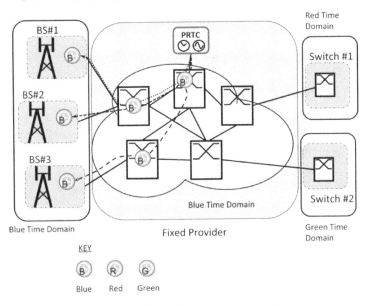

Figure 4.19. *Fixed provider provides managed service*

In this case, the fixed infrastructure is one timing domain. A defined synchronization service can be provided to the mobile operator base stations.

All the base stations are now traceable to the same timing domain, that is, the fixed provider's. In essence they all meet the same requirements, that is, traceability to a common timescale.

Advantages

– Architecture can be designed to support timing, not the entire architecture will need to support suitable packet clocks.

– Fully under one operator's control, easier to engineer performance to requirement.

– Fully under one operator's control for fault finding.

– Timing can be separated from mobile operator traffic using dedicated physical layer timing interface.

Disadvantages

– Requires some form of managed service/agreement between fixed and mobile operators.

– Requires the mobile operators to be comfortable and trust the fixed provider (performance can be enforced by defining SLA).

– Requires some investment by the fixed provider to deliver the timing reference. However, cost may be significantly less than that required in case 2 where multiple timing domains and flows across the network need to be supported and where the operational case may be complex.

– In the case where an MNO has to deal with many fixed line carrier operators, reliance on this model may not provide a solution to all requirements.

Phase/time synchronization analysis

All these cases have different advantages and disadvantages.

However, one key thing becomes clear regardless of ownership of the reference clock: the same requirement has to be met, that is, traceability to UTC. Different mechanisms of transport will create different levels of impairment and different sets of problems.

Case 1 in some respects is the easiest to implement although this does not reflect the reality of a multi-operator environment.

Separate instances of every operator passing timing through another operator's network will be unduly complex and may not be required. Technical solutions may exist (e.g. based on TCs), but are not discussed in detail in this chapter; it is a decision from the network operator to consider them. Anyway, this will require careful study to fully understand the relationships between mobile and fixed operators. This would suggest that case 2 should be considered very carefully and possibly avoided. It should also be noted that this model continues to propagate the model of timing transparency that may not be easy to meet with tight time/phase synchronization requirements.

Case 3 overcomes some of the challenges presented in case 2, whilst still providing a multi-operator environment. But this scenario may still create some challenges to some operators.

4.2.3.3.2. Example of FDD mobile networks with frequency requirements only

When only frequency synchronization is required, such as in some FDD mobile systems, the previous analysis may slightly differ in some aspects.

First, because meeting the frequency synchronization requirement of FDD mobile systems is much simpler compared with the stringent phase/time synchronization requirement of TDD mobile systems, this leaves room for timing reference degradation across the Fixed Line when using a timing transparent approach. The "not-too-stringent" FDD requirements explain why PDH timing transparency over SDH has been quite popular in the past.

However, as explained earlier, when moving to packet networks, the situation may be less simple. Transparently carrying the frequency synchronization reference of the mobile operator across the carrier operator network may not always be possible, for instance, depending on the technology used (SyncE transparency is possible only in specific cases, such as OTN, wavelength division multiplexing (WDM) and dark fiber transport), or may lead to unacceptable degradation of the reference, for instance, when using IEEE 1588 end-to-end solution if PDV is too high.

Therefore, although the problem is less critical than for the distribution of accurate phase/time, using fixed operator's timing reference can certainly be considered as the safest solution in terms of performance even when only frequency synchronization is required.

But this scenario may again still create some challenges for some operators. This is the reason why using IEEE 1588 end-to-end across carrier operator networks has been quite common in the recent years, although it does not provide formal guarantees of performance in most of the cases due to the lack of SLA related to PDV.

4.3. Deployment of timing solutions for frequency synchronization needs

This section discusses how to deploy synchronization networks addressing frequency synchronization needs. In particular, two technologies are analyzed: Synchronous Ethernet (see section 4.3.2) and IEEE 1588 end-to-end (see section 4.3.3).

When a network operator has to design and then deploy a synchronization network architecture supporting telecommunication network operations, several considerations should be taken into account in order to choose the most suitable technology or set of technologies. Combination of solutions is indeed relevant in some synchronization network architectures.

The selection of suitable technologies may depend on the following parameters, for instance:

– Is there an SDH-based synchronization network already deployed?

– Are the packet network nodes already deployed SyncE capable?

– Is a transmission network like OTN crossed? If so, does it support SyncE timing transparency?

– Is the network fully owned or not? If the network is not fully owned, is it possible to rely on a synchronization service based on physical layer technology from the operator delivering the connectivity?

When the answers to these questions are all "yes," it is likely to be a good sign that relying on physical layer techniques like SyncE should be possible. Otherwise, relying on IEEE 1588 end-to-end may be an alternative option, possibly in combination with physical layer techniques.

These aspects are further explained in the next sections.

4.3.1. *Overview of synchronization solutions for frequency needs*

Before discussing how to consider the deployment of some technologies, it is worth introducing them with a short comparison. The details of these technologies have been presented in Chapter 2. Table 4.2 summarizes the three main types of solutions that can be considered to address frequency needs.

As can be seen in Table 4.2, advantages and drawbacks are different depending on the solution considered. A trade-off has, therefore, to be found for different parameters, such as cost versus performance of the solution, applicability of some systems (e.g. some countries might have issues with GPS and not with network-based solutions). Long-term maintenance of the solution must also be considered.

As network contexts may be different, the design of the synchronization network architecture has to account for the specificities of the deployment. The objectives of the following sections are to guide the reader through the usual deployment steps of two major recent technologies: SyncE and IEEE 1588 end-to-end.

Technology	Type of sync.	Pros	Cons
Physical methods (e.g. SyncE)	Only frequency	– Very good frequency quality (no impact of PDV) – Straightforward integration in the existing SDH-based synchronization networks – Low-cost solution when anticipated in equipment design	– All the network nodes of the timing chain must recover the reference (link-by-link solution), upgrade of equipment may be required – Sync network design is required (e.g. to avoid timing loops) – Timing transparency is not always possible
Packet-based methods "end-to-end" (e.g. IEEE 1588 end-to-end)	Mainly frequency (accurate phase/time is hard to achieve)	– Transparency to the networks (timing is carried as data traffic) – Suitable for frequency delivery in some cases (e.g. controlled PDV, slave embedded in the base station or in a device co-located with the base station)	– Performance/timing quality impacted by PDV – Hard guarantees are difficult to provide for applications with stringent frequency requirements (solution mainly applicable for mobile networks) – Engineering rules to be respected to limit PDV – Bandwidth consumption and stabilization period
GNSS (e.g. GPS, Galileo)	Frequency and phase/time	– Excellent performance/timing quality achievable (e.g. 100 ns accuracy)	– Installation cost – Antenna infrastructure – Indoor configurations – Jamming, air interferences – Lack of commercial guarantee/SLA of GPS

Table 4.2. *Overview and comparison of synchronization solutions for frequency*

4.3.2. *Synchronous Ethernet deployments*

This section discusses how to approach SyncE-based synchronization networks from the deployment perspective as usually faced by network operators in practice.

This section, in particular, discusses what would be the reasons for choosing SyncE technology and the usual deployment steps.

4.3.2.1. *Rationale for Synchronous Ethernet deployments*

As explained in Chapter 2, SyncE is a physical layer timing distribution technique whose principles are widely based on SDH synchronization networks. In particular, special care has been taken during the development of the standards defining the SyncE technology so that it would be fully compatible and interworkable with an SDH-based synchronization network.

Therefore, when the synchronization network to be built corresponds to a migration scenario from a TDM network to a packet-based network, it is natural to consider SyncE technique.

Indeed, SyncE has several advantages:

– It is a fully standardized and mature technology, with principles that are well known by the operators who are used to operating physical layer solutions such as an SDH-based synchronization network (including testing procedures, deployment steps, troubleshooting, etc.).

– It allows a straightforward integration in existing SDH-based synchronization networks, in particular, step-by-step migration scenarios can be considered (i.e. replacement of TDM nodes by SyncE nodes as needed).

– The synchronization quality of this solution is not impacted by the PDV that is observed on the packet networks because it works at the physical layer (therefore, below the packet layer).

Some conditions need, however, to be met for using SyncE:

1) The entire packet network where timing is expected to be delivered must be SyncE capable: this is due to the fact that SyncE is a link-by-link technology, where each node involved in the timing distribution chain receives, processes, and retransmits the clock.

Note: the entire telecommunication network does not need to be SyncE capable because first, this technology can interwork with SDH networks and second, some packet nodes may not participate in the timing distribution chain.

2) SyncE may be more difficult to be deployed in scenarios where the network is not fully owned because the physical layer is not under the control of the operator in this case.

Note: as mentioned earlier, some operators may, however, propose either a synchronization service as part of the connectivity offer (e.g. based on SyncE signal or another physical layer synchronization signal) or leased line offers allowing timing transparency to SyncE (e.g. based on OTN transport).

As a summary, SyncE can be considered as a valid choice in general for fully owned networks. However, as mentioned earlier, this technology covers only frequency synchronization needs and not phase/time synchronization needs.

4.3.2.2. Architectures typically considered with Synchronous Ethernet deployments

SyncE is an attractive solution especially when used as a solution for TDM links replacement in an already deployed synchronization network. Obviously, it can also be deployed in a totally new network (e.g. green field deployments).

Figure 4.20 illustrates a typical migration scenario consisting of replacing a TDM-based network by a packet network and for which SyncE is introduced to support the migration.

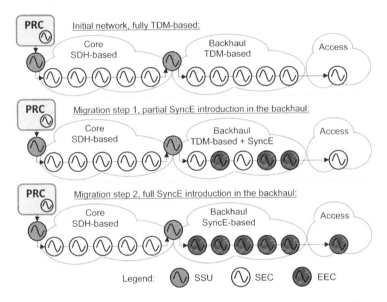

Figure 4.20. *Migration scenario from TDM network to packet network using Synchronous Ethernet*

This type of scenario is very common in mobile networks where, in general, the backhaul migrates toward packet-based technologies because the operator faces capacity upgrade needs at controlled costs. The migration toward SyncE starts therefore from the backhaul and propagates sometimes to the core network, which is in general impacted later (the core network is very often still based on SDH technologies in many cases). Similar migration steps as shown in Figure 4.20 may apply to the core network.

Figure 4.20 clearly illustrates the key property of SyncE, which is considered as a great advantage for the network operators: it allows smooth integration within an existing SDH-based synchronization network with the same architectural design and reusing some equipment already deployed, such as Synchronization Supply Units (SSU) or SDH Equipment Clocks (SEC).

Moreover, the different migration steps in Figure 4.20 also clearly show that interworking with SDH nodes embedding an SEC is not an issue even when mixing SEC and EEC in the same chain. This statement is true for the transport of the frequency reference and also for the channels carrying the Synchronization Status Message (SSM) associated with the reference (the Ethernet Synchronization Messaging Channel (ESMC) defined for SyncE has been designed so that it fully interworks with the SSM bits included in the SDH header).

Figure 4.20 also highlights an important point that needs to be considered when deploying SyncE solutions: all the nodes in the synchronization distribution chain must support specific SyncE functions. These functions differ depending on the type of node considered (Figure 4.20 shows only SEC/EEC functions):

– for transport nodes, for instance IP routers or Ethernet switches, but also for some transmission nodes, like SDH switches: a Synchronous Ethernet Equipment Clock (EEC) or an SDH Equipment Clock (SEC) is required; actually, the specifications of EEC and SEC are equivalent;

– for some transmission techniques such as OTN: an EEC is not required, but specific functions must be supported to ensure the timing transparency to SyncE (this point will be detailed later in section 4.3.2.3).

In addition, SyncE can be used to synchronize any type of access technologies: fixed broadband access (e.g. DSL and PON) or mobile access (e.g. base stations).

As mentioned earlier, not all the network needs to support SyncE, it is only required that a synchronization path (and possibly a backup path) be built from the PRC to the end equipment that needs synchronization (e.g. the base stations in mobile networks).

The synchronization paths can rely on various networks, including existing SDH or TDM-based networks, which are not necessarily always used for transporting the data delivered to the end nodes that are synchronized. Indeed, it must be kept in mind that a synchronization network is an overlay network and therefore can rely on different types of networks, synchronization and connectivity are not necessarily being linked together here.

IP routers are in general connected together via a transmission network. If this transmission network is an SDH network (or, more generally, a network built over a synchronous physical layer), then this transmission network can be used to deliver synchronization. This can be achieved without, in general, requiring the IP nodes to support SyncE (except sometimes for the final node), as illustrated in Figure 4.21 in the "core" segment (which is based on Packet Over SDH (POS) in this example).

If this transmission network is based on Wavelength Division Multiplexing (WDM) or OTN, then it does not normally have the capability to carry and deliver synchronization itself; generally, it may only have the capability to carry transparently the timing reference of their client signals (see section 4.3.2.3). In this case, the IP nodes need in general to support SyncE in order to provide the WDM/OTN transmission network with synchronous client signals. These considerations are illustrated in Figure 4.21, which shows that an SDH-based network can deliver timing reference to a SyncE network.

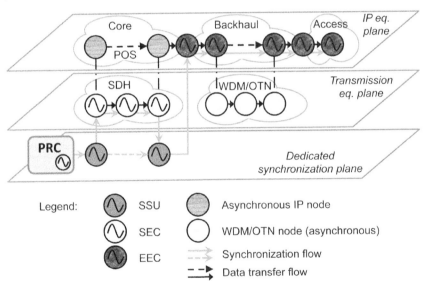

Figure 4.21. *Example of synchronization flow carried over multiple planes*

In this example, the SSU just after the PRC synchronizes an SDH network belonging to the transmission equipment plane and which supports the IP core network. The IP core network is asynchronous here because the SDH network is in charge of carrying the synchronization reference. Possibly, this SDH network providing the reference to the backhaul segment could have been totally independent of the IP core network (i.e. not supporting the IP core network transmission).

At the output of the SDH network, the synchronization reference is then filtered by a second SSU and is provided to a SyncE chain belonging to the IP backhaul network. This SyncE chain is partly carried over an OTN connection in the transmission equipment plane, which is asynchronous, but capable of carrying the SyncE client signals transparently. At the end of the chain, the timing reference is delivered in the access to the end application.

In this example, it can be observed that the IP core network does not need to support SyncE functions. This is represented in Figure 4.22, showing the connection between two end equipment. Of course, this scenario is only possible when an existing synchronization network can provide the timing reference to the backhaul network, for instance, the SDH network of Figure 4.21. This may be the case for already deployed networks. However, assuming that SDH networks could be replaced in the future by packet-based networks, the existing synchronization networks shown in Figure 4.22 might disappear. New architectures permitting us to deliver the synchronization down to the edge of the network will then need to be designed. Options based on physical layer timing could be either to build a SyncE core network, or to deploy a distributed PRC architecture, where PRCs are located closer to the edge of the network (and therefore, their number would be higher than when centralized).

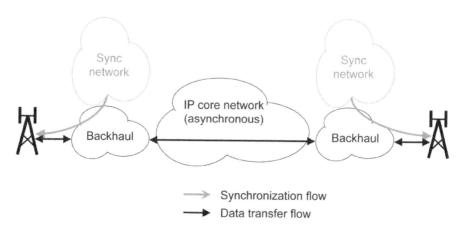

Figure 4.22. *Example of asynchronous IP core network*

Finally, it is important to repeat, as explained in Chapters 2 and 3, that the same architectural rules defined in the ITU-T Recommendation G.803 [G.803] apply for SyncE and SDH-based synchronization networks. Therefore, the same restrictions regarding the maximum number of clocks in tandem must be considered also for SyncE deployments:

– not more than 10 SSUs in tandem;

– not more than 20 EECs or SECs between two SSUs (this rule must be followed for EEC clocks and also when EEC and SEC clocks are mixed in a chain);

– in the same synchronization path, not more than 60 SECs or EECs in total.

4.3.2.3. *Deploying Synchronous Ethernet over Optical Transport Networks*

SyncE signals can be transported over transmission networks, such as OTN. Depending on the type of transmission technology, timing transparency may or may not be feasible. A proper design of the synchronization network must account for the different transmission techniques considered and whether or not they can offer timing transparency.

In order to better understand this concept here, let us give a simple example. A client signal carried over a transmission network is characterized by an inherent rhythm which corresponds to the rhythm of the symbols transmitted at the layer 1 of the OSI model. For instance, the frequency accuracy of a SyncE signal must be within +/– 4.6 ppm; it is however expected that a SyncE signal be traceable to a PRC under normal conditions. Timing transparency refers to the capability to carry this client signal such that the inherent rhythm characterizing it is conserved with a reasonable and controlled generation of noise. For example, a SyncE signal traceable to a PRC is carried over OTN with timing transparency when the output SyncE signal exiting the OTN network and rebuilt by the last OTN node is also PRC traceable; there is no timing transparency in case the PRC traceability is lost and the SyncE signal becomes a normal (asynchronous) Ethernet signal at the output of the OTN network.

The notion of timing transparency ensured by the transmission layer is not a new concept at all; indeed, PDH signals were already carried over the PDH hierarchy by ensuring timing transparency using justification techniques. Similar justification techniques were also introduced in the SDH hierarchy in order to allow timing transparency (in particular, PDH and SDH client signals can be transported through an SDH network with timing transparency, although it is recommended in general not to use PDH signals when carried through SDH to synchronize end applications requiring an accurate frequency reference because of the noise added to them due to the possible pointer adjustment events, see Chapter 2, and instead to favor the timing reference carried by the SDH network itself).

In the year 2000, when the OTN technology had been defined by the ITU-T, discussions ensued in order to decide if this technology should be in charge of directly carrying synchronization or not, that is, if the actual OTN layer should be synchronized. It was decided to define the OTN layer similarly to the PDH hierarchy, in the sense that it does not directly carry synchronization (the OTN transport signals are asynchronous), but allows timing transparency to client signals. Based on this decision, appropriate OTN mappings have been defined in order to allow timing transparency to the SDH client signals; it is the role of the SDH client signals to carry synchronization down to the end application, not of the OTN network, which should simply ensure timing transparency.

Figure 4.23 illustrates this discussion with an example of two SDH client signals synchronized from different frequency sources, f1 and f2, and carried over the same OTN network, which ensures timing transparency to these signals.

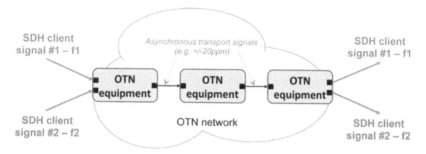

Figure 4.23. *Examples of SDH client signals carried with timing transparency over OTN*

With regard to traditional asynchronous Ethernet client signals, initial OTN mappings defined by ITU-T, such as Generic Framing Procedure – Framed (GFP-F), did not consider timing transparency. Indeed, these Ethernet signals being asynchronous, there was no need to ensure timing transparency; the only concern was to ensure transparency of the data. Things changed when SyncE was defined by ITU-T; some work was led in order to define how to carry these synchronous client signals over OTN. New OTN mappings allowing timing transparency, such as Generic Mapping Procedure (GMP) combined with Timing Transparent Transcoding (TTT)), were then introduced for SyncE signals (similarly to those used for SDH client signals). When using these mappings, the inherent synchronization reference carried as part of a SyncE signal is not lost through the OTN network. Simulations have been run to study the effect of these new mappings on a synchronization network (see details in the ITU-T Recommendation G.8251 [G.8251]). Now, SyncE is assumed to play the same role played by SDH signals: it is the role of the SyncE client signals to carry synchronization down to the end application, not of the OTN network, which should simply ensure timing transparency.

One final aspect to be considered with SyncE deployments over OTN networks is the transmission of the ESMC frames carrying the SSM information. When timing transparent OTN mapping is used, the ESMC frames must be transmitted together with the synchronization reference of the client signal and not discarded when entering the OTN network. However, when a non-timing transparent OTN mapping is used, the ESMC frames are not assumed to be forwarded by the OTN network and should be discarded because the Ethernet signal at the output of the OTN network is not synchronous anymore. It should be noted that some old OTN mappings designed for traditional asynchronous Ethernet signals may not discard the ESMC frames. Again, a careful design of the synchronization network should account for these aspects.

Figure 4.24 illustrates this discussion with two examples: in the first case, the SyncE client signal is carried with timing transparency over the OTN network and the ESMC frames are forwarded, while in the second case, timing transparency is not ensured by the OTN network for the SyncE signal and the ESMC frames are discarded.

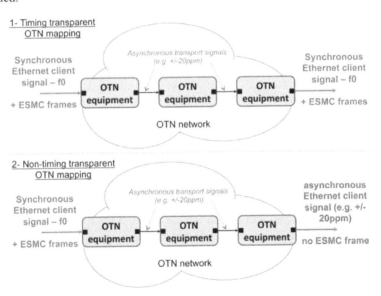

Figure 4.24. *Examples of Synchronous Ethernet client signals carried with and without timing transparency over OTN*

To conclude, carrying SyncE over an underlying transport infrastructure requires us to verify if that layer is transparent to timing. In certain cases, even though the technology is capable of transparent timing (as is the case with OTN) the specific mappings may not allow timing transparency.

Transmission technique \ Client signal	PDH	SDH	Ethernet/packet physical layer	OTN
PDH	OK	OK (Note 1)	Not OK (Note 2)	OK (but undefined)
SDH	OK	OK	Not OK (Note 2)	OK
Synchronous Ethernet	Not OK (Note 3)	Not OK (Note 4)	Not OK (Note 2)	OK (Note 5)
OTN	Not applicable	Not applicable	Not applicable	OK

Note 1: it is, however, not recommended in general to use PDH signals carried through SDH to synchronize end applications requiring an accurate frequency reference (i.e. requiring a synchronization interface as defined in the ITU-T Recommendation G.823 [G.823]), the reference carried by SDH should be used instead. Mobile applications may however tolerate PDH transport over SDH.

Note 2: other techniques involving higher layers than the physical layer (e.g. adaptive clock recovery, differential clock recovery) may, however, be considered to provide timing transparency at the "packet layer"; in general the noise added to the original signal with such techniques is much higher compared to traditional justification techniques. In case the transmission technique is SyncE, it can be synchronized in general using the reference of the input client signal (it does not, however, provide timing transparency *per se* because only a single timing reference can then be carried)

Note 3: some systems (e.g. MW) may allow carrying Ethernet signal over PDH frames, but do not ensure in general timing transparency; instead, it is the role of the PDH physical (radio) layer to carry the timing reference in this kind of system (some implementations allow synchronizing the PDH frames using an input SyncE reference signal).

Note 4: this case corresponds to POS techniques, which do not ensure timing transparency in general; instead, it is the role of the SDH layer to carry the timing reference in this kind of system (some implementations allow synchronizing the SDH layer using an input SyncE reference signal).

Note 5: timing transparency of SyncE is only possible when using appropriate mappings (e.g. GMP + TTT); old mappings defined for normal asynchronous Ethernet client signals are not appropriate and do not offer timing transparency.

Table 4.3. *Summary of transmission techniques providing timing transparency at the physical layer depending on the client signals as generally observed in current network equipment*

Table 4.3 summarizes the scenarios where timing transparency at the physical layer (with justification techniques) is possible, depending on the client signals and transmission techniques considered. Further details about timing transparency concepts and applicability on these various technologies can be found in Chapter 3. Note that Table 4.3 does not consider the case of timing transparency at the "packet layer" or the case where the transmission layer is synchronized by the client signal (some of the notes at the end of the table simply provide some indications).

4.3.2.4. Usual protection schemes with Synchronous Ethernet

As for an SDH-based network, SyncE mainly relies on the use of the SSM for most protection scenarios. This mechanism, which is carried as part of a specific channel called "Ethernet Synchronization Messaging Channel" (ESMC), is further detailed in Chapters 2 and 3. Essentially, it consists of changing the synchronization path when a failure is detected (i.e. when the quality of the source generating the synchronization reference degrades).

In addition to the SSM mechanism, more static schemes are also applicable to SyncE, similar to SDH:

– *static reference switching*: switching to a backup synchronization source may in some cases be based on static criteria, such as loss of the reference, or degraded input signal (which assumes that the clock is able to monitor the quality of the input);

– *holdover*: a clock with a good oscillator quality, for instance in an SSU, can be used to continue delivering the timing reference to a synchronization path during failure conditions.

During the design of the network, the main task consists of ensuring proper planning of the synchronization network, which should consist of properly configuring the SyncE equipment for protection (e.g. SSM configuration) while avoiding timing loops. Indeed, it is important to remember that the SSM mechanism is an efficient, very simple, although not fully automatic mechanism and, therefore, does not eliminate the creation of potential timing loops. Careful synchronization planning is required to properly design the protection of a synchronization network with the SSM in order not to create synchronization loops, which must be avoided. Such timing loops can indeed lead to substantial degradation of the timing reference delivered to end applications.

This network planning task can be done manually, or with the assistance of tools that can study the network topology in order to anticipate the consequences of network failures, and possibly make proposals for network configuration. Manual intervention is also, in general, necessary due to the lack of integration between the Network Management System (NMS) and Sync Management System. It is recommended to check the network configuration obtained when a tool is used to assist the design of the network, especially for large networks, as well as ensuring consistency between network and synchronization topologies after network updates (new equipment must, for instance, be reflected in the synchronization plane). Further details related to the SSM configuration are provided in Chapter 3.

After the deployment, a synchronization network can be monitored using traditional techniques used in SDH-based networks. For instance, the timing quality of a SyncE ring can be monitored by an SSU, which can raise alarms in case the wander recorded exceeds some predefined thresholds. The management aspects of a synchronization network are further detailed in Chapter 5.

It must be mentioned that some implementations may require ESMC frames to be transmitted together with the SyncE signal in order to consider the input link as a valid timing reference. In other words, some SyncE-capable network nodes may reject a SyncE link if associated ESMC frames are not transmitted (although in the ITU-T Recommendation G.8264 [G.8264], a note specifies that it should be possible to disable the SSM/ESMC process in the receiving equipment).

Such implementations interpret a lack of ESMC frames as an indication that the sending node does not support SyncE and, therefore, as a potential mistake in the synchronization network planning. The idea is to prevent possible confusion between a traditional asynchronous Ethernet link and a SyncE link.

However, this behavior may cause limitations, since in practice it is known that some equipment may implement SyncE with reduced functionalities: for instance, a valid SyncE signal may be sent without associated ESMC frames (this is not compliant with the standard, but it exists). In such cases, deployment issues may happen; hence an operator should consider carefully this aspect during the deployments and check that the ESMC mechanism is properly implemented at both sides of a link, or that it can be deactivated.

4.3.2.5. Summary of main aspects of Synchronous Ethernet deployments

Previous sections have introduced SyncE-based synchronization networks from the view of network operators in charge of deploying and operating them.

In summary, the following points must be kept in mind when considering SyncE deployments:

– SyncE fully interworks with SDH-based synchronization networks, which allows progressive introduction in existing networks.

– SyncE is a layer 1 mechanism, not impacted by PDV.

– SyncE provides only frequency synchronization.

– The full chain composing the synchronization path must be SyncE compatible (EEC, SEC or timing transparency); otherwise, the timing reference is lost.

– SyncE deployments in scenarios where the network is not fully owned should be considered carefully (for instance, timing transparency does apply only on specific cases).

– An appropriate timing transparent mapping must be used when carrying SyncE over OTN.

– Network planning should be done carefully in order to avoid timing loops.

– Support for the ESMC is important and should be verified.

The following ITU-T standards are related to SyncE: [G.8261], [G.8262], [G.8264] and [G.781].

4.3.3. *IEEE 1588 end-to-end deployments*

This section discusses how to approach IEEE 1558 end-to-end synchronization networks from the deployment perspective, as it is usually faced by network operators in practice.

This section, in particular, discusses what the reasons would be for choosing IEEE 1588 end-to-end technology and the usual deployment steps.

4.3.3.1. *Rationale for IEEE 1588 end-to-end deployments*

As explained in Chapter 2, IEEE 1588 end-to-end is a packet-based method, whose principles are widely based on adaptive clock recovery technique. Although similar adaptive clock recovery techniques have been specified during the Asynchronous Transfer Mode (ATM) times, they were not really used widely in practice in telecommunication networks (since the SDH layer was in charge of carrying timing in most of the cases). For some leased lines requiring timing transparency, adaptive mode was used, but it should be noted that differential method was also favored when a timing reference was available, provided by the SDH layer. Therefore, adaptive clock recovery techniques can be considered newer than the physical-based methods that are instead based on more established concepts.

The IEEE 1588 end-to-end technology has been defined when the delivery of synchronization over packet networks has started to become an issue, due to the fact that traditional Ethernet links were asynchronous. IEEE 1588 end-to-end is, in particular, useful for the cases where the use of physical synchronization methods is not possible. As an introduction, a short history about this technique may be useful to appreciate its subtleties.

4.3.3.1.1. Notion of PTP profiles

The first point to be noted is that, as explained in Chapter 2, the PTP protocol has been defined by the IEEE, but it is not a protocol dedicated to telecommunication networks. In particular, version 1 has not been considered by the telecom industry and was rather focused on automation in small networks (e.g. in LANs). Version 2 [IEEE 1588™-2008], which introduces the concept of PTP profiles, contains new features useful to telecommunication networks.

A first telecom profile corresponding to an "end-to-end" deployment scenario has been defined by the ITU-T in the Recommendations G.8265 [G.8265] and G.8265.1 [G.8265.1] to address frequency synchronization needs only; this section discusses this particular PTP profile and the related deployment scenario.

A second telecom profile corresponding to a "node-to-node" deployment scenario, and usually referred to as "full timing support", is currently under definition by the ITU-T in the future Recommendations G.8275 and G.8275.1 to address accurate phase/time synchronization needs; sections 2.5.1 and 4.3.4 briefly introduce this second PTP profile.

Study of a potential third telecom profile for the transport of phase/time synchronization, including IEEE 1588 unaware equipment, and usually referred to as "partial timing support", has been proposed in ITU-T but the technical discussions to define it precisely have just started; therefore, this chapter does not consider it.

4.3.3.1.2. "End-to-end" versus "full timing support" PTP deployments

The second point to be noted is that the original version of IEEE 1588 protocol (version 1, 2002) did not assume "end-to-end" deployments where the nodes in the middle of the network do not support PTP functions such as BCs (PTP-unaware nodes), but instead deployments where such functions are present, in a "node-to-node" architecture. This was due to the general target of distributing sub-microsecond time synchronization. Although PTP version 2 allows "end-to-end" deployments, high accuracy can generally be guaranteed only when these PTP hardware functions are present; otherwise, the protocol suffers from the classical network impairments, such as PDV.

As mentioned earlier, "end-to-end" deployments were foreseen as useful mainly for the cases where physical layer solutions were not possible (e.g. replacement of

PDH connections by Ethernet connections and cases of Ethernet-based leased lines), because this solution provides timing transparency at the "packet layer" (rather than at the physical layer as discussed earlier for SyncE deployments). It was, therefore, seen as very useful for helping the migration toward Ethernet.

Obviously, this key advantage of timing transparency at "packet layer" is lost when "node-to-node" architecture is considered, since some relevant hardware needs then to be implemented in the intermediate network nodes. In this case, for frequency synchronization needs only, the use of SyncE is not expected to be more complex. It is mainly the reason why the IEEE 1588 full timing support architecture is considered for very accurate phase/time synchronization needs, as it will be introduced later in section 4.3.4, and not really for pure frequency synchronization needs.

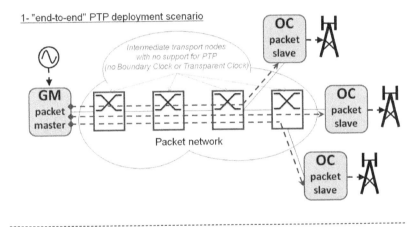

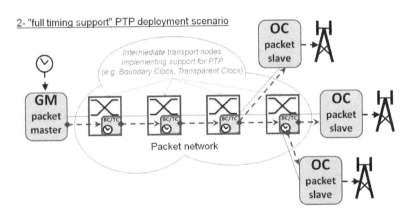

Figure 4.25. *"End-to-end" vs. "full timing support" PTP deployments*

Figure 4.25 reminds us of the major differences between "end-to-end" and "full timing support" PTP deployments. It can be observed, in particular, that when the "end-to-end" scenario is used, the PTP flows go through the queues of the intermediate transport nodes, leading to PDV generation (and therefore a degradation of the timing reference), while when the "full timing support" scenario is used, the PTP flows are processed (and even terminated when BCs are used) before going through the queues, which strongly limits the generation of PDV. The purpose of the PTP support in intermediate transport nodes (BC/TC) is, therefore, to avoid PDV generation and asymmetry due to background traffic load.

4.3.3.1.3. *Providing guarantees to an "Over-The-Top" type of solution*

Understanding the fundamental differences between "end-to-end" and "full timing support" deployment scenarios is really crucial, and led to a lot of confusion originally in the industry of telecommunications when the work on PTP started; it was sometimes thought that without the relevant hardware present in the intermediate nodes between the PTP master and the PTP slave, very high accuracy (such as sub-microsecond accuracy, and even sub-nanosecond accuracy) could be achievable.

These assumptions were originally supported by early results based on laboratory testing, which showed very good performance for this technology when using "end-to-end" deployment scenarios. However, the PTP slave clocks were not necessarily fully challenged, and the testing environment did not take into account all possible network conditions.

In practice, the PDV generated by real live packet networks could be significantly higher and more difficult to be filtered than those used in these original tests, and could cause problems for some PTP slave clock implementations, especially when highly accurate phase/time synchronization was targeted.

To give the reader a better idea of the issue, the amplitude of PDV generated by packet networks is typically in the order of several milliseconds, even tens or hundreds of milliseconds in some cases. Moreover, precisely anticipating the PDV that will be generated by a packet network is a very complex task, because it depends on many parameters (e.g. types of equipment, configuration of the equipment, size of the network, type of traffic carried on the network, load characteristics, etc.).

There are various techniques to reduce this amount of noise in the PTP slave, such as the use of packet preselection, but the major issue is that the preselection algorithms are not necessarily efficient in all the cases encountered in live networks. In other words, it is always possible to find a network that will generate a PDV pattern breaking even the most intelligent preselection algorithms. This is particularly an issue when accurate phase/time synchronization is targeted, the problem being far less complex for the not too stringent frequency synchronization needs of mobile networks, assuming that the packet slave clock is properly implemented.

We are discussing here the most interesting aspect of this IEEE 1588 end-to-end solution: it can be considered that IEEE 1588 end-to-end has initially been proposed by some PTP slave vendors as a kind of "Over-The-Top" (OTT) solution. Indeed, the PTP slave algorithm was assumed to be able to adapt to the network impairments in a flexible and intelligent way, similar to OTT applications delivered to the end users over the Internet. It was (and still is) considered by the PTP slave vendors that the intelligence of the PTP slaves is a key differentiator of their solutions (the filtering algorithm being not defined in the PTP standard), and therefore they are not willing in general to provide very much detail. Hence, slave designs are proprietary for all the aspects related to packet preselection algorithms, and differences are, in general, observed between implementations.

However, synchronization is totally different from an end user OTT service and cannot be considered the same way, or treated in a best effort manner. Indeed, synchronization is critical for the proper operations of the transmission networks, and therefore cannot suffer from random performance. It is of high importance for the network operator that the solution provides guarantees and meets the relevant performance requirements. Two aspects are required for this:

– to determine the amount of noise that the network can generate, in terms of PDV (indeed, PDV is the major type of impairment impacting PTP slave performance);

– to understand the amount of noise that can be tolerated and filtered by the PTP slave clock while meeting a given performance objective for the reference signal delivered at the output of the PTP slave clock.

This overall discussion explains why complex discussions occurred at ITU-T before defining applicable PDV network limits and PDV noise tolerance of packet slave clocks. It should be mentioned that some aspects are still under discussion.

Figure 4.26 aims at highlighting the main points that need to be considered when deploying IEEE 1588 end-to-end. These aspects are detailed in the following sections.

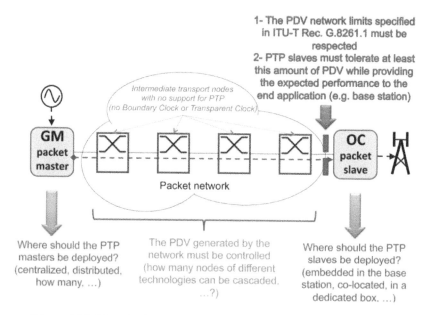

Figure 4.26. *Main challenges related to IEEE 1588 end-to-end deployments*

4.3.3.1.4. IEEE 1588 end-to-end deployments for mobile networks

Now that the history of this technology has been introduced, it is important to remember that IEEE 1588 end-to-end is a key building block of future synchronization networks when addressing specific scenarios. In particular, when synchronizing mobile base stations, which have quite relaxed requirements related to frequency synchronization, this technique is quite appropriate, and can be considered as providing a fair enough level of guarantee when properly deployed.

Using IEEE 1588 end-to-end to provide frequency synchronization to base stations has indeed become quite a common solution. Following are the main reasons that led some operators to consider this solution:

– IEEE 1588 end-to-end allows timing transparency at "packet layer" as explained earlier, which is very attractive when physical layer solutions such as SyncE are not possible.

– Although IEEE 1588 end-to-end only covers frequency synchronization needs today, some operators believe that deploying this solution might help simplifying the migration path toward solutions covering also phase/time synchronization needs. This aspect is less obvious than the previous aspect, and will be discussed in section 4.4.3.

Regarding the first point (timing transparency), it is generally observed that IEEE 1588 end-to-end is considered in particular when multiple operators are involved in the mobile backhauling segment (the network that connects the base station to the core mobile network), typically when leased lines are used. In general, IEEE 1588 end-to-end is considered when a migration from PDH leased lines toward Ethernet leased lines is performed, with an IEEE 1588 flow generated and terminated by the mobile operator, as shown in Figure 4.27.

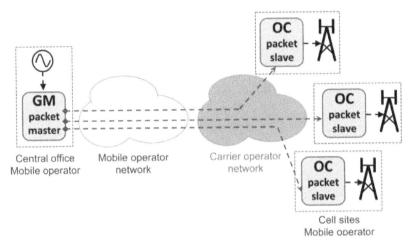

Figure 4.27. *Example of IEEE 1588 end-to-end typical deployment in mobile networks*

Although quite common, this type of deployment of IEEE 1588 end-to-end over a network that is not fully owned by the same network operator has associated technical challenges. As it will be further explained in the next sections, the main parameter impacting the quality of the frequency synchronization recovered by the PTP packet slave clock is the PDV generated by the network. Anticipating the PDV of a network for timing applications is not an easy task; this task becomes even more difficult when the network is not owned by the mobile operator, and sometimes even not known at all in terms of its size, technologies used, load, etc.

Typically, at the time of writing this book, no SLA related to PDV, defined in a way that would be useful for an adaptive clock recovery solution, was proposed in general to form part of the carriers' current Ethernet leased lines offers. This implies in practice no formal guarantee that the PTP packet slave clock will meet the appropriate performance. Some work is ongoing in standards to try to define some sort of SLA based on the PDV network limits defined by ITU-T, but this task implies some complexity, mainly due to the necessary budgeting of the overall end-to-end noise budget.

Indeed, the overall PDV to be filtered by the PTP slave clock is the combination of the PDV generated by the different segments of the connection between the PTP packet master clock and the PTP packet slave clock. Some segments may belong to the mobile operator, as shown in Figure 4.27; therefore, any theoretical SLA between the two operators should be only a portion of the overall PDV that the PTP packet slave clock can filter. Allocating PDV budget to different segments of the network is not an easy task, and has not been studied by ITU-T, which considered the entire connection as a unique segment when PDV network limits have been defined in the Recommendation G.8261.1 [G.8261.1].

It is important to bear in mind these technical aspects when deploying IEEE 1588 end-to-end in a multioperators environment; although many deployments do not raise particular issues, for example when the level of load of the networks is low, or when the PTP packet slave clock is tolerant to high PDV.

4.3.3.2. *Architectures typically considered with IEEE 1588 end-to-end deployments*

As already mentioned, IEEE 1588 end-to-end is an attractive solution especially when physical layer methods cannot be used, for example in some cases involving leased lines. This section presents the main architectural aspects to be considered for IEEE 1588 end-to-end deployments.

It is sometimes assumed that IEEE 1588 end-to-end will run over a packet network without any real view on the implementation of the packet network, its architecture, scale, etc. This is, in reality, not the case and networks will have different issues that affect performance and implementation. Clearly, these will affect end-to-end performance and scalability. To be able to develop a scalable architecture for IEEE 1588 end-to-end deployment, well-defined architectures need to be developed for master/slave deployment scenarios.

4.3.3.2.1. Options for PTP packet master clocks deployments

Properly positioning the PTP packet master clocks is an important aspect of the IEEE 1588 end-to-end solution. When several locations are possible, it is highly recommended that the various options are studied so that an appropriate choice can be made.

The first point to be considered is that a PTP packet master clock needs to receive an input frequency reference that is traceable to a PRC. Relying on a GNSS solution is possible, but requires a clear view to the sky for the antenna, which is not always feasible depending on the location. GNSS also has other drawbacks, as discussed in sections 4.2.2.1 and 4.4.1, and in Chapter 2. Therefore, quite a common alternative solution consists of using an input reference coming from a physical layer synchronization network (SDH or SyncE-based synchronization network), typically a reference delivered by an SSU.

The second point to be considered is how close to the edge of the network the masters should be deployed. This question is a trade-off between cost and performance: indeed deploying masters closer to the edge of the network implies a reduction of the number of hops of the master–slave connections, and therefore also a reduction of the PDV generated by the network, but it also implies in general the deployment of a higher number of masters.

Regarding the second point, a number of scenarios can be envisaged that are described for the purposes of discussion as a "centralized master", "distributed master" and "edge sited external master".

In all these scenarios, protection paths and redundancy of supply are ignored for simplicity, in all cases the PTP packet masters are traceable to a PRC and hence, by implication, the PTP slaves. These scenarios are described and discussed in turn in the following sections.

1) *Centralized master*

This scenario has a single PTP master deployed in a central location of the packet network (typically the core, or the switch site in case of mobile backhaul). In this scenario, the location of the PTP master may lead to differences in topology and length of network traversed (see Figure 4.28). In Figure 4.28, PTP slave (S2) may therefore have a worse level of performance than PTP slave (S1) as there are more hops to PTP slave (S2). These differences will increase as the scale and topology of the network change. The centralized master would typically be applicable to a small network both in scale and reach.

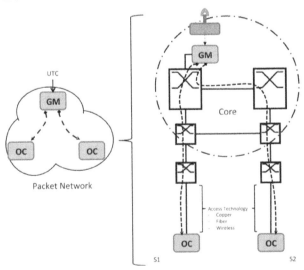

Figure 4.28. *Simple centralized master scenario*

2) *Distributed master*

This scenario has multiple distributed PTP masters deployed in the core/metro of the packet network. It is assumed that the PTP masters chosen will be the closest in terms of network hops and geographically closest to the PTP slave; therefore, the size of the master–slave connection is generally reduced in this scenario compared to the previous scenario. In the distributed master scenario, the PTP master may be located to architect out differences in topology (switch/router path) length thus equalizing network traversed in terms of number of hops (see Figure 4.29). Assuming the network performs consistently, PTP slave (S1) may therefore have a similar level of performance to PTP slave (S2) when considering the number of hops. The distributed master scenario would typically be applicable to a larger network both in scale and reach.

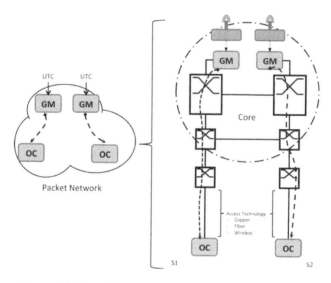

Figure 4.29. *Distributed master scenario for larger networks*

3) *Edge sited external master*

This scenario has PTP masters deployed in an edge location of the packet network, and may typically occur in a multi-operator/user situation. In the edge sited external master scenario, the PTP master is owned by operator A and in the center of their network but will typically be located toward the edge of the transit operator's network, that is operator B. This is illustrated in Figure 4.30. In this case, PTP slave S2 is closer to the PTP master than PTP slave S1 (in terms of hops) and may therefore have a higher level of performance. These differences will increase as the scale and topology of the network change. Variations will also be introduced dependent on path the PTP flow takes through the network, for example aggregation of traffic (and also PTP flows) further back into the core may allow the general transport architecture to scale better but may result in decreasing performance as more hops are introduced.

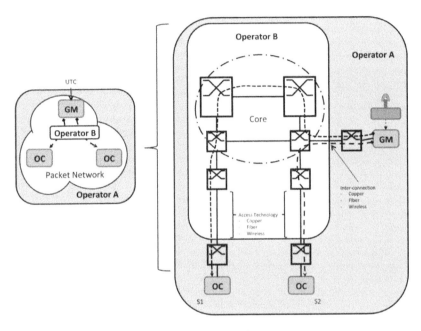

Figure 4.30. *Edge sited external master scenario*

Another issue with edge sited external master architectures is that performance requirements at the PTP slave may be difficult to achieve in all cases. Especially when the PTP flows go through another operator's network. Operator A has indeed no control over the architecture of operator B's network from a topology perspective, and by the nature of the location of the PTP master, the PTP flows may be routed using the shortest path or using a longer path. This is quite clearly an issue and shows some of the topology challenges when an operator does not have full control of the transport network.

Advantages/disadvantages of these scenarios are summarized in Table 4.4.

	Centralized master	Distributed master	Edge sited external master
Applicable scenario	Single network/operator domain	Single network/operator domain	Location in other operator/network domain/enterprise
Network scale	Smaller networks	Larger networks	Small or larger networks as applicable
Number of slaves supported	Will be limited by size/scale of network	Slaves partitioned to appropriate master	Will be limited by size/scale of network/transit network
Performance	Will drop as network grows	Flexibility to partition slaves to appropriate master	Will drop as size of transit network increases
Architecture	May be limited by technology/hop count	Flexibility to overcome hop count	May create architecture limitations in network, for example hop count
Cost	Limited (small number of masters)	Higher (number of masters to be deployed grows as the network)	Limited (small number of masters)

Table 4.4. *Issues surrounding location of IEEE 1588 end-to-end master in relation to location and type of network when full control on the transport network does not exist*

Protection of the PTP flow from the PTP master to the PTP slave and the impact of failure is also an important issue and needs to be considered in light of the underlying architecture used. This will be discussed in section 4.3.3.4.

4.3.3.2.2. Options for PTP packet slave clocks deployments

Positioning properly the PTP slave clocks is also another important aspect of the IEEE 1588 end-to-end solution. Several deployment options are possible for the PTP slaves, as shown in Figure 4.31.

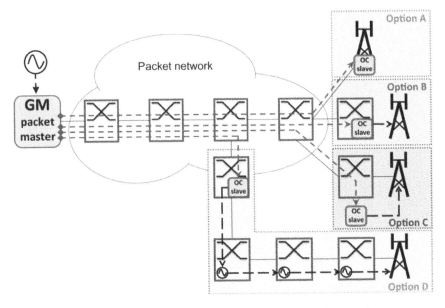

Figure 4.31. *Deployment options for the PTP packet slave clocks*

Assuming a deployment for a mobile network, the options depicted in Figure 4.31 are discussed hereafter. A network operator that decides to deploy the IEEE 1588 end-to-end solution should be aware of the pros and cons of each option before making a decision. Note: some options may not be possible in some cases, for example in case of legacy base stations, option A may not be available.

Option A – PTP packet slave clock embedded in the base station

In option A, the PTP packet slave clock is directly integrated inside the end application, for example inside the base station; therefore, the PTP communication is terminated directly at the end application, and the reference is delivered to the base station without the need for a physical layer interface between the PTP packet slave clock and the end application. The integration of a PTP packet slave clock in base stations is now very common, especially for the next generation of mobile networks such as LTE. The performance to be met at the output of the system containing the PTP packet slave clock is the air interface requirement (e.g. ±50 ppb accuracy and stability).

– Pros: the PTP packet slave clock is integrated inside a device normally having a good oscillator quality (indeed, base stations require good oscillators in order to generate a stable air interface radio signal), which is in general very useful to recover the frequency timing reference properly.

– Cons: the PTP packet slave clock is difficult to change totally in case of performance issues, since it is integrated in network equipment (software upgrades are however possible). Moreover, in case the backhaul network and the mobile base station are owned by separate operators, there is no demarcation point where the PTP flow is terminated; this might raise some performance problems in some cases, as discussed earlier in section 4.2.

Option B – PTP packet slave clock embedded within a network transport node co-located with the base station (e.g. Cell Site Gateway)

In option B, the PTP packet slave clock is integrated inside the last network node connecting the base station, typically an equipment called "Cell Site Gateway" (CSG); this last node then delivers the reference directly to the base station using a physical layer interface between the PTP packet slave clock and the end application (e.g. 2.048 MHz, 2.048 Mbit/s, SyncE, etc.). The integration of a PTP packet slave clock in CSG devices is also quite common. The CSG can act as a demarcation device in case of multiple operators' networks (e.g. if the PTP communication is handled only by the carrier operator). The performance to be met at the output of the system containing the PTP packet slave clock corresponds to the wander network limits defined in [G.8261.1], Figure 4 (note: these limits correspond to the Maximum Time Interval Error (MTIE) network limits defined in [G.823] for a traffic interface extended with a +16 ppb long-term slope).

– Pros: the PTP packet slave clock is integrated inside a network equipment (there is no additional box).

– Cons: the PTP packet slave clock is difficult to change totally in case of performance issues, since it is integrated in network equipment (software upgrades are however possible). Deploying a second external PTP packet slave clock (option C below) is however possible in case of performance issues.

Option C – PTP packet slave clock inside an external box dedicated to synchronization

In option C, the PTP packet slave clock is deployed in a dedicated box; this device then delivers the reference directly to the base station using a physical layer interface between the PTP packet slave clock and the end application (e.g. 2.048 MHz, 2.048 Mbit/s, SyncE, etc.). This dedicated box can act as a demarcation device for synchronization in case of multiple operators' networks (e.g. if the PTP communication is handled only by the carrier operator). The performance to be met at the output of the system containing the PTP packet slave clock corresponds to the wander network limits defined in [G.8261.1], Figure 4 (note: these limits correspond to the MTIE network limits defined in [G.823] for a traffic interface extended with a +16 ppb long-term slope).

– Pros: the PTP packet slave clock is easier to change in case of performance issues.

– Cons: the dedicated box is an additional equipment to be deployed, maintained, etc.

Option D – PTP packet slave clock within a network transport node not co-located with the base station (e.g. DSLAM, MW, etc.)

In option D, the PTP packet slave clock is integrated inside a network node upstream from the base station, typically in a Digital Subscriber Line Access Multiplexer (DSLAM), an Optical Line Termination (OLT) or a MW node; this upstream node delivers then the reference to the base station via physical layer methods, such as SyncE, along a chain of clocks in tandem.

In practice, it means that the PTP packet slave clock has to deliver a physical layer timing signal (e.g. 2.048 MHz, 2.048 Mbit/s, SyncE, etc.) with sufficiently good quality so that this signal is accepted by the next nodes. The relevant quality of such signal has not been studied by ITU-T so far.

– Pros: this type of solution might allow the avoidance of using IEEE 1588 end-to-end solution over a segment that is noisy in terms of PDV (e.g. DSL link, MW links, etc.), in the case where this segment supports a physical layer solution, such as SyncE. Moreover, in the case where the backhaul network and the mobile base station are owned by separate operators, the network transport node may be used as a demarcation point where the PTP flow is terminated (e.g. if the PTP communication is handled only by the carrier operator).

– Cons: the performance aspects of this deployment scenario have not been studied by ITU-T; the current specification of the PTP packet slave clock in [G.8263] may not be applicable to this scenario. Potential performance issues have to be considered, because the PTP packet slave clock must deliver a reference signal which can be tolerated by the clocks downstream (note: the signal delivered by a PTP packet slave clock may exceed the minimum tolerance of an EEC). Although this scenario may have interest and could be applicable in some specific cases, any potential deployment with this option needs to be considered very carefully, with associated testing of the final chain to ensure noise tolerance compatibility.

4.3.3.2.3. Interactions between IEEE 1588 end-to-end and physical layer methods (e.g. SyncE)

Interactions between IEEE 1588 end-to-end and physical layer methods such as SyncE are obviously possible, and even quite common. They have been standardized by ITU-T as explained in Chapter 2. These interactions occur at both ends of the PTP communication, that is, at the PTP packet master clock and at the PTP packet slave clock, as illustrated in Figure 4.32.

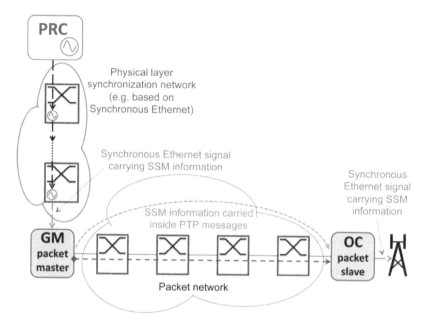

Figure 4.32. *Example of interactions between Synchronous Ethernet and IEEE 1588 end-to-end*

The following aspects should be underlined in Figure 4.32:

– As mentioned earlier, a PTP packet master clock needs to receive an input frequency reference traceable to a PRC; in general, this reference can be delivered by a physical layer synchronization network, which can be based on SyncE nodes (or on SDH nodes, or on a mix between SyncE and SDH).

– This SyncE input reference carries also traceability information via the SSM in the ESMC (or from the SDH frame in case of SDH input reference); the PTP packet master clock can select the source to be used based on this information (in case multiple sources are present), and the relevant traceability information is inserted in the PTP messages using the mapping between SSM Quality Levels (QLs) and the PTP clockClass attribute defined in [G.8265.1], section 6.7.3.1, Table 1 (see Chapter 2).

– The SSM QL information is then transported transparently through the PTP communication segment in the PTP messages.

– The PTP packet slave clock receives the frequency reference and the traceability information via the PTP messages; when this slave is external to the end application, like in Figure 4.32, it delivers the reference to the end application (e.g. base station) using a physical layer frequency signal (e.g. a SyncE or an E1 signal). This physical layer frequency signal may carry the original SSM QL information received by the PTP packet master clock.

4.3.3.3. *Performance aspects of IEEE 1588 end-to-end deployments*

IEEE 1588 end-to-end safe deployments require studying properly the performance aspects of the entire system, which includes the PTP packet master clock and the PTP packet slave clock, but also the network between these two elements. This section presents the main performance aspects to be considered for IEEE 1588 end-to-end deployments (Note: the PTP packet master clock will not be discussed in this section, although it is an important piece of the solution).

4.3.3.3.1. Anticipating the PDV generated by a packet network

As already mentioned several times, the main network impairment impacting the performance of the IEEE 1588 end-to-end solution is the PDV generated by the network, also called packet jitter.

There are well-known techniques used to anticipate the jitter of a packet network, for instance for real-time services such as voice. While some of these techniques can also be applied to synchronization flows, as it is discussed in this section, it should be mentioned that the criteria to be used to analyze the results strongly differs from traditional real-time end user services.

Indeed, in general, performance objectives over IP networks for real-time services such as voice are specified such that a jitter in the order of a few milliseconds has to be respected for almost all the packets, for example 99.9% of the packets must not exceed a few milliseconds of jitter.

For synchronization flows, it is the opposite: performance objectives are specified such that a very small amplitude of PDV, in the order of tens of microseconds, has to be respected, but it concerns only a very small portion of the packets, in the order of a few percents (for instance [G.8261.1] specifies as the PDV network limits for the Hypothetical Reference Model 1 (HRM-1) that the 1% of fastest PTP packets must not exceed 150 µs of PDV amplitude). This is due to the fact that PTP packet slave clocks perform in general packet selection; for instance, only the fastest packets are selected and used to recover the frequency reference.

To meet these very stringent PDV requirements related to synchronization flows, a specific engineering of the network needs to be setup: for instance, the synchronization flows are in general marked with the highest priority, and positioned in the highest priority queue. It is also recommended to avoid carrying large packets such as jumbo frames over the links where PTP messages are transported. Some useful information regarding expected engineering can be found in the description of the HRMs defined in [G.8261.1], section 7.

Determining precisely the jitter that a packet network will produce, especially at the levels required by synchronization flows, is a very difficult task. Indeed, many parameters enter into the equation, and the problem is made even more complex by the fact that the design of network equipment such as switches or routers may strongly differ from one to another (for instance software-based routers do not experience the same performance as hardware-based routers).

However, anticipating approximately the jitter that a packet network may produce is an achievable task. Identifying the network equipment producing excessive PDV is also feasible. While it could be done in theory via simulations only, network PDV estimation based on measurements, performed on the network equipment that will be deployed in the network, provides in general more accurate results, due to the variety of the equipment designs. This type of testing can be done, for instance: over a live network, over a chain of network equipment assumed to be representative of the network or over a single network equipment.

When the PDV generation testing is done over a single network equipment, it is then possible to anticipate the amount of jitter that should be observed over a chain composed of several similar devices, under certain assumptions. For instance, assuming that the sources of noise/delay measured for one equipment are independent of those of the other equipment, the measured statistical PDV distribution can be convolved in order to estimate the combined statistical PDV distribution produced by a chain of network equipment, as explained in Figure 4.33.

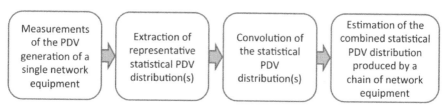

Figure 4.33. *Estimation of the PDV of a chain of network equipment based on measurements of the PDV generated by a single network equipment*

The difficulty of this type of test lies in being able to stress the relevant sources of noise/delay inside the equipment, in order to obtain meaningful results. In general, a network equipment is stressed by loading the device with one or several traffic generators. This topic is still under discussion at ITU-T, the objective being to develop a methodology for testing the PDV generation of a single network equipment. One of the interests of this work is to allow identifying the network equipment producing excessive PDV.

As mentioned earlier, when multiple operators are involved in the PTP communication path, it might be difficult to anticipate the PDV generated by the network, since operators are seldom sharing detailed information about their network infrastructure.

Introduction to JESS metric and examples

Jitter Estimation by Statistical Study (JESS) metric has been defined to analyze the PDV generated by a network equipment. This mathematical tool aims at determining the number of nodes that can be cascaded between a PTP packet master clock and a PTP packet slave clock. This metric typically applies to a PDV generation measurement performed on a single network node.

This metric has been presented to ITU-T in September 2011 [JOB 11b]; it consists of studying the maximum number of nodes that can be cascaded while meeting a certain percentage objective of timing packets (e.g. 1% for the HRM-1 of [G.8261.1]) within a floor delay window of a given size (e.g. 150 µs for the HRM-1 of [G.8261.1]). A floor delay window is defined here as the delay window starting at the smallest measured delay (minimum delay).

JESS estimates the PDV accumulation using the mathematical operation of convolution: the PDV measured on a single node is convolved by itself n times in order to estimate the PDV accumulated over $n + 1$ nodes. As the convolution operation describes the Probability Density Function (PDF) of the sum of independent random variables, JESS assumes that the PDV noises produced the network nodes (i.e. the PDV histograms extracted from the measurements) are independent. This assumption is however not always correct in real networks; JESS is anyway considered as providing a conservative estimation of the jitter produced by the network when high levels of load are applied on the network equipment on which the PDV is collected.

The mathematical definitions of this metric (JESS and JESS-w) are provided in Appendix 2 of this book and in [JOB 11b].

JESS and JESS-w metrics have been used to obtain PDV accumulation results that have been presented to ITU-T when [G.8261.1] has been developed. These results, supported by additional PDV measurements over various chains of network nodes, were the basis of the definition of the HRM-1 PDV network limits in [G.8261.1].

Some examples of results presented to ITU-T in [JOB 11c] are provided in Figures 4.34 and 4.35, in order to illustrate the use of JESS metric. The original PDV distributions were collected during PDV generation testing performed on IP routers equipped with 1 and 10 Gbit/s Ethernet interfaces.

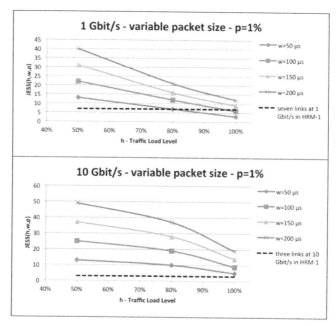

Figure 4.34. *JESS metric estimations on a router with 1 and 10 Gbit/s Ethernet interfaces*

The following aspects should be highlighted in Figure 4.34:

– Unlike curves providing information about the level of noise such as Time Interval Error (TIE) or MTIE, for which "the lower, the better", JESS estimation provides information about the maximum number of nodes that can be cascaded for a given histogram; hence "the higher, the better" with JESS metric curves.

– More precisely, Figure 4.34 shows that, when the objective is to ensure at least 1% of timing packets within a floor delay window of 150 µs (curves with triangles), 31 and 37 nodes can be, respectively, cascaded for 1 and 10 Gbit/s interfaces if not more than 50% of load is applied on the tested equipment, while only 16 and 28 nodes can be, respectively, cascaded for 1 and 10 Gbit/s interfaces if up to 80% of load is applied on the tested equipment.

– The decreasing curves in Figure 4.34 show therefore the degradation of the maximum number of nodes estimation when increasing the traffic load on the tested equipment.

– Larger floor delay windows allows more nodes to be cascaded, but larger percentage objectives ($p > 1\%$) would allow fewer nodes to be cascaded (not shown in the figure).

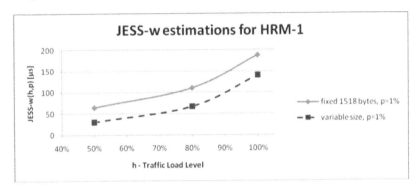

Figure 4.35. *JESS-w estimations for the HRM-1 of [G.8261.1]*

The results shown in Figure 4.35 led to choose the value of 150 μs for the PDV network limits of the HRM-1 of [G.8261.1]. Two types of background data traffic applied on the routers are compared: the first traffic is based only on large packets with fixed size (1518 bytes) and second traffic is based on variable packet size (from 64 to 1518 bytes). The two curves in Figure 4.35 show how the size of the floor delay window that contains the 1% of fastest packets increases when increasing the level of load on the network equipment. It has been considered that the case of variable packet size (dotted curve) was more representative of the networks, and it has been decided, based on a request from operators, to consider the highest level of load (i.e. 100%) in order not to put restrictions on the load that the operator can apply on the network. Again, these results were supported by additional measurements on various platforms, using different types of chains of network nodes and of background data traffic.

This type of methodology based on JESS metric can be used for network planning purposes, for instance in order to anticipate the PDV generated by a packet network and determine relevant locations for PTP packet master clocks and PTP packet slave clocks.

Synchronization Design and Deployments   217

Usage of PDV metrics to characterize an entire packet network

The characterization of a full network in terms of PDV can be done using some of the PDV metrics defined in the ITU-T Recommendation G.8260 [G.8260], Appendix I. The definition of a PDV metric to be used as a pass/fail criterion has been a complex task in ITU-T; this is the reason why many proposals are defined in Appendix I.

It should be mentioned that, at the time that this book was written, Floor Packet Percent (FPP) is the only PDV metric coming from [G.8260] that is used in a normative way by ITU-T to define the PDV network limits in [G.8261.1] applicable to IEEE 1588 end-to-end; any characterization of a packet network should therefore use at least this metric.

Further details about PDV measurements over an entire network and about the definition and usage of the FPP metric are provided in Chapter 7.

4.3.3.3.2. Compliancy to the ITU-T PTP telecom profile and PDV tolerance of the PTP packet slave clock

Two main aspects must be covered when studying the performance of the PTP packet master and PTP packet slave clocks:

– The compliancy to the PTP protocol [IEEE 1588™-2008] and to the IEEE 1588 end-to-end telecom profile defined by ITU-T in [G.8265] and [G.8265.1] – this is a key aspect to be verified, applicable to the PTP packet master and PTP packet slave clocks, in order to ensure interoperability between the implementations.

– The PDV tolerance of the PTP packet slave clock – it consists of ensuring that the PTP slave can tolerate at least the PDV network limits defined in [G.8261.1]. It applies to the PTP packet slave clock only. Indeed, while the evaluation of the PTP packet master clock function is not too complex, since this equipment basically converts a physical layer reference into a packet timing flow composed of PTP messages, the main complexity of the IEEE 1588 end-to-end system resides in the PTP packet slave clock, which has to filter the PDV generated by the network in order to recover a clock with an appropriate quality.

Testing the compliancy to the PTP protocol and ITU-T telecom profile of a PTP packet master or packet slave clock and PDV tolerance testing of a PTP packet slave clock will be detailed in Chapter 7.

4.3.3.4. *Usual protection schemes with IEEE 1588 end-to-end*

Originally, [IEEE 1588™-2008] defines a protection mechanism called Best Master Clock Algorithm (BMCA), which is detailed in Chapter 2. Chapter 3 also provides information about architectures and protection schemes applicable to

IEEE 1588 end-to-end. The BMCA mechanism consists essentially of creating and maintaining a tree between the best PTP clock of the PTP synchronization network and the other PTP clocks. The default BMCA hence avoids timing loops.

When the PTP telecom profile defined in [G.8265.1] has been developed by the ITU-T, it was decided not to rely on the default BMCA specified in [IEEE 1588™-2008] , but instead to remain in-line with traditional telecommunications practices and with the principles of SDH and SyncE selection mechanism defined in the ITU-T Recommendation G.781 [G.781]. An alternate BMCA has been specified for this, as also detailed in Chapter 2. This alternate BMCA allows having multiple PTP packet master clocks active at the same time, so that a PTP packet slave clock can listen to several active PTP packet master clocks at the same time, in order to permit PTP master redundancy schemes.

Therefore, the protection schemes related to the IEEE 1588 end-to-end solution can be considered as very similar to the schemes used for SDH and SyncE-based synchronization networks. They can, moreover, interwork with such physical layer frequency networks, as explained in section 4.3.3.2.3, since the SSM QL information can be transferred over the PTP communication path.

The PTP packet slave clock receives SSM QL information in the PTP clockClass attribute from each potential PTP master, and has a local priority configured for each potential PTP master; the PTP master is selected based on these two parameters, in a similar way as defined in [G.781]. Essentially, each PTP connection between each PTP packet master clock and the PTP packet slave clock can be considered as a logical link carrying (noisy) synchronization reference, similarly to an SEC or EEC clock having multiple input links carrying synchronization reference.

It should be mentioned that there is no risk of timing loops inside the PTP communication path with the IEEE 1588 end-to-end solution, since there is no PTP clock between the PTP packet master clock and the PTP packet slave clock with this architecture.

In addition to this alternate BMCA, other protection schemes are also possible when using IEEE 1588 end-to-end. For instance, the PTP packet slave clock can enter the holdover state during failure in case no backup PTP packet master clock would be available.

Finally, it should be mentioned that monitoring the performance of the IEEE 1588 end-to-end solution is an important challenge for the network operator. Two main aspects can be monitored: the output performance of the PTP packet slave clock and the network PDV. Both require having access to a frequency reference source at the edge of the packet network, which is not possible permanently in general (otherwise, the use of IEEE 1588 end-to-end would be useless). The use of the internal reference recovered by the PTP packet slave clock is clearly not an ideal solution, since this reference is based on the PTP communication, and may therefore suffer from inaccuracies; this reference might not allow detecting all possible problems. It may, however, be used to detect short-term impairments.

It is therefore important to keep in mind that on-site monitoring and troubleshooting might be required in case performance issues are faced after the deployment of such solution.

4.3.3.5. Summary of main aspects of IEEE 1588 end-to-end deployments

Previous sections have introduced IEEE 1588 end-to-end synchronization networks from the view of network operators in charge of deploying and operating them.

In summary, the following points must be kept in mind when considering IEEE 1588 end-to-end deployments:

– IEEE 1588 end-to-end does not provide as good a performance as IEEE 1588 full timing support or SyncE, because it is impacted by the PDV of the network.

– IEEE 1588 end-to-end is considered as useful when physical methods such as SyncE are not possible (a typical example is when leased lines are used).

– IEEE 1588 end-to-end finds relevant use cases, in general, to support mobile applications.

– IEEE 1588 end-to-end has been designed in order to allow a certain degree of interworking with physical layer based synchronization networks (e.g. SyncE); protection schemes are quite similar.

– The positioning of PTP packet master clock and PTP packet slave clock in the network should be studied carefully, depending on the network topology.

– Studying the PDV generated by the network is highly recommended before deploying IEEE 1588 end-to-end solution, in order to verify that it does not exceed what can be tolerated by the PTP packet slave clock.

– Testing the PDV tolerance of the PTP packet slave clock is also an important item when considering IEEE 1588 end-to-end deployments.

– Testing the compliancy to the PTP protocol and telecom profile for the PTP packet master clock and PTP packet slave clock is also very important to ensure interoperability.

– Accurately monitoring the performance of IEEE 1588 end-to-end solution after deployment may be challenging.

The following IEEE and ITU-T standards are related to IEEE 1588 end-to-end: [IEEE 1588™-2008], [G.8265], [G.8265.1], [G.8261.1] and [G.8263].

4.4. Deployment of timing solutions for accurate phase/time synchronization needs

This section discusses how to envisage the deployment of synchronization networks addressing accurate phase/time synchronization needs, required, for instance, by some mobile technologies, as explained in Chapter 1. In particular, two synchronization technologies are analyzed: GNSS, such as GPS and Galileo, and IEEE 1588 full timing support.

At the time this book was written, the standardization of the IEEE 1588 full timing support technology was still ongoing; therefore, only a general introduction to this solution will be provided in this section (essentially, the main architectural aspects and associated challenges). Other solutions based on so-called "partial timing support" (see short description in section 4.4.2 and in Chapter 2) have been proposed in standards, but not yet studied; they are therefore not presented in this section.

Before discussing how to consider the deployment of these technologies, it is worth introducing them with a short comparison. Table 4.5 summarizes the two main types of solutions that can be considered to address accurate phase/time needs. Other off-air solutions, introduced in section 4.2.2.2 and in Chapter 2, might also be applicable in some cases, but they are not detailed in this section.

As explained in Chapter 3, the major driver leading potentially to mix these two technologies is essentially the optimization of the cost of the overall synchronization solution, for instance by limiting the number of GNSS receivers to be deployed. Securing the GNSS receivers from potential failures (e.g. air interferences) is also an important driver when designing an architecture for accurate phase/time synchronization delivery.

One very basic solution to address the needs for accurate phase/time of mobile networks might simply be to deploy one GNSS receiver per cell site, and not rely at all on the IEEE 1588 full timing support solution. SyncE could still be used with this architecture to secure the GNSS receivers against radio signal failures, by providing enhanced phase/time holdover.

An alternative, potentially attractive, architecture could be to move the GNSS receivers higher in the network, for instance in a central office, and to distribute phase/time synchronization using IEEE 1588 full timing support over a network providing hardware support for PTP. A combination with SyncE can also be used with this alternative architecture to secure the GNSS receivers against radio signal failures by providing enhanced phase/time holdover. As it will be discussed, this architecture has also some drawbacks to be considered.

Technology	Type of sync.	Pros	Cons
Packets based methods "node-to-node" (e.g. IEEE 1588 full timing support)	Frequency and phase/time	– Hardware support from the network should remove the impact of PDV and asymmetry inside the network nodes – Combination with Synchronous Ethernet is possible (e.g. for phase/time holdover)	– All the network nodes of the timing chain must implement hardware timing support (e.g. PTP Boundary Clock or Transparent Clock) – Link asymmetry issue not solved by PTP, manual calibration required today – Very tight budgeting of noise sources
GNSS (e.g. GPS, Galileo)	Frequency and phase/time	– Excellent performance/timing quality is achievable (e.g. 100 ns accuracy) – Should play a significant role in a phase/time distribution architecture (e.g. as Primary Reference Time Clocks)	– Installation cost – Antenna infrastructure – Indoor configurations – Jamming, air interferences – Lack of commercial guarantee/SLA of GPS

Table 4.5. *Overview and comparison of synchronization solutions for accurate phase/time*

4.4.1. *GNSS deployments and associated issues*

Although originally designed mainly for positioning purposes, GNSS, such as the GPS in the United States or future Galileo in Europe, can also be useful as a source of synchronization. Indeed, the GNSS receivers receive the timing reference generated by the satellites, which is based on atomic clocks. The GPS system provides, for instance, traceability to the GPS time, which is based on the "Temps Atomique International" [International Atomic Time] (TAI), and therefore related to UTC. Hence, these systems can be used to accurately synchronize in phase and time a telecommunication network. The details of this type of technology are introduced in Chapter 2, section 2.7, and also discussed in section 4.2.2.1.

It can be considered that the use of GNSS systems is a relevant solution in many cases to addressing accurate phase/time synchronization needs. Indeed, this solution does not require any synchronization function to be supported inside the telecommunication network, since the phase/time reference is delivered outside the network, using radio signals. It simplifies the design of the telecommunication network and avoids additional costs related to the implementation of dedicated synchronization functions in the network equipment. Note that GNSS systems are also the main technology envisaged to generate the primary phase/time synchronization reference when using IEEE 1588 full timing support solution (GNSS-based PRTC).

However, there are drawbacks that have to be borne in mind related to the use of such satellite systems:

– *Installation cost*: to receive the GNSS radio signals, it is necessary to deploy a GNSS receiver equipped with a relevant antenna. This type of installation may imply a non-negligible associated cost, especially in some locations where antenna installation is not easy and where, for instance, a long cable between the antenna and the receiver could be required. A specific engineering should be considered when high performance is targeted; for instance, the antenna cable needs to be properly calibrated. Moreover, some GNSS receivers may be expensive due to the use of high-quality oscillators in some designs.

– *Antenna infrastructure*: as discussed, the antenna infrastructure requires specific engineering. As an example, it is in general recommended to deploy the antenna with a clear view to the sky in order to receive as best as possible the GNSS signals, and minimize the various radio impairments.

– *Indoor configurations*: For some network equipment deployed indoors, the use of GNSS solution may become difficult to consider, and sometimes not feasible. It is, in particular, not always trivial or even possible to install an antenna on the roof of some buildings and a cable to connect it to the receiver or to the network equipment. The utilization of window antenna is in general not recommended because of the multipath issue.

– *Jamming, air interferences*: there are known examples of interferences with GPS radio signals, including for instance the United States where the use of GPS solutions is more developed than in Europe. Some interferences may be intentional, while others may be unintentional. As an example, it is quite easy to have access to low-cost GPS jammers (even if using such devices is usually illegal), some of them are designed simply to be plugged onto a cigarette lighter in a car. In any synchronization network design, this risk should not be underestimated, and appropriate protection schemes or backup systems should be considered. GNSS vulnerabilities are further discussed in Chapter 6 (section 6.2).

– *Lack of commercial guarantee/SLA of GPS system*: as far as the GPS system is concerned, it should be recalled that this system is free of charge, in terms of the usage of the GNSS radio signals, but does not provide any minimum guarantee of availability or of quality. Usually, this system is very reliable, but there are some examples where the service has been temporarily stopped in some areas of the world, or sometimes where the civil signals have been intentionally degraded. In this respect, the Galileo system is expected to provide some improvement, since commercial services with a certain commitment in terms of availability and quality of service are planned to be offered.

To mitigate these issues related to the use of GNSS systems, various protection schemes can be envisaged; they may include, for instance, phase/time holdover, switching between different GNSS systems or switching between different off-air systems. These options are discussed hereafter. Other information is provided in Chapters 2 and 6.

Concept of phase/time holdover and associated alternatives

The concept of phase/time holdover corresponds to maintaining locally in the GNSS receiver the phase/time reference during a certain period when the GNSS radio signals are not available or are degraded. Phase/time holdover is based on the use of a frequency reference to continue driving the local timescale in the GNSS receiver when the input time reference (GNSS signals) of the system cannot be used. There are two main alternatives to providing phase/time holdover.

First, phase/time holdover can be based on the holdover of a local oscillator with good quality, that is, the oscillator is not locked to an external frequency reference. It should be mentioned that, because of the tight requirements related to some mobile technologies (e.g. sub-microsecond phase/time accuracy), relying on the holdover of a local oscillator does not allow in general for a long holdover period; this approach is rather appropriate for limited outage periods (e.g. a few minutes) or during protection switching between different phase/time references (e.g. between different GNSS systems).

Second, phase/time holdover can be based on an external accurate and stable frequency reference, such as a reference delivered by a SyncE synchronization network, synchronizing a local oscillator in the GNSS receiver. The ITU-T Recommendation G.8272 [G.8272] describes a functional model of a PRTC based on GNSS, which can implement such external frequency input. At the time of writing this book, studies were still ongoing at ITU-T in order to determine how such combination could be used, and the related performance (i.e. the phase/time holdover period). In particular, the wander accumulated along the SyncE chain should be controlled to a low enough level, or filtered properly, in order to allow such combination. With an appropriate design and engineering of the SyncE network, it is expected that such a solution could allow for relatively long holdover periods (e.g. several hours).

Switching between different GNSS systems

Another option to mitigate the potential outages of a GNSS system is to rely on multiple GNSS systems. For instance, a GNSS receiver can be designed to receive timing information from multiple GNSS systems, for example GPS, Galileo and GLONASS. If one of the GNSS systems fails, then the receiver may switch to another one. It is also possible that the receiver listens to all the GNSS systems at the same time and produces a combination of the various timing references (ensemble).

It should be mentioned that this protection scheme is not robust against strong radio interferences that might affect all the GNSS radio signals at the same time.

Switching between different off-air systems

Another option to mitigate the potential outages of a GNSS system is to use backup radio systems in addition to GNSS systems, such as eLORAN/LORAN-C. However, such systems have limited regional availability, as highlighted in Chapter 2, and are today not really used for synchronizing telecommunication networks. A GNSS receiver could, for instance, also integrate a receiver to these additional radio systems, or several receivers could be deployed. If the GNSS system fails, the other radio system could be used as a backup.

Again, it should be mentioned that this protection scheme is not robust against strong radio interferences that might affect all the radio signals at the same time, although interfering with LORAN-C signals is in general more difficult than with GNSS signals, due to the high transmission power of the LORAN-C signals.

The GNSS receivers act as PRTC. It is possible to deploy such functions at different places in a telecommunication network; they can be more or less centralized in the network, and sometimes can be combined with the IEEE 1588 full timing support solution.

Figure 4.36 shows an example of an architecture where a GNSS receiver is deployed in each cell site to synchronize mobile base stations. These receivers are secured by an input SyncE signal in order to provide phase/time holdover during potential periods of GNSS failures.

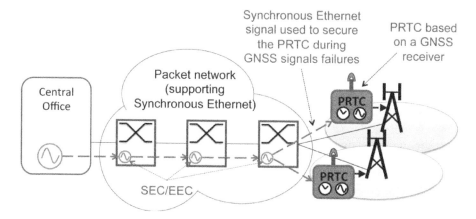

Figure 4.36. *GNSS receivers deployed at each cell site and secured with SyncE*

To optimize the cost of the phase/time synchronization solution, a network operator could be interested in trying to centralize more the PRTC functions based on GNSS than in Figure 4.36, and share them for multiple cell sites. In this case, the distribution of phase/time synchronization reference from the PRTC to the mobile base stations can rely on the IEEE 1588 full timing support solution, as it will be described in the next section.

4.4.2. *IEEE 1588 full timing support deployments*

One of the main drivers for considering phase/time distribution by the telecommunication network instead of a solution purely based on GNSS is to minimize the number of GNSS systems to be deployed, and try to centralize them so that these PRTC functions would be shared. Figure 4.37 shows one example of such architecture.

4.4.2.1. Rationale for considering IEEE 1588 full timing support compared to a partial timing support deployment scenario

The IEEE 1588 full timing support solution is generally considered as a solution technically viable to deliver very accurate phase/time synchronization, although at the time when this book was written, the standardization of such solution was still ongoing at ITU-T.

Section 4.3.3.1.2 discussed the major differences between the IEEE 1588 end-to-end and full timing support deployment scenarios; Figure 4.25 has illustrated what is meant by IEEE 1588 full timing support: it corresponds to a deployment scenario where all the nodes in the middle of the network between the PTP packet master clock and the PTP packet slave clock do support PTP functions such as BCs and TCs (PTP-aware nodes). As explained, the purpose of these hardware functions is to mitigate network impairments such as PDV and asymmetry inside the equipment.

Relying on a full PTP timing support from the network, where all the nodes support PTP hardware functions, can be considered as a more robust solution compared to a deployment scenario with partial timing support, where not all the nodes support such PTP hardware functions. In the case of partial timing support, the PDV generated by the network is not removed. To achieve high performance, this PDV has to be filtered, and the asymmetry that may be generated by the network nodes has to be compensated. Various testing results on telecommunication platforms show that achieving high performance with such a solution could be difficult, even with smart packet selection algorithm, due to the high amplitude of PDV generated by telecommunication equipment, and especially due to the variable asymmetry that is generally observed. At the time when this book was written, the partial support case was still an item under study at ITU-T; the architecture and performance associated with such approach were still unknown.

Obviously, similar to SyncE, the major drawback of the IEEE 1588 full timing support solution is that it requires all the nodes from the network between the PTP packet master clock and the PTP packet slave clock to support some PTP hardware function, which might in some cases imply the need to replace or upgrade some network equipment. A new hardware may indeed be required in some cases to support the BC or TC functions, and to implement hardware time-stamping properly (i.e. at the PHY level). However, while a progressive introduction of SyncE was possible due to the interworking with existing SDH-based synchronization networks (see section 4.3.2.2), it should be mentioned that building a fully PTP-aware network with such PTP functions is expected to be rather a "one-step" process, since there is, in general, no possibility to interwork with an old solution for phase/time synchronization distribution by the network already deployed. The progressive introduction of PTP hardware functions is, therefore, not as easy as SyncE. Supporting BC or TC in the network nodes may, however, be anticipated by the network operator in order to prepare the network for future synchronization needs.

These aspects need to be taken into account during the decision to deploy the IEEE 1588 full timing support solution, and also later during the design of the synchronization network. Indeed, various locations could, in general, be envisaged for the PRTC functions (i.e. GNSS receivers), and whether or not to support BC and TC functions in the network nodes is obviously one parameter to be considered in deciding if the PRTCs should be centralized or not. It could be possible, for instance, to deploy the PRTC functions quite close to the edge of the network in order to limit the number of network equipment having to support PTP hardware functions.

Figure 4.37 illustrates the IEEE 1588 full timing support deployment scenario. In this example, the IEEE 1588 full timing support solution is combined with SyncE. The next section will discuss this technical option.

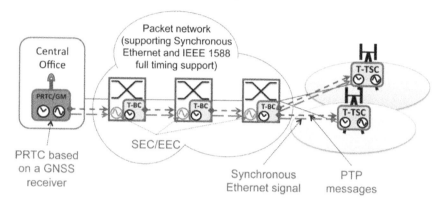

Figure 4.37. *GNSS receivers deployed at a central office with phase/time being delivered to the cell sites using IEEE 1588 full timing support combined with Synchronous Ethernet*

4.4.2.2. *Combined Synchronous Ethernet and IEEE 1588 full timing support deployments*

During the development of the IEEE 1588 full timing support solution, the possibility to combine this solution with SyncE has been considered from the beginning. Indeed, such a combination was considered as providing valuable technical advantages.

Having access to a stable and accurate frequency reference can, indeed, help the recovery of accurate phase/time synchronization. For instance, such a combination may allow in some cases: the reduction of the PTP packet rate, designs based on cheaper oscillators, and a faster start-up and convergence for the phase/time recovery.

But probably, the main advantage of such combination is related to phase/time holdover, as described in section 4.4.1 previously. Phase/time holdover can, indeed, also be applied to equipment other than the PRTC/GNSS receivers. As an example, a SyncE signal could be delivered to a mobile base station receiving also PTP messages, as shown in Figure 4.37; this signal can be used to maintain the phase/time local reference of the base station in case PTP messages would no longer be received. Different causes might lead to this situation of phase/time holdover, with various related periods, for instance: reference switching between different PTP packet master clocks (short phase/time holdover period is expected, in general, in this case, since the network should be reconfigured automatically), failure of a PTP packet master clock with no backup (long phase/time holdover period is expected in this case, since on-site operations might be required), etc. The combination with SyncE is expected to help extend the phase/time holdover period, therefore to be useful for long periods.

There are also associated challenges when combining IEEE 1588 full timing support and SyncE: as far as ITU-T has considered this combination when this book was written, frequency and phase/time planes were assumed to be maintained independently. It means, for instance, that the frequency plane (SyncE) maintains its own topology independent of the topology of the phase/time plane (PTP), which can lead to situations where the topologies are different (it is called "non-congruency" in ITU-T terminology, see also Chapter 3). It should be understood that maintaining two synchronization planes independently may in some cases lead to additional operational complexity: as an example, it means that a network equipment might receive on two separate ports frequency synchronization via SyncE and phase/time synchronization via PTP messages. Debates about the complexity of such assumptions were still ongoing at ITU-T when this book was written.

Also, simulations have shown that SyncE rearrangements (e.g. in ring topologies) can cause the accumulation of phase/time error in the PTP phase/time plane. Discussions were still ongoing at ITU-T when this book was written, but various mitigation solutions had been identified to minimize this behavior.

Obviously, it is also possible to rely on the IEEE 1588 full timing support solution without the combination with SyncE. In this case, frequency is delivered to the various PTP functions (e.g. BCs) directly via the PTP messages. This option is, of course, also studied by the ITU-T; at the time of writing this book, simulation activity was starting. It should be mentioned that the advantages provided by SyncE are lost in this case (holdover, or redundancy, has to be provided by some other means).

4.4.2.3. *Link asymmetry issue and possible solutions*

Another important parameter to be considered when making decisions about the location of the PRTCs is the issue of link asymmetry. Typically, PRTC functions may not be deployed in a very centralized location in some synchronization network designs in order to avoid having to carry PTP over uncalibrated links with potentially high asymmetry. Actually, this point is probably a major technical issue that could lead to excessive costs, making the IEEE 1588 full timing support irrelevant in some cases, economically speaking.

An optical link connecting together two network nodes such as routers is in most cases composed of two separate optical fibers, one per direction. When these fibers do not have the same length, it leads to link asymmetry, which impacts the performance of IEEE 1588 full timing support if this asymmetry is not compensated: it is, in general, considered that a difference of 1 m between the two fibers leads to roughly 2.5 ns of time error (considering the information is transferred at the speed of 5 ns/m in optical fibers, and that the accumulated error with the PTP protocol is half the asymmetry). This static time error in the worst case accumulates in an additive way along the IEEE 1588 full timing support chain. Connections over copper or MW links may suffer from similar asymmetry impairments.

The link asymmetry issue may, in particular, be very problematic when a transmission system, such as WDM and OTN, is inserted between the two network nodes. Indeed, in some cases, especially when the distance between the two sites to be connected is long, Dispersion Compensating Fibers (DCF) can be inserted in the optical system to compensate the chromatic dispersion of the system. DCFs that are designed to compensate for the same amount of chromatic dispersion do not necessarily have precisely the same fiber length; important differences, in the order of several hundreds of meters or more, have been recorded. It is shown in Figure 4.38.

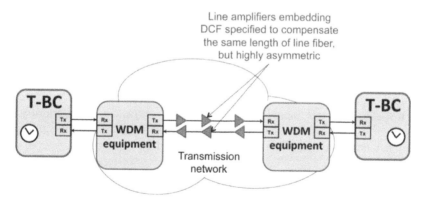

Figure 4.38. *Link asymmetry may be caused by the use of asymmetric DCF*

One possible way to mitigate this general problem could be using monofiber links; with this technology, information is transmitted over the same fiber in both directions using a different wavelength per direction. However, this type of links is not widely used today and is at the moment not as efficient as links based on two separate fibers, especially for long distances. This is why most transmission links are based on two separate fibers, which may suffer from asymmetry problems due to DCF.

Therefore, the only practical solution today is to measure the delay of each fiber separately in order to determine the asymmetry of the link, and to compensate for it during the computation of the PTP time stamps. It is important to mention that the PTP protocol does not itself allow automatically calibrating the link asymmetry. On-site measurements are generally necessary, using GPS receivers or Optical Time-Domain Reflectometer (OTDR) systems, which imply potentially important costs, especially if the number of links to be calibrated is high.

At the time of writing this book, research was still ongoing with regard to automatic asymmetry calibration and compensation mechanisms inside the network equipment; such features might imply new hardware.

4.4.2.4. *IEEE 1588 full timing support over OTN*

In addition to link asymmetry issues related to DCF, the use of IEEE 1588 full timing support over OTN is also a topic where challenges have been identified, and regarding which debates were also ongoing at ITU-T at the time of writing this book. Note that the elements provided hereafter are, therefore, initial considerations only, since standardization activity was still ongoing.

Several options have been identified [JOB 11a] to carry phase/time synchronization over OTN using PTP (some of them being proprietary). They can be classified into two main categories:

1) *Transparent transport of PTP client*: it corresponds to the traditional use of OTN as transport layer of a client signal, where timing transparency (for phase/time synchronization here) is expected from OTN. No additional specific PTP hardware functions inside the OTN equipment (e.g. BC or TC) are necessary to process the PTP messages with this approach, since the PTP messages are carried transparently; however, strict control of asymmetry generated by the OTN buffers is required, which may not be compatible with all generations of OTN equipment.

2) *"Time-synchronized" OTN*: it corresponds to a hop-by-hop synchronization distribution system where each OTN node receives a phase/time reference, for instance, via PTP messages, processes them and sends the phase/time reference to the next OTN node. Specific PTP hardware functions inside the OTN equipment (e.g. BC or TC) are necessary to process the PTP messages with this approach. Several options have been proposed with regard to the channel inside which the PTP messages can be transported hop-by-hop over the OTN segment (see, for instance, options 2.1 and 2.2 described below). Note that this "time-synchronized" OTN approach does not imply a "frequency-synchronized" OTN at the physical layer: it is expected that OTN remains an asynchronous transport layer, providing timing transparency (via justification techniques) to the client signals when necessary.

In practice, three main options have been presented to the ITU-T; they are referred to below as options 1, 2.1 and 2.2, and will be briefly introduced with a short comparison.

Option 1: transparent transport of the PTP messages within the client signals

Figure 4.39 illustrates option 1 where PTP messages are carried inside the client signal (e.g. Ethernet client signal) transparently over OTN.

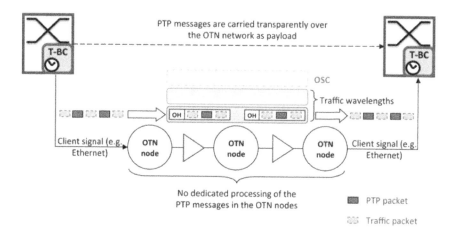

Figure 4.39. *Transparent transport of the PTP messages within the client signals (option 1)*

Some key points should be underlined with option 1:

– No PTP hardware functions (e.g. Telecom-BCs) are necessary in the OTN nodes and in the line amplifiers with option 1.

– Unlike options 2.1 and 2.2, PTP messages always stay inside the client signal and are never extracted and reinserted with option 1. This is in line with the current architectural positioning of OTN as a transport layer.

– Proper calibration of the link asymmetry in the OTN connection due to the DCFs and to the line fiber is required with option 1.

– Strict control of asymmetry generated by the OTN buffers is required with option 1.

Option 2.1: transport of the PTP messages in the OTN "in-band", for instance, within the OTN overhead

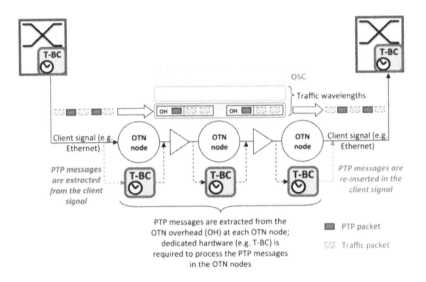

Figure 4.40. *Transport of the PTP messages "in-band" in the OTN overhead (option 2.1)*

Some key points should be underlined with option 2.1:

– PTP hardware functions (e.g. Telecom-BCs) are necessary in the OTN nodes with option 2.1, but not in the line amplifiers.

– PTP messages are extracted from the client signal when entering the OTN connection, and are potentially reinserted inside the client signal when exiting OTN with option 2.1. This is not in-line with the current architectural positioning of OTN as a transport layer, since OTN becomes fully part of the phase/time synchronization network, with PTP clocks in the OTN nodes processing the PTP messages.

– Proper calibration of the link asymmetry in the OTN connection due to the DCFs and line fiber is required with option 2.1.

Option 2.2: transport of the PTP messages in the OTN "out-of-band", for instance, within the Optical Supervisory Channel (OSC).

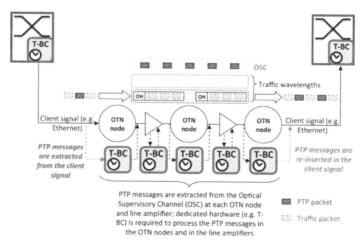

Figure 4.41. *Transport of the PTP messages "out-of-band" in the Optical Supervisory Channel (option 2.2)*

Some key points should be underlined with option 2.2:

– PTP hardware functions (e.g. Telecom-BCs) are necessary in the OTN nodes and in the line amplifiers with option 2.2.

– PTP messages are extracted from the client signal when entering the OTN connection, and are potentially reinserted inside the client signal when exiting OTN with option 2.2. This is again not inline with the current architectural positioning of OTN as a transport layer, since OTN becomes here fully part of the phase/time synchronization network, with PTP clocks in the OTN nodes processing the PTP messages.

– Proper calibration of the link asymmetry in the OTN connection due to the line fiber is required with option 2.2. However, the DCFs should not create link asymmetry with option 2.2 because they are by-passed.

These different options were still under study at the ITU-T when this book was written. The third option 2.2 based on the OSC was considered very difficult to be standardized, although proprietary industrial implementations exist, because the OSC channel itself is not standardized. Some testing results were reported also for options 1 and 2.1, showing technical feasibility of the solution in some cases, but also a certain amount of noise and asymmetry for the option 1 with some OTN equipment design. The interworking of such OTN mechanism with the IEEE full timing support telecom profile under definition at the ITU-T is also an important point still under study.

4.4.2.5. *Mechanisms to carry phase/time over access technologies and interworking with IEEE 1588 full timing support*

Similarly to the transport of phase/time synchronization over OTN, the use of IEEE 1588 full timing support together with certain technologies considered in the access of a telecommunication network, or close to the access (e.g. first mile), was still a subject under study when this book was written, with identified challenges. Typically, the transport of accurate phase/time over PON, DSL or MW links, for example, requires specific functions to be supported in the involved network nodes.

Some of these technologies, such as PON or DSL, have defined specific technology-oriented mechanisms to transport accurate phase/time synchronization across these links that are not based directly on PTP, but on a mechanism inherent to the technology.

Interworking between these specific mechanisms and IEEE 1588 full timing support is required; it is also a subject under study at the ITU-T, and some proposals have been made, such as the concept of "distributed Boundary Clock", corresponding to some of the BC functions split into separate network elements, for example OLT and Optical Network Unit (ONU) for PON systems. This is not considered however as a full BC *per se*, functionally speaking, but rather a PTP master function and a PTP slave function, and a conversion between both mechanisms (technology specific and IEEE 1588 full timing support), as illustrated in Figure 4.42.

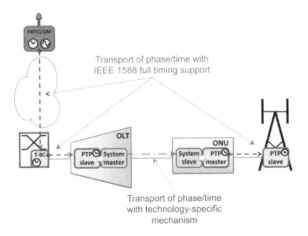

Figure 4.42. *Interworking between IEEE 1588 full timing support and technology-specific phase/time transport mechanism*

Synchronization Design and Deployments   235

With regard to MW systems, some of them tend to rely on the use of TC defined in PTP, although using the BC type of support is also fully applicable to MW links. The use of TC is also envisaged in some simple network equipment, such as 2-ports demarcation devices.

It should be mentioned that some issues related to the use of TC in a telecommunications environment have been identified and were still under study when this book was written. The most important issue that has been identified is the layer violation problem, as described in Chapter 2, due to the fact that the TC function may imply in some cases the need to modify some part of the PTP payload beyond the layers that are normally under the responsibility of the network equipment. Models of TCs with no layer violation have been proposed to ITU-T.

4.4.2.6. *Carrying accurate phase/time synchronization across multiple operator domains*

Carrying synchronization across multiple operator domains has always raised some issues. This was already the case when only frequency synchronization was required, although sometimes feasible for some applications having relaxed synchronization requirements. It becomes even more complex as far as accurate phase/time synchronization is required, with very strong requirements such as those required by some mobile networks.

In particular, as already discussed in sections 4.2.1 and 4.2.3.3, the concept of "timing transparency across another operator network" may become very difficult to achieve in practice for accurate phase/time. Moreover, end applications requiring real plesiochronous phase/time reference (i.e. phase/time reference not traceable to some form of UTC) are expected to be quite marginal, the key requirement being traceability to some form of UTC in most the cases, whoever provides the reference. Therefore, relying on its own reference for a mobile operator is technically not required in general; another carrier operator could deliver the reference as long as the quality of the synchronization reference delivered meets the expected requirements.

However, as discussed in sections 4.2.1 and 4.2.3.3, business and commercial aspects need also to be taken into account and might drive, in specific cases, the will to develop other solutions allowing phase/time transparency, despite the associated technical challenges highlighted in earlier sections.

Very likely, accurate phase/time synchronization needs will rely in most cases on a "synchronization as part of the service" paradigm when leased lines will be used, as is already the case with some frequency synchronization solutions. An alternative could be to deploy a GNSS receiver at the cell sites, as mentioned previously, in order to avoid having to carry phase/time synchronization across another operator network.

4.4.3. *Possible migration paths toward IEEE 1588 phase/time profile*

Network operators having deployed, as a first step, next-generation synchronization networks addressing frequency needs only, such as Synchronous Ethernet or IEEE 1588 end-to-end, are, in general, looking for possible migration paths toward solutions addressing future needs for accurate phase/time synchronization, in order to secure their investment. This section gives some insights about this subject.

With regard to network operators having deployed a SyncE network, it has been discussed in this chapter that SyncE could be useful also when phase/time synchronization is required, for instance, in order to provide enhanced phase/time holdover, and secure the synchronization network during the failures of GNSS systems. Therefore, a network operator having deployed SyncE could migrate toward:

– either an architecture where GNSS receivers are deployed at each cell site as shown in Figure 4.36, without using IEEE 1588 full timing support (and therefore without having to implement new PTP hardware functions in the network);

– or an architecture where IEEE 1588 full timing support is used and the GNSS receivers are positioned inside the network as shown in Figure 4.37 (in this case, the network equipment should be upgraded to support the necessary PTP hardware functions).

With regard to network operators having deployed an IEEE 1588 end-to-end solution, the logical migration path should be to add PTP hardware functions inside the network in order to build an IEEE 1588 full timing support architecture. This migration could be envisaged in two different ways:

– Directly toward an IEEE 1588 full timing support architecture by ensuring that all the network nodes implement PTP hardware function. This option is expected to provide guaranteed performance.

– By adding step by step the PTP hardware functions in the network equipment and relying temporarily on a partial timing support. As discussed earlier, there are identified challenges with the partial timing support approach, and the performance may significantly depend on the characteristics of the network, for example its size, its load, its asymmetry and types of PTP unaware devices. It is too early to discuss this approach as it is not yet defined.

4.5. Bibliography

[FER 08] FERRANT J.-L., GILSON M., JOBERT S., *et al.*, "Synchronous Ethernet: a method to transport synchronization", *IEEE Communications Magazine*, September 2008.

[FER 10] FERRANT J.-L., GILSON M., JOBERT S., *et al.*, "Development of the first IEEE 1588 telecom profile to address mobile backhaul needs", *IEEE Communications Magazine*, October 2010.

[G.781] ITU-T Recommendation G.781, Synchronization layer functions, 2008.

[G.803] ITU-T Recommendation G.803, Architecture of transport networks based on the synchronous digital hierarchy (SDH), 2000.

[G.823] ITU-T Recommendation G.823, The control of jitter and wander within digital networks which are based on the 2048 kbit/s hierarchy, 2000.

[G.8251] ITU-T Recommendation G.8251, The control of jitter and wander within the optical transport network (OTN), 2010.

[G.8260] ITU-T Recommendation G.8260, Definitions and terminology for synchronization in packet networks, 2012.

[G.8261.1] ITU-T Recommendation G.8261.1, Packet delay variation network limits applicable to packet-based methods (Frequency synchronization), 2012.

[G.8263] ITU-T Recommendation G.8263, Timing characteristics of packet-based equipment clocks, 2012.

[G.8264] ITU-T Recommendation G.8264, Distribution of timing information through packet networks, 2008.

[G.8265] ITU-T Recommendation G.8265, Architecture and requirements for packet-based frequency delivery, 2010.

[G.8265.1] ITU-T Recommendation G.8265.1, Precision time protocol telecom profile for frequency synchronization, 2010.

[G.8272] ITU-T Recommendation G.8272, Timing characteristics of primary reference time clocks, 2012.

[GIL 06] GILSON M., *Synchronisation in the Converging World, "Synchronous Ethernet – A Carriers Perspective"*, WSTS, 2006.

[GIL 07] GILSON M., *The Transport Infrastructure. "Carrier Scale Solutions – The Infrastructure Problem"*, ITSF, 2007.

[GIL 11] GILSON M., Multiple independent timing domains, COM 15 – C1388 – E, BT contribution to ITU-T SG15 Q13 Plenary Meeting, Geneva, February 2011.

[GOL 12] GOLDIN L., MONTINI L., "Impact of network equipment on packet delay variation in the context of packet-based timing transmission", *IEEE Communications Magazine*, October 2012.

[IEEE 1588™-2008] IEEE Std 1588™-2008, Standard for a Precision Clock Synchronization Protocol for Networked Measurement and Control Systems, 2008.

[JOB 11a] JOBERT S., LE ROUZIC E., BROCHIER N., Analysis of phase/time distribution over OTN networks, COM 15 – C1451 – E, France Télécom Orange contribution to ITU-T SG15 Plenary Meeting, Geneva, February 2011.

[JOB 11b] JOBERT S., BROCHIER N., TABOURE B., Introduction to Jitter Estimation by Statistical Study (JESS), a metric to analyze the PDV generated by a network equipment, WD38, France Télécom Orange contribution to ITU-T SG15 Q13 Interim Meeting, York, September 2011.

[JOB 11c] JOBERT S., TABOURE B., G.8261.1: proposal of PDV network limits for the HRM-1, WD69, France Télécom Orange contribution to ITU-T SG15 Q13 Interim Meeting, York, September 2011.

Chapter 5

Management and Monitoring of Synchronization Networks

5.1. Introduction

Up to this point in the book, we have discussed architectures, requirements and mechanisms used in the next-generation synchronization network. While these are generally considered new methods, operationally they need to be managed in a way that is consistent with the operation and management of the evolving telecommunication network. In many cases, the operation of a network is very cost sensitive and any added costs associated with managing a network will ultimately impinge on a carrier's ability to profit from the services that it offers. Management of the network on a technology agnostic basis will ultimately drive operational costs down through the "economies of scale" that would occur in the application of similar operational methods to different technologies.

A network, for the purposes of this book, represents a collection of individual devices that are interconnected for the purposes of allowing a user to send information between geographically distributed points. Between these two points, the information may traverse multiple separate components ("network elements"). Generally, multiple functions are needed before a user can transfer information. First, each individual network element associated with the service end points may need to be configured (or provisioned). Second, the end-to-end connection needs to be formed from the individual links between the network elements. Once the connection is set up, providing service and generating revenue, the focus changes to provide an uninterrupted flow of traffic. Effective and efficient monitoring is now critical.

For the purposes of connection setup, there may be different ways in which an end-to-end connection may be established. Historically, the first telephone exchanges were completely manual switchboards; these represent a form of manual connection setup. "Switched services" evolved to provide a signaled connection setup. For packet networks, the concept of a connection was removed and individual packets could be switched or routed through a network resulting in a "virtual circuit". The mechanism to set up connections could be either manual, via a network management system, or via a control plane (in the case of routed connections). While the discussion of these types of systems is beyond the scope of this book, the key point is that for synchronization networks connections are generally set up via the management system (the management plane).

5.2. Network management systems and the telecommunications management network (TMN)

Network management systems used in telecommunication networks generally follow the principles outlined in the M.3100 series of recommendations within the International Telecommunications Union (ITU-T). This is sometimes referred to, in general terms, as a telecommunications management network (TMN). The TMN includes various operating systems (OS) responsible for various management functions required by the TMN. Note that the terms "OS" and "network management systems" (NMS) are often used interchangeably.

The general architecture of the TMN is shown in Figure 5.1, which is an adaptation of Figure 7/M3010. This includes a dedicated data communications network (DCN) [G.7712] that allows communications between various components of the network (the network elements and control elements.) The DCN supports both the management network (management plane) and the control plane, if present. Note that for the purposes of synchronization, automatic connection setup using a control plane is not used as described in section 5.4. As with other aspects of telecommunication equipment, network management standards have been developed to facilitate the deployment of networks with equipment from multiple vendors. The individual boxes in Figure 5.1 may, therefore, be from different vendors.

Multiple operational support systems (OSS) may be present and cover different systems. For example, the OS covering the transport network (layer 1, optical transport network (OTN), or Synchronous Digital Hierarchy (SDH), for example) may be separated from a service layer OS (e.g. backhaul).

OS are also used as central collection points for mentoring information provided by the individual network elements. This is covered in greater detail in the sections below.

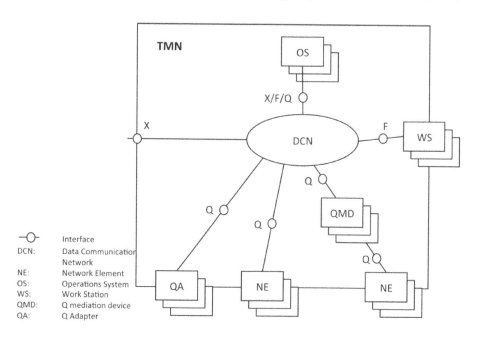

Figure 5.1. *Telecommunications management network framework, an adaptation of Figure 7/M.3010*

The functions provided by the network management systems include fault management, accounting management, configuration management, performance management and security management. These are often referred to as FCAPS. These are generic functional terms that are applicable to different service types. For some services, not all functions are required. For example collection of accounting or billing information by an NMS is generally required for call-based services and is therefore not required for management of the synchronization network. If timing as a service is provided, then the billing aspects are by contractual agreement (service-level agreement).

Networks will generally consist of multiple types of network elements depending on the functionality provided. For the synchronization network, the critical components may include the network clocks and transport equipment. The delivery of synchronization will involve all of these components, which may also operate at different levels within the general Open Systems Interconnection (OSI) stack either in the physical or in the packet layers. Providing the service of delivering synchronization will require the interaction of these components and since these must be considered to be geographically remote, centralized network management systems are used.

It is important to note that an alternative architecture for network management exists for use in some cases. The Simple Network Management Protocol (SNMP) is defined for use in managing Internet protocol (IP) networks. Due to its relative simplicity, SNMP may be seen in some management systems used for controlling and managing SSU or BITS clocks. In the frequency synchronization network, this is generally not an issue, as the management of the synchronization network is often run by separate organizations within an operating company. Unlike traffic connections, the number of synchronization links that need to be configured is far less than that required for network elements managed for the purposes of carrying traffic. If necessary, the TMN framework has the facility to interoperate via the Q adapter function shown in Figure 5.1.

5.3. Synchronization Network management: the synchronization plan and protection

The primary goal of the synchronization network is to allow the distribution of a timing signal (frequency only or frequency, phase and time). In general, the network will be geographically dispersed and will consist of network elements and the necessary transmission facilities. From the perspective of the network, the timing signal is a form of information that has to be managed. Synchronization network management, while similar to payload traffic management, may impose different requirements on the network management systems. In particular, the protection schemes used within the synchronization network are slightly different from traffic protection. Where traffic protection is concentrated on the protection of a path or circuit between two points, the broadcast nature of the synchronization information impacts multiple connections and needs to be coordinated at the network level to the overall network synchronization plan that may be developed by the network operator.

While the overall objective is to deploy a network that is always capable of transporting information, the reality is that networks are subject to unforeseen failures, the most common of which are accidental cable cuts. Failures may also be equipment related, but in many cases, the design of the equipment itself will attempt to minimize the impacts of equipment-related failures. Often, equipment will contain redundant components that may be automatically switched and minimize any outage events.

Network protection mechanisms are used to provide similar redundancy but at the network level. As with the case of equipment protection, the use of network level protection can minimize the impacts of network failures.

All protection mechanisms aim to minimize the service disruption due to network failures by allowing information to travel over alternative pathways. There are different ways of providing these paths and these will have a different impact on the network, either in protection performance or in terms of network resource utilization (e.g. cost). End-to-end protection mechanisms are generally referred to as "path based" mechanisms and generally consume the most network resources since two duplicate connections are required to be established between protected endpoints. "Hit-less" traffic protection may be provided by sending fully duplicated streams of information. Other methods for improving network utilization also exist that include line protection (where only the line section is protected) and protection at the packet layer.

Network protection mechanisms are well known for traffic protection, in particular for TDM systems. For example, in TDM systems such as SONET/SDH, it supports different types of protection mechanisms (1+1, 1:1) or topologies (linear or ring) and may operate to protect only a segment (e.g. line) or an entire path. Each mechanism meets the overall requirement of increasing the network availability under failure conditions but offers different capabilities, for example, 1+1 protection provides error-free protection by the transmission of two identical bit streams, in which case, the downstream network element will choose the appropriate path, 1:1 (one-for-one) protection offers the ability to carry extra traffic on the protection line, and 1:N (one-for-N) provides the ability to protect multiple lines with a single protection line.

Fundamentally synchronization network protection is similar to traffic protection. In most cases, redundant synchronization links are provided. One working synchronization reference can be supported by multiple backup references. Using similar terminology for traffic protection, this could be considered an "N:1" system, with N backup paths supporting one working path. However, one of the key differences between traffic protection and synchronization network protection is the scope of protection. In traffic protection, the aim is to protect paths between two points. Redundant paths are provided for the transfer of information. A second difference is that while traffic protection is generally revertive operation, that is, traffic is returned to the original working path upon clearance of the triggering fault condition, synchronization protection can be either revertive or non-revertive. Since protection events may cause phase transients in the synchronization network, using non-revertive protection reduces the number of transients in the network.

For synchronization protection, the goal is to have each network element protected, which generally means that each network clock will be provisioned with redundant timing links. However, the nature of the synchronization links is fundamentally different from that of the traffic in two key ways; for traffic, information carried between the two end points has significance only between the two end points. With synchronization, the information has network-wide significance and can be distributed to other nodes. This results in synchronization information being "broadcast" from a network element. Using SDH or synchronous Ethernet as an example, all links emanating from a network element are timed from the same local clock and can, therefore, be used as synchronization links. This can result in a tree-line distribution topology. Second, although the target is to provide a timing path from a PRC to a network element, network elements along the path can substitute the synchronization information to accommodate for failures along the path. This, however, can result in a different timing trail as shown in Figure 5.2.

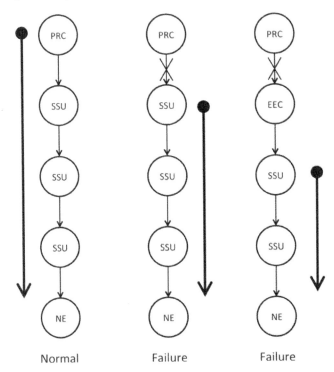

Figure 5.2. *Network topology dependency on clock type under failure conditions*

While not an issue in simple chains, it is important that the impact of changes to the synchronization network be understood at the planning stage. This is most critical with the use of more complex topologies, including ring topologies and tree-like distribution, which can lead to the creation of timing loops – a potentially damaging situation where the input of the clock may, in fact, be driven by its output – if topology changes due to protection are not well understood. Avoiding timing loops has been greatly reduced with the advent of synchronization status messages (see Chapter 2); however, careful deployment is still required. As the network grows, it is important that the overall topology of the synchronization network is well understood so that additions to the network can be made in an orderly manner. Since degradation of synchronization performance occurs with the increasing number of network elements (clocks), it is important that protection events be planned to also limit the length of the clock chain.

The overall structure of the synchronization network is often referred to as the synchronization plan. In a network deployment, the synchronization plan should be developed to accommodate not only network additions but also changes in topology due to network protection events. In general, robust networks are those with relatively simple network synchronization plans.

5.4. Provisioning and setup: manual versus automatic

Some functions that are necessary for the network to operate, for example connection setup, can be provided either by a network management system or by a control plane. Voice calls (either mobile or fixed line) are an example of services that are set up by signaling through a control plane. Private line services have traditionally been set up by provisioning actions through the management plane but may also be set up via a control plane.

As noted in section 5.3, the synchronization plan is critical to the operation of the network. The synchronization plan will be developed to understand the impacts of failures and result in deterministic behavior in the case of network failures. The network management system must allow the implementation of the synchronization plan, which may mean that segments of the network may be treated differently. Other than the setup of protection, synchronization management generally precludes automatic actions as these cannot be implemented to address the wide range of behavior that may be required to implement the synchronization plan. While this is true for layer 1 synchronization (SDH or synchronous Ethernet), it also applies to the IEEE 1588 telecom profile for frequency.

The basic IEEE 1588 standard defines an automatic protection mechanism (the best master clock algorithm, see Chapter 2). However, due to the need to interwork with the existing synchronization network, the ITU-T chose protection based on G.781. Since the carrier develops the synchronization plan to fit the precise requirements of their network, provisioned protection is also required for the IEEE 1588 within the telecom profile. The need for predictability is generally viewed as more important than having a "plug and play" system.

5.5. Monitoring functions

Monitoring of service quality has been considered an integral part of the development and deployment of modern telecommunications networks.

A critical aspect of network management involves understanding the overall health of the network. Monitoring of service has long been considered an integral and critical part of the development and deployment of modern telecommunications networks. Indeed, the modern telecommunications network relies on the ability to self-monitor and take action when necessary. In some cases, this may be required for the assessment of the SLA. Monitoring involves accessing, collecting and analyzing measurement information at end points and points within the network. In many cases, the monitored information is also used to trigger protection events. Monitoring may be direct or indirect. For example, direct detection of bit errors is generally possible through the use of parity bit calculations. If a transmitter calculates the parity of a bit stream (e.g. parity calculated over a frame or cell) and inserts the result within a dedicated field within the data, the receiver can directly detect the presence of bit errors by performing the same calculation on the received data. A difference in result between the calculation performed at the receiver and the information inserted by the transmitter is a direct indication that bit errors have been introduced into the transmission path.

Monitoring synchronization introduces an interesting challenge since measurement of synchronization signal performance requires the presence of an accurate reference. In many cases, this reference is not available, and thus direct monitoring methods cannot be used. Indirect methods may be used (S1 byte and SSM content) for frequency; however, these can still only partially monitor the performance as they cannot assess the noise levels (MTIE/TDEV). However, in normal operating environments, the need to monitor frequency is not required for physical layer frequency distribution (SDH or EEC) since the noise accumulation can be controlled with proper network design.

The network management interfaces to a network element, the Q interface in Figure 5.1 above, which can be viewed in more detail as shown in Figure 5.3. Here, information is segregated into two general types: control and provisioning information and event /monitoring information.

Information needed to monitor the network is typically defined as part of the OAM functionality of the transport system and is intended to allow a direct indication of the capability of the network to carry a service. Typically, information is contained within the transport overhead that can allow statistical collection of information such as bit errors. Alarm reporting is also provided to give an immediate indication of critical events such as the triggering of a protection event, which could be due to some type of failure (equipment or cable cut, for example). In most cases, the information provided over this interface is detected directly from the network element, for example signal loss, while in other cases, it may be derived from information contained within the OAM overhead (e.g. bit errors based on FEC calculation). In many cases, the performance monitoring is a self-contained function, as it is directly measureable or derivable from the embedded OAM defined for the transport system. Some variation may exist between different technologies, but the essentials are generally applicable.

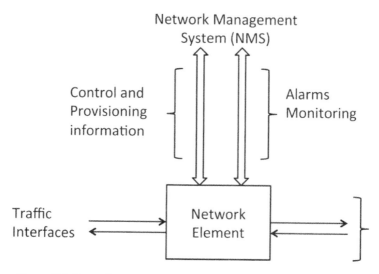

Figure 5.3. *Logical network management interfaces to a network element*

Monitoring of synchronization performance poses a challenging problem. For general transport systems, the health of the network, and the capability to carry traffic, can be determined from information provided by the network elements without any external equipment. The NMS system can collect all the information and thus provide a picture not only of specific connections but also of the entire network.

For synchronization distribution, performance monitoring generally provides a relative indication of the heath of the synchronization signal. Without a high-quality reference – the distribution of such a reference is the function of the synchronization network – only relative performance can be monitored by the network element itself. This, however, can still provide the operator with valuable information.

Synchronization status messages are one such measure that can be used to measure the relative health of the synchronization network. Although not ubiquitous for SDH, the use of messages is mandatory for synchronous Ethernet. Thus, each network element has an indication of the intended quality of the signal it receives. Changes in SSM levels are reported via the management interfaces (see G.781) and are, in fact, a trigger for reference switching. An example of the information exchanged between the network element and the network management system is shown in Figure 5.4, which is an adaptation of Figure A.1/G.781. This shows the G.781 synchronization selection function that is used for frequency synchronization (SDH, synchronous Ethernet and the IEEE 1588 for telecom). The management information shown in the figure is identified in selection algorithm itself with a prefix "MI". The information associated with the transfer of data is the characteristic information of a layer (CI) – see Chapter 4.

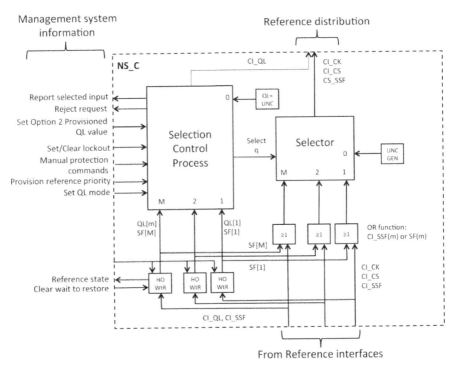

Figure 5.4. *Monitoring information provided by the G.781 synchronization selection function, an adaptation of Figure A.1/G.781*

For packet IEEE 1588, the SSM value is also carried. However, as a packet bit stream, additional aspects need to be monitored and reported. Specifically, loss of messages at the packet layer needs to be reported and alarmed, as this may indicate a potential failure in the ability to deliver synchronization.

For frequency distribution over networks that do not process PTP messages at every node, the quality level of the recovered reference may be dependent on the levels of packet delay variation produced by the network. In these cases, it may be possible to utilize probes within the network to get an understanding of the performance output in these cases. Note also that packet loss will also impact performance.

As discussed in Chapter 2, the distribution of time is also sensitive to the packet delay variation. While this can be controlled and monitored, network asymmetry can result in a static error. Again, careful design of the network at the deployment stage is required. Once this is done, strategic placement of probes to measure PDV can provide a useful indication of the relative performance delivered by the synchronization network.

By collecting information at the NMS, proactive maintenance may be performed. Since packet timing is sensitive to traffic load, that is additional packet clocks place additional constraints on the packet master clock, the NMS is able to correlate lost timing messages and warn of potential server overload.

5.6. Management issues in wireless backhaul

The application of next-generation synchronization to the wireless backhaul case presents new challenges. Historically, network timing distribution was a function provided only within a single carrier domain. Wireless backhaul, as a driving service for next-generation synchronization, requires distribution of timing toward the edge of the network. In some situations, the carrier in order to offer a service ubiquitously over a large geographic area may have to lease network capacity from another carrier. In this case, the ability to monitor the heath of the network may be restricted. Network design has to assume that the management of the network may be fragmented as shown in Figure 5.5. Although one operator provides the service, monitoring the service at the midpoints is limited. In this particular example, the operator responsible for the end-to-end service does not have management/control access to the intermediate network.

Another option that can also exist and may need to be considered is when the service operator may require leased equipment at the edge of the network, as in the case of a shared radio access. In cases where full access is not provided, contractual arrangements are required.

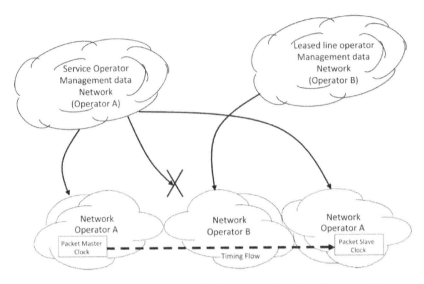

Figure 5.5. *Wireless backhaul over multiple operator domains. Operator A has no management access to operator B's network*

5.7. Network OS integration: M.3000 versus SNMP

So far, this chapter has discussed network management aspects in relatively general terms. Deployment in a live network requires consideration of the type of operational environment present and the capability provided by the chosen technology. The level of integration of the various functions may also need to be considered.

In general, large telecommunications networks have to manage multiple technologies (e.g. packet and TDM) in a multidomain environment, while enterprise-based systems (e.g. local area network (LAN) technologies) may have different requirements and will thus use management systems based on different standards. Deployments may need to consider these differences. For example, in the case of PTP, the IEEE 1588 standard fully specifies network management aspects applicable for LAN-based systems. Special network management messages have been defined that provide an "in-band" management channel. In certain applications, the use of the management channel defined within the IEEE 1588 standard may be suitable, but in other cases, notably when systems are embedded into the existing equipment, the operator must be aware of that this may be inconsistent with their existing systems. In the telecommunication environment, management is typically provided by dedicated management interfaces, the so-called "north-south" interface, as previously illustrated in Figure 5.3. Aspects related to network element management may not be possible in certain implementations due to this restriction.

Integration of different technologies may imply the need to manage the devices in a specific manner. The case of the boundary clock providing frequency syntonization presents such a case. Management of the physical layer aspects is solely in the realm of telecommunication management systems, which effectively forces similar processes on the PTP aspects.

While the above discussion demonstrates that in some cases the use of PTP management is supplanted by telecom-based management, there are cases where the opposite may be true. Consider the BITS clocks found within networks. These clocks have specialized roles within the network and, in many cases, the management aspects of these devices may be considered relatively simple in comparison to management systems for some existing telecommunication network elements involved with the transport of user traffic. For these network clocks, the choice of management systems based on the use of "Enterprise" management standards is fully justified. However, this now presents a situation where the management of the master may be under the control of one type of management system, while the management of the slave may be under a different system.

Network management systems have been developed to address specific application environments. For the purposes of this book, network management systems can be categorized as applying to either the broad telecommunication environment or to the enterprise environment. The capabilities of the network management systems reflect the needs of the market for which they are targeted Broadly speaking, enterprise environments generally require less functionality to manage. For example billing, something required in the telecommunication environment, is not needed in an enterprise environment.

Network management standards also reflect this. Many IP data network devices that were originally intended for enterprise or LAN application are based on SNMP, intended to carry management messages within the IP network on an in-band basis. This is substantially different from telecommunication applications where a separate DCN is used for carrying management messages. Furthermore, in the telecommunication environment, it is assumed that separate management specific interfaces are present.

The frequency-based synchronization network is typically composed of dedicated synchronization network elements working closely with transport network elements. Network management principles are based on telecommunications applications. The introduction of synchronous Ethernet is also consistent with this as Ethernet management standards follow the same principles. In fact, both Ethernet and SDH/SONET follow similar network management principles (based on OSI) and, therefore, are supported by the same management system. The Telemanagement Forum (TMF) Multi-Technology Operations System Interface

(MTOSI) is an example of a system capable of supporting management actions related to SDH/SONET and Ethernet.

The incorporation of the IEEE 1588 as a technique to transfer timing represents a departure from the traditional SDH/SONET and synchronous Ethernet. SDH/SONET and synchronous Ethernet provided transport of timing as an overlay but the use of the IEEE 1588 means that synchronization management now needs to be extended in some cases to include functionality previously only required for the management of payload data traffic, as timing information is now also carried together with user data.

From the perspective of management systems, we have noted above that synchronous Ethernet and traditional transport (SDH/SONET) may be supported by the same network management system. When the IEEE 1588 is introduced into a network, the choice of the underlying packet network may dictate the type of the network management system. In some cases, this may result in the physical synchronization network being managed with one type of network management system, while the packet synchronization network is run with a separate network management system. For example, enterprise IP networks typically use SNMP management. This would result in the potential of having to access two network management systems, one for the IP portion based on SNMP and one for the Synchronous Ethernet portion based on the traditional telecommunication network management. For frequency synchronization, this, while inconvenient, can be managed. In the case of time distribution, some operators wish to use synchronous Ethernet to provide support to boundary clocks. In this case, both the physical and packet layers must be managed by the same NM system.

Meeting the stringent performance requirements for time distribution will require coordination among multiple layers. In some cases, this coordination may only be possible at the network level and thus integration of physical layer and packet layer network management is crucial.

5.8. Bibliography

[IEEE 1588] IEEE 802.1500-2008, Standard for a Precision Clock Synchronization Protocol for Networked Measurement and Control Systems, IEEE Instrumentation and Measurement Society, July 2008.

[IEEE 802.1] IEEE 802.1D-2004, IEEE standard for local and metropolitan area networks, Media Access Control (MAC) Bridges, IEEE Computer Society, June 2004.

[G.8264] ITU-T Recommendation G.8264/Y.1364 (2008), Distribution of timing information through packet networks, ITU-T, 2008.

[G. 781] ITU-T Recommendation G.781, Synchronization layer functions, ITU-T, 2008.

[G. 7712] ITU-T Recommendation G.7712, Architecture and specification of data communication network, ITU-T, 2010.

[M.3000] ITU-T Recommendation M.3000 (2005), Overview of TMN Recommendations, ITU-T, 2005.

[M.3010] ITU-T Recommendation M.3000 (2000), Principles for a telecommunications management network, ITU-T, 2000.

[RFC 3411] RFC3411, An Architecture for describing simple Network Management Protocol (SNMP) Management Frameworks, 2002.

[TS32.102] 3GPP Technical Specification 32.102 Digital cellular Telecommunications system (Phase 2+); Universal Mobile Telecommunications System (UMTS); LTE; Telecommunication management, 2012.

Chapter 6

Security Aspects Impacting Synchronization

This chapter introduces possible security threats that can specifically impair synchronization services. It discusses some attacks and security techniques that may pertain to synchronization and help limit vulnerabilities and it considers overall risk management in the context of synchronization distribution. However, it is not intended to detail or to be exhaustive on these topics that are more precisely covered in literature partially referenced in the chapter.

Security is a broad subject covering multiple areas that can interact. Synchronization can then be a subset of various security aspects.

Section 6.1 introduces the main concepts of security, section 6.2 covers the vulnerabilities, the possible threats, attacks and measures to mitigate those attacks on timing sources, section 6.3 covers the same topics but for timing transmission over a network and section 6.4 addresses some related security management aspects.

6.1. Security and synchronization

In the context of synchronization in telecom environment, different segments in the timing distribution path lead to some specific possible threats and thus are various areas of investigation.

This chapter, in particular, considers the potential risks related to the timing source and the transmission and reception of a timing signal, as depicted in Figure 6.1. Options are considered to reduce vulnerabilities, detect potential threats, mitigate attacks and limit timing service impairments.

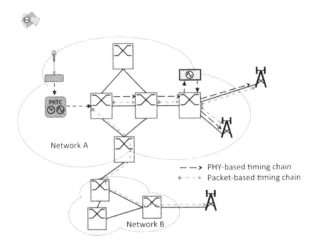

Figure 6.1. *Studied timing security scope*

Security should be viewed from the following perspectives:

– where the operator is providing a timing service for their purpose;

– where the operator is providing a timing service to customer;

– where the timing is transferred over a third-party operator.

For the first case where the operator is providing timing to its own network, generally it is assumed that the operator already has security aspects well controlled as part of its general operations plan. In this case, security of timing will generally follow existing security rules established by the operator. The operator will still need to be cognizant of purely synchronization threats such as jamming (as discussed later in this chapter).

Generally, when an operator provides a timing service, this service would have to be considered safe and trusted. Assuming that the operator's timing sources (reference signal and packet master) are known, identified and located in the operator network, the utilization of security mechanisms specific to timing might be unnecessary. It would be under operator responsibility to assess risks associated with its timing services and implement proper security features.

A customer of this timing service might contemplate a mitigation or protection mechanism against corrupted timing. However, this could lead to a more expensive solution such as provisioning a distributed synchronization model (see Chapter 4) for monitoring the operator's timing service or for using it as a primary timing source (the operator timing service could then be used as backup).

In the last case, where the transmission of a packet-based timing signal over a third-party operator does not provide timing service, further attention is needed. The carrier operator might be considered safe because the operator provides a transport service (e.g. Virtual Private Network (VPN) or "virtual circuit"). That is, carrier or metro Ethernet operator infrastructure could be considered a closed and safe environment assuming that proper and usual mechanisms to protect from external and internal attacks are correctly implemented. As for its other traffic transmitted over this network, the customer might not consider it necessary to utilize specific security mechanisms.

However, if such confidence cannot be given, the timing information (and probably other data) would have to be secured.

It is thus worth reviewing the possible threats, vulnerabilities, types of attack and mitigation options for the timing distribution, including the synchronization plane (the signal) and control plane (e.g. for building the hierarchy).

6.1.1. *Terminology used in security*

Security of timing synchronization is becoming an important topic, especially since timing information flows over multiple operator domains and now includes the packet layer. While this book is primarily intended to describe how new synchronization technologies are used within the evolving network, it is necessary to review some concepts and define the scope or meaning of usual security terms as they are used in this chapter.

The term "security" covers multiple concerns. It can be viewed from different perspectives: from technical to social, from configuration to organization. A first important remark is that synchronization security should not be taken independently. It has to be part of a wider risk management strategy. In the context of packet-based and next-generation synchronization networks, synchronization security management would particularly relate to the computer or information security strategy of the organization owning the distribution and/or the end application. Hence, this chapter mainly adopts computing security assertions and, from a computing system perspective, the IETF RFC 4949 [SEC 07] is a good reference.

However, some aspects in synchronization distribution, particularly from timing source perspective, may go beyond the boundary of the computing environment.

This chapter limits the discussion of security to vulnerabilities, threats and attacks. Acknowledging those terms might have nuances as definitions come from international, government and commercial organizations (e.g. ISO 270001/2 or 27005, IETF RFC 2828, Committee on National Security Systems of United States or The Open Group), the reader is invited to take into account, in this chapter, the meaning of following terms.

A *threat* is a possible danger to a service. It represents a potential source of harm to either part or to the entire timing domain. In larger sense, this is a "circumstance or event", categorized by type and origin. In computing systems, this can refer to the intruder, relating to a motivated threat agent capable of exploiting vulnerability. There are various origins of threat: some natural, some human driven, some external and some internal to the affected service.

Vulnerability usually refers to a weakness or a flaw. It can be technical but can also be human. Some definitions consider this is a weakness that can be intentionally exploited by an agent (a threat) through a vector, a target (or victim) of an attack. Some others define vulnerability as a probability of not being able to resist a threat action. A vulnerability becomes a potential target for an attack.

An *attack* is an intentional, intelligent or malicious act to acquire, modify information or to disrupt/impair a service. An attacker would use tools to access a vulnerability for specific results. This is a threat action conducted with some intention that describes the objective of the attack.

Hence, natural threat or unintentional cause of a threat affecting vulnerability is not an attack but can impact a service the same way. This is particularly true for radio-based timing sources.

Threats, vulnerabilities and attacks may be classified based on a number of categories. These may refer to their type, an action or its possible consequence, the affected community, the intended objective and possible severity. Security incidents may be passive or active and could impact hardware or software systems or other network components.

Timing domains can have various vulnerabilities related to physical signals (e.g. radio or electrical) or devices (e.g. clock, switch and link) and to packet and protocol. From a threat perspective, this chapter focuses on three areas for the elements or the segments of the synchronization chain:

– availability;

– integrity;

– authentication.

Some threat actions would attempt compromising confidentiality. A peculiarity of synchronization is that confidentiality of time, for instance exposure or interception of time stamp, is not yet referenced as a concern because the timing

Security Aspects Impacting Synchronization   259

"data" are considered public information (e.g. Universal Time Coordinated (UTC)). We might use a specific timescale as a security mechanism and then might protect this information. Moreover, beyond the signal itself, further information, for example pertaining to a timing device or to the hierarchy, can be eavesdropped upon. However, as of today, no user case has been presented requiring the hiding of timing-related information – either sync or control plane when they are distinct.

6.1.2. *Synchronization in network security ensemble*

As noted earlier, synchronization security should not be taken as a silo. Many aspects of network, computing or information security are directly relevant to synchronization security particularly because of utilization of packet-based methods. For instance, security incidents on any device such as:

– the deployment of improper security mechanisms allowing access or intrusion to link or device;

– the deployment of improper or faulty software;

– a misconfiguration, the lack of knowledge or of training;

– or an improper human behavior (social engineering), and performed on an equipment directly or indirectly involved in the synchronization distribution can compromise the security of the synchronization service.

For instance, tapping can compromise short or long haul cabling, copper or fiber. There are various options but they always assume an access to the media.

Passive and active wiretappings allow eavesdropping. Active wiretapping could also allow modifying, redirecting, recording, eliminating, replaying and delaying the timing signal. Copper wiretapping is not easy (particularly to remain discreet, e.g. due to resulting impedance changes) but possible by direct connection to the wires, by induction or by active relay. Wiretapping optical has a higher level of complexity but is also possible by clamping and, with appropriate tool, permits both passive and active wiretapping. Inserting a device on the copper (such as by active relay) or fiber path is direct but would trigger a link failure that can be easily detected and thus can trigger log and alarm.

Any code error in software can create vulnerability. For instance, one version of Network Time Protocol (NTP) public code had a security vulnerability that permitted issuing denial-of-service (DoS) attacks or allowed unauthorized access. The code has been fixed. In case of Precision Time Protocol (PTP), if servo algorithms are implementation dependent, there are commercial PTP protocol stacks as well as a public PTP daemon. But public or commercial codes can have vulnerability. Such vulnerability remains until it has been detected, corrected and that the fixed version is implemented. Vulnerability in the operating system (OS) supporting the timing code might also be leveraged.

Any threat on availability (e.g. DoS), on privacy or integrity (e.g. middle man attack and intrusion) and on confidentiality (e.g. misuse, access to the data, for instance through break-ins using identity spoofing) in the system (network, device, OS, etc.) can be exploited to attack timing service. This can be in the form of indirect attacks impairing support of synchronization (in such a case, the timing service might not be the target but a collateral victim) or of direct attack(s) to synchronization distribution functions. A threat agent willing to specifically attack the network timing service would often have to first exploit network vulnerability (e.g. to insert or control a device). Examples of availability threat that can end as (unintentional) DoS attack have been described on the Internet, for example as NTP server misuse and abuse.

Malicious access to equipment (timing or network equipment) is not specific to synchronization. Any threat for accessing a device implied in some way, being in the path of the timing signal, is one step threatening synchronization transfer. It is thus critical to first assess security of the network supporting the timing service.

For some applications such as for mobile wireless radio synchronization, the security aspect may not be limited to one timing domain and dependency between timing domains may exist.

Synchronization is hierarchical from reference down to application (e.g. radio interface of wireless mobile base stations). As depicted in Figure 6.2, if synchronization in one domain is corrupted, it may impair the synchronization purpose. It actually depends on the scope and objective of the synchronization distribution as well as on available mitigations.

The consequence of a timing source attack might not be noticeable. If the synchronization scope is strictly bounded to slaves synchronized to this same reference master, they all refer to same signal and the applications would use the same reference. Hence, a deviation of the reference would just be shared. This may be the case of phase synchronization in one timing domain. However, if slaves refer to distinct masters, in same or distinct management domains, synchronization may fail.

In Figure 6.2, timing domain B is assumed to have compromised timing distribution (whatever the victim and threat effect) and its synchronization reference provided to the base station is corrupted. In this domain, the application requirement (e.g. phase alignment) might not be achieved, impacting the expected behavior. Consequences may be limited such as degraded radio service to the user equipment but might impact critical service such as emergency location service.

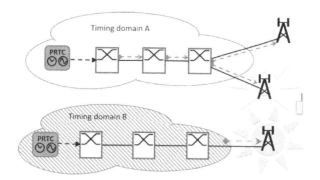

Figure 6.2. *Timing corruption in one timing domain*

The above discussion should be borne in mind particularly when synchronized end systems sharing the same application requirement are split among distinct timing domains. The application could thus depend on the security of multiple timing domains. Mitigation might involve that each timing domain monitors the neighboring timing domain, however this is not currently being studied in standards and thus may not be universally possible.

6.2. Security of the timing source

Transmission of a timing signal involves multiple network elements. Each network element is assumed to degrade the original timing signal as it traverses the network. The characteristics of the equipment can be defined and constrained so that the degradation of the signal is controlled to a degree that is acceptable to the end application. This, however, assumes that the source is correct and operating within its performance targets. Hence, the security of the source (availability, integrity and authentication) should be a concern particularly because this timing signal distribution could be out of the control of the operator.

Nevertheless, the protection of the distribution of a signal (e.g. the Global Navigation Satellite System (GNSS) infrastructure as described in section 2.7) received by network timing source as depicted in Figure 6.3 is out of the scope of this chapter. That is, the vulnerabilities, possible threats and their impact of the transmission of a signal up to the operator's timing source are not discussed. This section covers the security of the timing source equipment and of received GNSS timing signal.

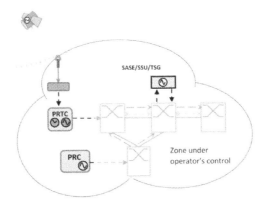

Figure 6.3. *Area controlled by operator (GNSS infrastructure – here the user segment is outside the area)*

As shown in Figure 6.3, timing source equipment term encompasses:

– GNSS receiver (standalone or embedded in another device) or other reception device (see section 2.7 for alternate sources) and associated installation parts;

– dedicated frequency devices (i.e. Primary Reference Clock (PRC)/ Primary Reference Source PRS) and Synchronization Supply Unit (SSU)/ Stand Alone Synchronization Equipment (SASE)/ Timing Signal Generator (TSG));

– dedicated time (and frequency) device (i.e. PRTC).

Timing source device threats include:

– physical access to the device;

– break-ins (virtual access) to the device;

– compromising (disrupting or modifying) signal to/from the device.

6.2.1. *Access security to device*

Access to any device or device location is a security item that is not specific to timing. Physical access can be verified by video or biometry systems. Such verification can prevent physical damage or intervention on device environment such as electrical power and air conditioning. Identification and authorization for accessing the device must be implemented, as for any network equipment considered as sensible to a service, with use of passwords, cryptography, authentication, authorization and accounting (AAA).

Assuming the installation is correct, verified and maintained appropriately (e.g. the antenna has a clear view of the sky with minimal multipath distortion and the equipment is properly commissioned and calibrated for fixed offsets such as antenna cable length and cable amplifiers), threats could be reduced. Protection from some events such as lighting may require redundant antennas. However, despite the best efforts to protect antennas to maintain service availability, unforeseen incidents may occur. One remarkable and anecdotic threat is the known "shooting-target exercise" on Global Positioning System (GPS) antennas, a threat that is difficult to prevent.

Cabling is the access to the signal from or to the equipment. Logging any pluggable cable connection and disconnection would allow tracking possible insertion of third-party equipment. However, this often means physically accessing the device location (e.g. office, rack or building roof in case of GNSS antenna). The main vulnerability is thus on the radio signal, particularly from GNSS.

6.2.2. *GNSS signal vulnerability*

In a nutshell, GNSS (i.e. satellite-based) timing references are free and accurate. The satellite and control segments of the GNSS are designed to be very reliable ensuring high availability of the GNSS signals. But as mentioned in section 2.7, the signal provided to the receiver, that is between the space segment and the user segment, may be very vulnerable to accidental or intentional interference. As a highly reliable system, the "always on" nature may have resulted in increased dependence on GPS. An often-quoted description of GPS (from Dr Marc A. Weiss) provides a nice summary of the issue:

> Main advantage of GPS signal is its reliability.
>
> Main vulnerability of GPS signal is its reliability.

As pointed out for many years by US organizations such as GPS.gov ([GPS 98]) and the US Army (e.g. [BEN 05]), over reliance on GPS (and in GNSS in general) is a public source of threat. [BEN 05] quotes:

> GPS provides many benefits to civilian users. It is vulnerable, however, to interference and other disruptions that can have harmful consequences. GPS users must ensure that adequate independent backup systems or procedures can be used when needed.
>
> Source: Interagency GPS Executive Board, 02/2001 [GPS 01]

The main cause of vulnerability of GNSS is that the signal strength from the constellation segment is purposely designed very weak so as not to disturb other signals. To provide a perspective, a 100 W bulb lamp is 10^{18} times more powerful than a GPS signal received at an antenna. Normally, a GPS signal cannot be seen with a spectrum analyzer, and detecting it would be considered abnormal. Because of the low signal strength, receivers attempt to seek the known frequency of the signal. The fact that frequency is known and that the structure of the public (civil) signal is also known is obviously another vulnerability.

Multiple reports or studies on GPS vulnerability have been written. To start, readings of [RAN 95], [VOL 01], [PAP 08], [PNT 10], [RAO 11] and [AIR 12] are suggested.

Because of weak signal strength, GNSS user segments are vulnerable to unintentional and intentional disruptions at the signal level, covering small to extensive areas, for durations from minutes to days. To study the threats, attacks and mitigations, let us consider a redundant synchronization network configuration as shown in Figure 6.4. Multiple receivers with holdover capability based on atomic clocks would protect them from localized disruption. In the figure, one of the receivers is equipped with a redundant antenna.

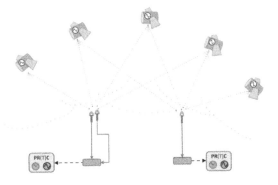

Figure 6.4. *Redundant sources of timing using GNSS signal*

A list of GNSS threats would include:

– Natural phenomenon affecting the signal and leading to performance degradation. Examples of such phenomena are solar activities (flares, often cyclical storms) that can affect the Earth magnetosphere including the ionosphere that the GPS signal traverses. GPS reception issues are reported during aurora borealis events. Considering Figure 6.4, all GNSS can suffer from same disturbance. GNSS satellites can be hardened to protect against solar storms, for example as occurred during a cyclical peak in 2012–2013, but the signal itself cannot be protected.

– Blocked signal reception. For best performance, GPS antennas need to be installed with clear access to the sky as shown in Figure 6.4, thus allowing the receiver to track as many satellites as possible and avoiding multipath reception problems. With reduced vision, such as in urban canyons, the receiver can have degraded performance. Proper installation should limit such situations but natural events (e.g. snow) might still perturb the reception.

– Spectrum competition and interference with other radio frequency (RF) systems. A strongly debated example happened in 2011–2012 in the United States: the deployment of LightSquared base stations would have blinded GPS receivers due to close frequency and high power. After months of debate, the spectrum license to provide such a service was revoked by the Federal Communications Commission (FCC).

– System anomalies and failures in the GNSS signal. An anomaly in the control or satellite segment, such as a clock failure in a satellite, can mislead a receiver. There are means to mitigate those errors but which involve more sophisticated receivers with optional utilization of other systems (e.g. space-based augmentation system (SBAS), see section 2.7) as discussed in the following section.

– Human factors. Such threats can be unintentional or intentional, that is hostile. These are as follows:

- Jamming is a voluntary interference of the signal that can affect, unintentionally or intentionally, GNSS signal reception.

- Meaconing, counterfeiting or spoofing are intentional threat actions, described in section 6.2.2.2.

It is worth noting that human factors are now considered the main threat to GNSS signal availability.

6.2.2.1. *Jamming*

There can be deliberate or inadvertent jamming. Unintentional GNSS outages are increasingly common, and this includes jamming, for instance due to illicit handheld personal jammer, such as $20 "anti-spy", "anti-tracker" GPS blockers pluggable into car's lighter socket, or environmental effects, such as *aurora borealis*, actually due to solar storms. Jamming impacts the availability of the timing source and, when intentional, should be considered as DoS attack. Preventing the GNSS signals from reaching the receiver (shielding) would also generate a DoS.

Jamming techniques are well known and it is claimed that a jammer can be built easily. Limited but highly disturbing jams are regularly reported due to availability of small individual jammers available from Internet. The objective of utilizing such jammers varies in their signal type (e.g. broadband or narrowband covering particular GNSS frequency or all GNSS frequency bands) and in performance. Their emitting power ranges from milliwatt for local, "domestic" usage, such as so-called "personal privacy devices" (PPDs) as shown in [PNT 10], to hundred of watts versions and even to megawatt for military application. Power less than milliwatt is sufficient to provide hazardous and misleading information (1 W jammer can create a 20 km area outage).

If high power jammers tend to lead to intentional jamming (e.g. North Korea jamming South Korea) with large coverage area (e.g. 300 km), the prevalence of low power and low-cost "privacy" GPS jammers and their availability increases as the utilization of GPS for tracking and location finding (e.g. truck float, car theft and terrorist) grows. Jamming in some military operations is intentionally performed by armies, to protect from remotely controlled improvised explosive device (IED). Privacy jammers are now the main sources of unintentional but sometimes very disturbing jamming for critical civil operations (e.g. in airport or maritime traffic). Some governments regularly organize exercises (e.g. US JAMfest or UK's Ofcom) to evaluate impact and counter measure to jamming. Such exercise sometimes leads to disruption (e.g. on maritime navigation) or is detected by a remote receiver used to control GPS signal.

Figure 6.5 illustrates a jamming situation. As depicted, the receiver on the left is the victim of a jamming condition. Depending on the location of its two antennas and of the power and location of the jammer, the receiver might become ineffective. Considering a low-power jammer, a timing source located kilometers away (as the one at the right) from the other receivers would not be affected. However, in case of intentional threat, an attack may be orchestrated and target different locations simultaneously.

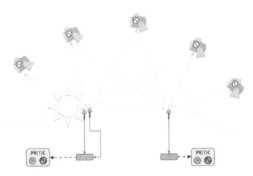

Figure 6.5. *Jamming of GNSS receiver*

Laws in many countries forbid the use, or in some case possession, of jammers. For example, in the United States, "Jammers are illegal to market, sell, or use in the

United States" (source: [USG]) while in Australia, a 2004 declaration from the Australian Communications Authority (ACMA) "prohibits the operation or supply, or possession for the purpose of operation or supply of a RNSS (Radio navigation satellite service) jamming device" (source: [ACM]). In Europe, the Electronic Communications Committee (ECC) within the European Conference of Postal and Telecommunications (CEPT) administrations, adopted Recommendations (03)04 and (04)01 [ECC 04] giving no legitimate civil use of various jammers (see also [ECO 99]).

The existence of laws still does not prevent the acquisition and utilization of jammers. Jamming presents an interference signal to the receiver. In some cases, it is possible to detect the presence of a jammer. However, this can be difficult due to various aspects, such as possible low signal strength of the jamming signal or if the jammer is not stationary (mobile). Without locating the source, it is difficult to determine if the signal is intentional or not. Other possible disturbances include natural phenomena (as previously noted), multipath or other electromagnetic sources. Research projects in the United Kingdom (GNSS Availability Accuracy Reliability and Integrity Assessment for Timing and Navigation (GAARDIAN) and SENTINEL – which leverages GAARDIAN infrastructure) developed nationwide detection capabilities, detection and report of interference events to allow discrimination between natural, accidental or deliberate situations. This might provide the tools necessary to effectively enforce antijamming laws.

Jamming conditions as discussed previously can be detected. It is possible that jammer location (JLOC) systems allow detecting the location of a jamming source. However, detection is not generally a simple function that can be implemented in all GNSS receivers. Despite detection, the effect of jamming is still present, although it can be mitigated to a certain degree as discussed in section 6.2.3.

6.2.2.2. *Signal counterfeiting*

Jamming of radio signals has been used for many years and devices have been produced with relatively little skill. Jamming, as discussed in the previous section, can generally be detected. GNSS counterfeiting, where a purposely false signal is generated, is relatively recent and requires more sophisticated devices and knowledge.

Following are the key aspects of counterfeiting threat:

– It is an intentional action in nature, hence reflects an attack.

– It is difficult to detect (i.e. it will not raise an alarm as jamming).

– There is no real way to mitigate it in current receivers.

Meaconing and spoofing are two counterfeiting threats. Sometimes both are referred as spoofing but the scenarios can be segmented into spoofing on-air (as shown in Figure 6.7), simulator spoofing and replay spoofing (also known as meaconing, as shown in Figure 6.6). These provide false information to the receiver by the creation of a GNSS signal.

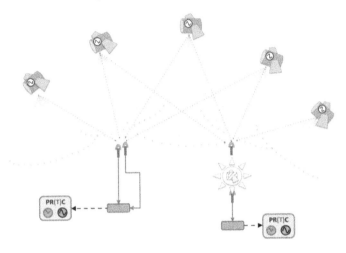

Figure 6.6. *Meaconing scenario*

One key reason for such a threat is that GNSS public codes (e.g. the C/A code for GPS) are available (so, well known) in order to build GNSS commercial equipment. GPS spoofing has been demonstrated [WAR 02, HUM 08, HUM 12] and can start with GNSS simulators forging fake navigation data. Authentication (if possible) would allow protecting the integrity of the information.

Meaconing is a disruption of a GNSS signal requiring first the recording of a valid signal. A meaconing attack would reuse real navigation data from one or multiple satellites, delaying and rebroadcasting the signal(s). Note that accidental meaconing might happen due to malfunctioning or inappropriate installation, for example the antenna. With meaconing, authentication of the signal would not be sufficient as even authenticated signals can be recorded and replayed.

Spoofing is a more sophisticated technique. Unlike jamming it targets an individual device. But, if implemented poorly, the spoofing source could simply jam the receiver. Figure 6.7 shows one possible spoofing example. Here, the spoofer needs to synchronize its false emitted signal to the signal received from a satellite by the victim device (1), and by providing a signal with higher power (2) tends to cause the victim to lock onto the false signal (3).

Security Aspects Impacting Synchronization 269

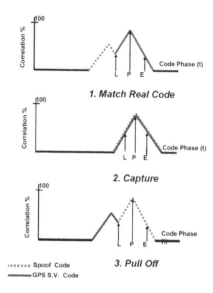

Figure 6.7. *Spoofing mechanism ([CAR 01])*

Research on spoofing countermeasures has been underway for a decade, but the subject is evolving. For instance, spoofing mitigation proposal from [VOL 01] is discussed but defeated in [HUM 08].

Anti-spoofing techniques evolve and solutions are being proposed ([LED 10] and [DAN 12] are two examples) and could make GNSS receivers more able to detect spoofing. But such solutions are not implemented in commercial off-the-shelf (COTS) products. Note that one technique ([LED 10]) expects other RF signals for combination, such as SBAS, LBAS or Loran signals.

As discussed, spoofing would target one device. Moreover, a single spoofer will generate from the same (fixed or moving) location signals assumed to be sent from multiple satellites. One method to mitigate spoofing (or make it more difficult), provided that the receiver (the left receiver on Figure 6.8) is able to perform differential carrier phase measurements between the antennas, would be to utilize multiple antennas with direct sky view (i.e. avoiding multipath). Coordinated spoofers can, however, bypass this protection.

Counterfeiting attacks address integrity and authentication vulnerability of the GNSS. As mentioned in section 2.7, public codes from available GNSS do not provide authentication services and have limited integrity information. Some options exist to overcome current limitations and more are expected to be available in the future.

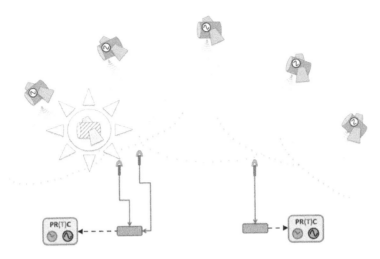

Figure 6.8. *Spoofing scenario*

6.2.3. Protecting and mitigating from compromise signal

Integrity of the information and authentication of the source of information are the main elements to protect from a spoofing attack.

As of 2013, authentication services are not available from the GPS and GLONASS public services. For GPS, these are expected as part of its modernization program. Galileo and COMPASS will allow authentication in the future as an optional service. GPS "clear access" or Galileo "open service" would continue to be freely available but to unprotected data. It is important to remember that authentication does not help for jamming or other interference conditions.

One potential issue in using cryptography is about, as usual, the key distribution process. Spoofers may intercept the key and then generate "authenticated" data. More generally some fear that providing a public cryptographic service, with civil equipment holding confidential data, can compromise the system itself.

Until authentication service is commercially available, both from control, satellite segment and from user segment, and safe to use, protection of timing distribution would have to remain focused on integrity checking, detection and mitigation.

6.2.3.1. Protecting and detecting signal impairment

The integrity of a GNSS defines the level of reliability of the signal (and thus of its Position, Navigation and Timing (PNT) information) from each satellite.

As suggested by [PNT 10], receivers and antennas could be hardened against jamming, for example with controlled radiation pattern antennas. GNSS receiver's antennas with phased arrays can resist jamming, once the direction of the jammer is determined. Jam resistant antennas may become more widely used. However, legacy equipment may not have this capability. Techniques are being developed to detect spoofing, including techniques that can be incorporated into receivers.

Telecom GNSS receivers should include a receiver autonomous integrity monitoring (RAIM) function for signal integrity verification. RAIM monitors multiple satellites to detect faulty conditions and exclude one deviant signal. Among GNSS PNT receivers, some are specialized in providing precision timing in stationary applications (not nomadic). As described in section 2.7, after precise location of a stationary receiver using signals from a minimum of four satellites, a single satellite signal would be sufficient to maintain timing information. A specialized GNSS receiver for synchronization application would have to keep scrutinizing multiple satellites for RAIM function. Note that RAIM is ineffective for spoofing.

Tracking more satellites is thus a way to improve integrity verification and mitigate from individual satellite fault. As simulations show [HEW 04], a combination of multiple GNSS (i.e. a multi-GNSS receiver, see section 2.7) can also improve RAIM results. RAIM performance analysis in case of a GNSS cooperation scenario shows that integrity depends on the clock offset between GNSS and the number of satellites available in certain elevation mask (height position of satellite from any constellation) [HEW 05]. However, RAIM detects errors in the user segment at the receiver.

Galileo will be able to provide an integrity navigation (I/Nav) message supporting Galileo system integrity. GLONASS and GPS will also provide similar integrity assurance from their restoration and modernization programs with the introduction of new signals from new satellites and new control segments to support new satellite capabilities. In particular, the L5 frequency would improve the mitigation against perturbation.

Integrity messages would allow a satellite to announce that the information provided should not be tracked. Hence, this is a continuous control of the information broadcast by the satellite segment.

Today, the control and satellite segments can provide such information but the delay to report such a faulty state can be excessive for the receiver to not be compromised. Utilization of augmentation systems by a receiver, if available, can mitigate this limitation.

Augmentation systems allow checking such integrity and announcing the fault much faster. As mentioned in section 2.7, augmentation systems verify the signal from the GNSS satellites and provide integrity information to receivers using the augmentation systems service. Wide Area Augmentation System (WAAS) in North America has been developed to overcome the lack of precision due to the enablement of the selective availability (SA) in the public C/A code. Since SA has been disabled, WAAS keeps maintaining enhanced services, for instance allowing faster announcement of integrity for any satellite that the augmentation systems is tracking.

Augmentation systems will improve the reliability and the precision of the data sent by the GNSS, by reporting in real time on the GNSS signal quality and thus allowing local mitigation. However, they cannot overcome spoofing attacks.

6.2.3.2. *Mitigating from signal impairment*

The first step of mitigation is to be able to detect signal impairment.

A receiver should be able to detect signal interferences due to natural (e.g. obstacles to signal), intentional attack (e.g. jamming) and some signal integrity issues from satellite segment, either directly or by some assistance. Most of the recently proposed anti-spoofing techniques also focus on spoofing detection rather than on spoofing mitigation.

Timing-specialized GNSS receivers aim to provide stable frequency and "one pulse per second" (1PPS) outputs. To achieve this objective they host a stable source of frequency (either Oven-Controlled Crystal Oscillator (OCXO) or atomic) that can be used to provide holdover in the case of signal loss. This stable reference can allow detecting signal variations beyond a threshold and trigger an appropriate mitigation action. These actions may also be initiated after detecting integrity issues or a possible spoofing attack.

When detecting signal impairment, the receiver would generally enter into holdover mode. It is thus important to evaluate the holdover requirement for the time source. As an alternate to local holdover capability, the receiver device (e.g. the PRTC in Figures 6.1–6.7) might also be able to select an external PRC-traceable frequency source for maintaining time information (see Figure 4.37 in section 4.4 for example). The time source (PRTC) can also announce any change on reference signal condition to a packet master to further forward this information. This can trigger provisioned protection. Alternatively, the packet master might also become a boundary clock relaying the time from another packet master with a valid source. There are technical backup options. However, as underlined in previous sections, those options should be considered during the network engineering phase with the objective to support the application's requirements.

Because spoofing attacks would be difficult to detect before integrity and authentication services from new generation of GNSS become commercially available, alternative solutions should be provisioned to mitigate against GNSS source impairment. This would ensure some redundancy and increase timing reference availability.

Alternatives to GPS and GNSS are mentioned in section 2.7. As also discussed in this chapter, the use of a multi-GNSS receiver would make the system more resistant to spoofing, since the spoofer would have to implement multiple signals. However, this still does not prevent jamming all signals reaching a multi-GNSS receiver.

One drastic way to overcome this risk could be not to use GNSS at all and turn to alternate solution(s). Lower frequency and stronger signals such as Enhanced Long Range Aid to Navigation (eLORAN) are technically attractive as they also allow for better signal penetration in buildings. Although eLORAN is a solid alternative, jamming and spoofing are still possible [SHE 09].

Another alternative in the future could be the utilization of dedicated (dark) fiber or wavelength options to connect to time generation laboratories [EUR 12]. Even if this option is not available for widespread use in the public telecommunication network, it might still serve as a primary or backup source on a limited scale to a few operator locations. Note that, as discussed in section 6.3, a fiber can be compromised.

A less drastic option would be to rely on GNSS as primary source and use one or multiple alternative solutions as secondary sources. The most common option, mentioned already, would be to protect the GNSS time signal with holdover provided by an atomic frequency standard (see Figure 4.34 in section 4.4). The use of other radio solutions such as eLORAN is possible, but currently limited regionally. For instance, as of the end of 2012, the number of operational eLORAN stations in the United Kingdom was limited.

In short, to provide a secure timing source, we need to consider:

– *Availability*: combination of sources in terms of technology and/or location, network-wide backup strategy with atomic frequency for holdover transient period;

– *Integrity verification*: RAIM, augmentation systems, GNSS combination;

– *Authentication*: when GNSS authentication service becomes available.

To conclude on GNSS mitigation, the authors would like to share the concern expressed by Dr Weiss, with regard to the prevalence and impact of jamming and spoofing:

"I think the issue is that the market responds better to disasters than prevention. I think there is a parallel between GPS and Internet vulnerabilities historically. Early on, [with the Internet,] there were no viruses or other attacks, hence no defenses. I am concerned that with GPS there will be more active jamming and spoofing attacks as time goes on and both awareness and motivation for attacking increase".

6.3. Security of synchronization distribution

Now assuming the timing source (i.e. the PRC or PRTC and its reference signal) is secured, the signal needs to be transferred to the end systems that require timing.

This section studies the security aspects of the synchronization chain from the timing source to the end system, looking at each element of the timing chain in telecommunication network. The synchronization transmission models taken into consideration are:

– physical frequency distribution (which can be associated with any packet-based distribution mode);

– packet-based end-to-end distribution;

– packet-based distribution with full timing support;

– packet-based distribution with partial timing.

As discussed in section 6.1, common basic security aspects on devices and cables are assumed to apply to any distribution mode and to physical or packet-based signal transmission. Compromised devices pave the way to more pernicious attacks but for timing aspects there are differences when considering timing-enabled or non-timing-enabled network equipment. Indeed, attacks include environmental threats such as on temperature or power, as they are critical elements for crystal-based oscillators or phase-locked loop (PLL) behavior. But attacks can also access microcode to modify parameters that control temperature of those elements or any variable maintained by the timing system for hardware or software functions (e.g. PTP data sets).

The following two sections describe, respectively, physical frequency transmission and packet-based timing distribution.

6.3.1. *Security aspects of physical timing transmission*

The upper part of Figure 6.9 shows a network supporting Synchronous Ethernet with PRC and SSU representing typical physical layer frequency synchronization network architecture. The bottom node in this network might be timing aware. The distinct operator cloud at the very bottom of the figure depicts a third-party network connected to the SyncE-enabled network. As can be understood from previous chapters, physical layer distribution is unique in each operator network, that is each operator with physical frequency transfer is assumed to own at least one PRC, and would normally utilize the clock from the third-party operator only when using timing as a service, as discussed in section 4.2.

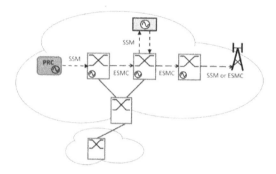

Figure 6.9. *Network with physical layer timing transfer*

The PRC and SSU, as well as the connections from PRC and SSU to network elements, should be secured and monitored. Network elements are network devices and their access is also secured. If the physical signal is compromised, the data transmission is affected. Attacking a node could further compromise the control plane. From the perspective of synchronization, this could allow false control information, such as clock selection and Synchronization Status Message (SSM) processes. In such a case, an attack on a single node becomes an attack on the full network.

An attack to the control plane may disturb all or part of the timing chain. Figure 6.10 shows some examples of threats, with compromised signal in synchronization and control planes illustrated in gray color. Different segments can be studied. Each enables operating an attack on the synchronization or control plane but their impact on the chain decreases as timing signal goes down in the hierarchy.

Careful synchronization planning and provisioning can mitigate most of the threats. For instance, in Figure 6.10, configuration of the list of timing sources can easily isolate internal or external nodes such as the base stations (segment C) or the gray nodes at the bottom. But controlling the network node would enable modifying this configuration.

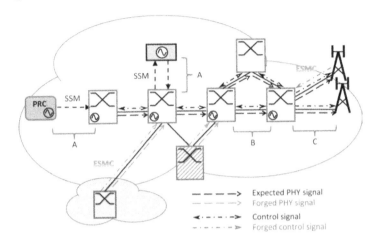

Figure 6.10. *Threats in physical layer timing transfer*

The link between two network elements transmits a frequency signal that can be synchronous such as with SyncE or SDH, but might also be a standard Ethernet link. By modifying the configuration of a synchronous node, the timing signal chain can be broken. For instance, an interface connected to non-synchronous node can be selected. Misconfiguration of a node has similar consequences. Tolerance limits specified in standard specifications should trigger signal failure at some point in the timing chain. The utilization of SSM or Ethernet Synchronization Messaging Channel (ESMC) and of QL-enabled clock selection could prevent this mistake, if wrong or fake information is not received.

Transmission of forged SSM or ESMC could impact the hierarchy of the timing chain. Attacking a segment A implies wiretapping local connections. Such a risk should be easily prevented. The consequence of such an attack could be limited, as the network hierarchy should switch to a backup source.

Modifying the QL-value in SSM or ESMC protocol data unit (PDU) in some configured path (e.g. segment B) may, however, generate a timing loop. Again, if node noise tolerance can overcome the issue, the impact can nevertheless last sufficiently long to impair the synchronization quality particularly if the physical layer frequency supports packet-based time synchronization. In such a case, the timing-packet servo might be able to alleviate this risk by verifying the frequency signal from packet-based signal.

The ESMC PDU can benefit through the use of MACsec [802.1 AE] integrity protection but it would not protect from the compromised node.

6.3.2. Security aspects of packet-based timing transmission

This section describes the security aspects of two main areas, the synchronization plane and the control plane. The synchronization plane transmits the timing signal while the control plane includes any functions setting and maintaining the synchronization hierarchy and distribution paths. A third area, the management plane, is not addressed here because, in the telecom environment, management would in most cases be performed out of band (e.g. Management Information Base (MIBs) and network management system/operational support system (NMS/OSS), see Chapter 5) and should already be part of the operators security policy.

Before going further, the authors strongly recommend reading draft-ietf-tictoc-security-requirements[1] [MIZ 13] from the IETF TICTOC working group. This document lists the security threats related to packet-based timing and

> "defines a set of security requirements for time synchronization protocols, focusing on the Precision Time Protocol (PTP) and the Network Time Protocol (NTP). This document also discusses the security impacts of time synchronization protocol practices, the time synchronization performance implications of external security practices, the dependencies between other security services and time synchronization".

As a community of timing and security experts is responsible for its development, the authors feel unnecessary to paraphrase this document. The approach taken in this section is to place the vulnerabilities and threats in the perspective of operator networks.

Figure 6.11 depicts main timing distribution scenarios[2]. The upper network shows operator network A implementing full timing support in some sections of its network (for instance, when precise time synchronization is required) and an end-to-end deployment model (e.g. for frequency or looser time synchronization). This operator is also using third-party networks B and C to connect its equipment. If the

---

1 At the time of the publication of this book, this document is still a draft but is expected to become an Informational RFC. Refer to http://www.ietf.org for the latest status of this document.

2 These scenarios do not distinguish cases by using physical frequency support or not. Such support, if available, owned or as third-party operator service, is trusted. See section 6.3.1 for some aspects of physical layer timing security.

operator A utilized timing service offered by network C, it will transmit its own timing signal over network B.

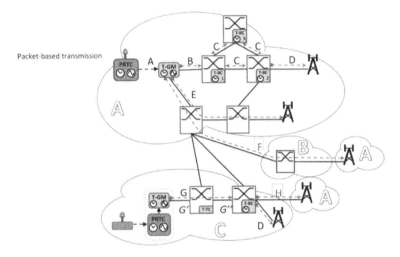

Figure 6.11. *Networks with packet-based timing transfer*

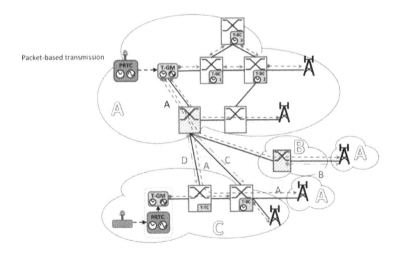

Figure 6.12. *External threats on packet-based timing transfer*

By following the timing path or flow, different segments or paths (A to H in Figures 6.11 and 6.12), various threats can be studied. Hereafter, the terms "reference master", "master", "master source" and "slave" are utilized for packet-based timing transfer with the following meaning:

– A *reference master* is the packet-based timing reference for a timing domain or hierarchy chain in a domain. This is usually the equipment receiving a reference signal from timing source (e.g. PRC, PRTC or Stratum server 0) and typically is a IEEE 1588 (telecom) grandmaster or NTP server of stratum level 1. Note: a *reference master* is a *master source*.

– A *master* is any device with at least one interface or port transmitting a packet-based timing signal, typically an IEEE 1588 (telecom) boundary clock or an NTP stratum server of level 1. Note: a *master* is a *master source*.

– *Master source* refers either to a *reference master* or to a *master*. *Master source* designates the nearest source of a packet-based timing signal.

– A *slave* is any device receiving packet-based timing signals and recovering a reference timing signal. A slave can be an IEEE 1588 (telecom) boundary clock, an IEEE 1588 ordinary clock, such as an ITU-T Recommendation G.8265.1 [G.8265.1] telecom slave, a Telecom Time Slave Clock (T-TSC), an NTP stratum server with a level greater than 1 or an NTP or Simple Network Time Protocol (SNTP) client.

– The (telecom) transparent clock, which does not recover or generate a clock signal, will be referred as T-TC when necessary.

Starting from the timing source, it is possible to threaten one or multiple signals (frequency, phase and or time) feeding a reference master (Figure 6.11, segment A). Risks are related to wiretapping thus relate to cable path, cable type, then type of interfaces (including physical and electrical characteristics) and communication protocol. Beyond the synchronization signal, control data can be spoofed. Such data from a timing source can be utilized by the reference master, then transmitted toward the packet-synchronization network. This vulnerability would not exist if the PRTC and reference master were in same device.

The interface from the timing source to reference master or first network element (e.g. synchronous Ethernet Equipment Clock (EEC)) is typically a dedicated timing interface (e.g. 2.048 MHz, 1PPS with time to live (TTL) or RS422 signal, Time of Day (ToD) with serial communication), connected and configured during initial deployment. Any unexpected disconnection (i.e. when not planned and managed by operational process) should be logged and trigger an alert as previously discussed. Because new timing interfaces can be Ethernet based (e.g. with SyncE and PTP), timing security aspects discussed in this chapter also applies.

Beyond this point, the security threats existing for current network protocols (at layer 2 and above) can apply to packet-based timing. The distinction would come from the information being modified, for example time stamps or timescale, control plane and effect of delay attack. Please refer to [MIZ 13] for further details.

A rogue master source (e.g. Telecom-GM (grandmaster) (T-GM) and Telecom-BC (boundary clock) (T-BC)) would attempt to replace the current and legitimate master source. To mitigate risk using the control plane, a known list of master sources can be configured (e.g. Figure 6.11, segments B, C, D, E, F and H). A list of master source unicast addresses, such as in G.8265.1 telecom slave or NTP server or peer association, and acceptable master table in IEEE 1588 when using multicast are examples of such control. Furthermore, port security (usually at MAC layer) and source/destination address filtering are simple security mechanisms that could be used before applying cryptography techniques requiring key management. With multicast, utilization of alternate multicast group addresses can provide further flexibility and control on destination addresses. However, a compromised master device may be utilized to simulate a reference master, as discussed later.

When communication between devices is not trusted (e.g. Figure 6.11, path F) or when the peer node can be masqueraded (e.g. node in network B impersonates the role of master source and masquerade timing flow − segment B in Figure 6.12), a slave should secure its communication paths from master sources, for synchronization and control signals. In some cases, depending on the processes defining the hierarchy or selecting the reference master, the master source should protect itself from slave control signals. Such security level implies authentication of the peer node, master or slave. For instance, NTP security applies, usually on server to secure from clients for client–server association, when peer associations would call for mutual authentication. For T-BCs that use dynamic port state establishment necessary for reacting to network rearrangement (see Chapter 4), securing communication with PTP peer port should be performed whatever the port state (e.g. Figure 6.11, segments C). Indeed, spoofed control information might destabilize the hierarchy. When possible, for instance when a fixed hierarchy is expected, defining statically the port state (i.e. forcing master or slave state) would prevent such risk (e.g. Figure 6.11, segments B, D, E, F, G or H).

When using T-TC, the network address of the master source might not be accessible for verification by the slave (Figure 6.11, segment G). This would be the case if the T-TC does not infringe the layer rules (see section 2.4.3). In such a case, for instance, the T-BC in network C (Figure 6.11, subsegment G″) or the base station of the network A that utilizes network C (segment A in Figure 6.12) cannot authenticate its T-GM.

Alternatively, the T-TC function may be disabled or instead, each link would have to be individually secured (Figure 6.11, subsegments G′ and G″). This second alternative could also be implemented for securing the PTP pdelay mechanism.

In a unique management domain, securing links between IEEE 1588 clocks is conceivable with symmetric cryptographic algorithms (e.g. within network A).

However, if the path utilizes third-party network equipment, asymmetric cryptography becomes recommended (two examples: Figure 6.11, segment H, and Figure 6.12[3], path A, using Telecom-Transparent Clock (T-TC) of network C) and introduces key management aspects. It also means that the T-TC device (actually, the network C) is being trusted because such node may introduce errors. Using the timing service from network C is an alternative (segment D', Figure 6.11).

Despite a communication path between master source and slave being secured for authentication and for integrity, it remains subject to other threats such as DoS, duplication or delay attacks (e.g. network B can delay path F, Figure 6.11 or replicate timing messages). Delay attack includes asymmetry attack (i.e. modifying delay in one direction to generate fixed asymmetry that the end node cannot correct without using external reference). Recall that a network node can generate high packet delay variation (PDV) and that its configuration can create a fixed offset in one direction, hence an asymmetry threat (e.g. gray nodes in Figure 6.12 using Network Processor Unit (NPU) [GOL 12]).

The gray-hatched network A node of Figure 6.12 helps to introduce two other outside threats. This node might be timing aware or not (case depicted in the figure). In both cases, if compromised, it gives an attacker the opportunity to manipulate the timing flows. The attacker might initiate new communication paths or hijack current timing initiated paths. It might spoof or impersonate the role of reference master, the master or the T-TC relative to slaves (e.g. T-BC path C Figure 6.12), to a T-TC (e.g. path D) or via a T-TC in network C. Authentication or integrity check in this case would not help.

To mitigate timing threats, a protection mechanism related to timing might be used. Holdover might mitigate instability in the control plane. Byzantine fault protection might be utilized at the slave but necessitates maintaining multiple communication paths with masters to allow selection of the "right" masters or discarding the "wrong" masters. NTP algorithms provide some level of protection by allowing multiple associations (and authentication) and combining timing signals (see section 2.6.3 and Figure 2.30). IEEE 1588 does not specify algorithms and security mechanisms.

[G.8265.1] provides a means to mitigate against PDV threats by the use of multiple reference masters, together with the use of a Packet Timing Signal Fail (PTSF) unusable signal. The precise definition of PTSF-unusable is not standardized (as of 2013), therefore utilization of this signal is implementation dependent.

---

3 In Figure 6.12, path A, which transfers timing signal from reference master A, bypasses the T-BC that is synchronized to network C reference master.

Another model has been proposed at IETF ([SHP 12])[4] where a slave compares signals from the same master source but sent over distinct network paths. Such techniques, using either multiple master sources or multiple network paths, aim to improve the quality of the recovered timing signal by the signal with optimum PDV characteristics discarding high PDV path and by combining signals. Hence, it could allow detecting and mitigating delay attacks that could affect one path. Note, these are also referred to as "multipath" techniques. However, the reader is cautioned that the use in this context is not the same as used in section 6.2 to describe degradation conditions on GNSS signals.

However, for efficient multipath, as for any path protection mechanism (e.g. MPLS or Internet Protocol (IP) Fast Reroute (FRR)), the utilization of the same segment (or link) for distinct paths should be avoided. Thus, multipath will be engineered for instance using specific MPLS pseudo wire (PW) or LSP for PTP [DAV 13][5].

To conclude, it is worth noting that authenticating communication paths in a full timing support design with T-BC or T-TC assistance would require attention. In such a scenario, authenticating with the reference master is not possible without introducing a specific or complex mechanism. NTP relies on an inductive authentication model called proventication[6]. With PTP, end-to-end or partial timing support scenarios would ease securing the path but would introduce a trade-off with regard to packet-based timing assistance.

6.4. Synchronization risk management

Before concluding this chapter, it is worth recalling few key points.

Security should be considered as an ensemble of techniques that aim to, but cannot prevent, every single threat. Perfect security does not exist. Because security mechanisms can be very expensive and beyond requirements, the goals are:

1) to minimize the vulnerability of a device or a protocol to attacks;

2) to limit the impact to the service (and its applications) from an attack;

3) to make any threatening action more difficult.

---

[4] At the time of publication of this book, this document is still a draft. Refer to http://www.ietf.org for the latest status of this document.
[5] At the time of publication of this book, this document is still a draft but is expected to become an international RFC. Refer to http://www.ietf.org for the latest status of this document
[6] "Proventication" is a term introduced by D. L. Mills for describing a chain of trust from stratum server 1 (see [NTP 10]).

As is the case with home intrusion, prevention is more about making the intrusion attempt more difficult (as any extra time increases the risk of being caught) than to actually prevent it. Because security can be expensive, the protection should remain proportional to the risk of service interruption or degradation.

Thus it is important to appropriately choose and implement a correct set of available security techniques, that is the security options will be balanced with the value of what is to be protected. This starts by studying vulnerabilities and threats, and analyzing the probability of a service being attacked, understanding the effects and consequences of an attack and defining their cost and related expenses in order to mitigate against them. The weakest element in a system defines the security strength of that system but security management is about cost-effectiveness.

As stated many times, appropriate network and device protection must be enforced and prioritized before considering optimizing security of timing distribution. Because of the importance of synchronization in the domain of application, to the architecture of the network and to security practices already in place, it might not be strictly necessary to implement security mechanisms that are specific to timing. As for other services, there is a need to clearly define a reasonable level of security. For timing, it would be necessary to articulate the benefit of security for a specific application and, if applicable, the benefit versus drawback for applying a particular security mechanism to the appropriate timing method. For instance, the application of cryptography could add uncertainty to the time stamping: for precise timing synchronization, the cryptography process will have to offer deterministic delays.

To minimize synchronization security administration and its performance impacts on some equipment, the application of security measures might be limited to certain areas of a timing domain. In a mixed application environment, that is when the same timing network supplies synchronization to distinct applications, only certain timing paths would have to be fully secured when others might not require any security. However, the latter must not become vulnerability points and a source of threats.

For integrity, authentication and, if necessary, encryption or non-repudiation, management of the symmetric or asymmetric keys should preferably already be in place (e.g. if IPsec [IPS 11] or MACsec are utilized). But key exchange protocols need time synchronization to define and maintain key validity duration. If the level of accuracy required for such dynamic management is quite loose compared to the timing needs of the mobile wireless applications, such circular interdependence must be acknowledged.

To conclude, security policy is a circular process that consists of repeated phases, which can include definition, analysis, classification, assessment, enforcement (remediating and mitigating) and verification phases. With carriers deploying critical timing distribution over new packet networks, it is becoming critical that carriers consider timing security as part of their overall security policy.

6.5. Bibliography

[ACM] Devices Prohibited by the ACMA (Australian Communications and Media Authority). Available at http://www.acma.gov.au/WEB/STANDARD/pc=PC_1296, 2004.

[AIR 12] AIRST M., "GPS network timing integrity", August 2012. Available at http://www.gps.gov/governance/advisory/meetings/2012-08/airst.pdf.

[BEN 05] BENSHOOF P., US Air Force 746th Test Squadron, "Civilian GPS Systems and its Potential Vulnerabilities", 2005. Available at http://www.gps.gov/cgsic/international/ 2005/prague/benshoof.ppt.

[CAR 01] CARROLL J., "Vulnerability assessment of the transportation infrastructure relying on the global positioning system", *DOT/OST Outreach Meeting*, October 2001. Available at http://www.navcen.uscg.gov/ppt/Volpe%20Slides.ppt.

[DAN 12] DANESHMAND S., JAFARNIA-JAHROMI A., BROUMANDAN A., *et al.*, "A low-complexity GPS anti-spoofing method using a multi-antenna array", ION GNSS, 2012. Available at http://www.ion.org/meetings/abstract.cfm?meetingID=38&pid= 421&t=B&s=3; http://plan.geomatics.ucalgary.ca/papers/iongnss2012_sdaneshmand_26sep12.pdf.

[DAV 13] DAVARI S., "Draft-ietf-tictoc-1588overmpls-04", IETF TICTOC WG, February 2013, Expires: August 2013. http://www.ietf.org

[ECC 04] ECC RECOMMENDATION (04)01, "With regard to forbidding the placing on the market and use of jammers in the CEPT member countries", February 2004. Available at http://www.erodocdb.dk/docs/doc98/official/pdf/Rec0401.pdf.

[ECO 99] EUROPEAN COMMISSION, ENTERPRISE and INDUSTRY, "Interpretation of the Directive 1999/5/EC, Jammers (item #37)", Radio and telecommunications terminal equipment (R&TTE); Available at http://ec.europa.eu/enterprise/sectors/rtte/documents/ interpretation/index_en.htm#h2-37.

[EUR 12] EURAMET, Accurate time/frequency comparison and dissemination throughoptical telecommunication networks, Publishable JRP Summary Report for JRP SIB02 NEAT-FT, March 2012; Available athttp://www.euramet.org/fileadmin/docs/ EMRP/JRP/JRP_Summaries_2011/SI_JRPs/SIB02_Publishable_JRP_Summary.pdf

[G.8265.1] ITU-T Recommendation G.8265.1, Precision time protocol telecom profile for frequency synchronization.

[GOL 12] GOLDIN L., MONTINI L., "Impact of network equipment on packet delay variation in the context of packet-based timing transmission", *IEEE Communication Magazine*, vol. 50, no. 10, pp. 152–158, 2012.

[GPS 98] GPS, Biennial Report to Congress on the Global Positioning System, 1998. Available at http://www.gps.gov/congress/reports/1998/biennial/.

[GPS 01] GPS, GPS Policy, Applications, Modernization, and International Cooperation, Interagency GPS Executive Board (IGEB), February 2001. Available at http://www.gps.gov/multimedia/ presentations/1997-2004/2001-02-northerneurope/main.ppt.

[HEW 04] GPS/GLONASS/GALILEO Integration: Separation of Outliers, 2004. Available at http://www.gmat.unsw.edu.au/snap/publications/hewitson_etal2004c.pdf.

[HEW 05] HEWITSON S., WANG J., "GNSS receiver autonomous integrity monitoring (RAIM) performance analysis", School of Surveying and Spatial Information Systems, The University of New South Wales, Australia, 2005. Available at http://www.gmat.unsw.edu.au/snap/publications/hewitson_etal2005a.pdf.

[HUM 08] HUMPHREYS T., LEDVINA B., PSIAKI M., *et al.*, "Assessing the spoofing threat: development of a portable GPS civilian spoofer", *ION GNSS Conference*, Savanna, GA, 16–19 September 2008. Available at http://108.167.174.48/~ ledvina/wp-content/uploads/2012/07/assessing_spoof_threat.pdf.

[HUM 12] HUMPHREYS T., "How to fool a GPS", TED Talks Video, 2012. Available at http://www.ted.com/talks/todd_humphreys_how_to_fool_a_gps.html.

[IPS 11] IP Security (IPsec) and Internet Key Exchange (IKE) Document Roadmap, Informational IETF RFC 6071, June 2011.

[JRP SIB02 NEAT-FT] "Accurate time/frequency comparison and dissemination through optical telecommunication networks", March 2012. Available at http://www.euramet.org/fileadmin/docs/EMRP/JRP/JRP_Summaries_2011/SI_JRPs/SIB02_Publishable_JRP_Summary.pdf.

[LED 10] LEDVINA B., BENCZE W., GALUSHA B., *et al.*, *An In-Line Anti-Spoofing Device for Legacy Civil GPS Receivers*, August 2004. Institute of Navigation ITM, San Diego, CA, 26 January 2010. Available at http://108.167.174.48/~ledvina/wp-content/uploads/2012/07/ inline_antispoofing.pdf.

[MIL 04] MILLS S.D., "NTP security model briefing slides", Network Time Synchronization Research Project, August 2004. Available at www.ece.udel.edu/~mills/database/brief/autokey/ autokey.pdf.

[MIZ 13] MIZRAHI T., "Draft-tictoc-security-requirements-04", IETF TICTOC WG, February 2013, Expires: August 2013. http://www.ietf.org

[NTP 10] "NTPv4 Autokey Specification", Informational IETF RFC 5906, June 2010.

[PAP 08] PAPADIMITRATOS P., JOVANOVIC A., "Protection and fundamental vulnerability of GNSS", *Proceedings of the International Workshop on Satellite and Space Communications* (IWSSC), EPFL Lausanne, 2008. Available at http://infoscience.epfl.ch/record/134639/files/IWSSC08-GNSS-sec.pdf.

[PNT 10] NATIONAL PNT ADVISORY BOARD, "Comments on jamming the global positioning system – a national security threat: recent events and potential cures", November 2010. Available at http://www.gla-rrnav.org/pdfs/interference_to_gps_v101_3_.pdf.

[RAN 95] RAND CORPORATION, "The Global Positioning System, assessing national policies", 1995. Available at http://www.rand.org/pubs/monograph_reports/MR614.html

[RAO 11] ROYAL ACADEMY OF ENGINEERING, "Global Navigation Space Systems: Reliance and Vulnerabilities", March 2011. Available at http://www.raeng.org.uk/news/publications/ list/reports/RAoE_Global_Navigation_Systems_Report.pdf.

[SEC 07] Informational IETF RFC 4949, "Internet Security Glossary, Version 2", August 2007.

[SHE 09] LO S., PETERSON B., ENGE P., "Assessing the security of a navigation system: a case study using enhanced loran", *European Navigation Conference GNSS*, Naples, Italy, May 2009. Available at http://waas.stanford.edu/~wwu/papers/gps/PDF/ LoENCGNSS09.pdf.

[SHP 12] SHPINER A., "Draft-shpiner-multi-path-synchronization-00, multi-path time synchronization", Experimental draft, IETF TICTOC WG, October 2012, Expires April 2013). http://www.ietf.org

[US GOV] "Information About GPS Jamming", Available at http://www.gps.gov/spectrum/jamming/.

[VOL 01] JOHN A. VOLPE NATIONAL TRANSPORTATION SYSTEMS CENTER, Vulnerability assessment of the transportation infrastructure relying on the global positioning system, Final Report, August 29 2001. Available at https://www.transportationresearch.gov/dotrc/pnt/Documents%20of%20Interest/Forms/DispForm.aspx?ID=25.

[WAR 02] WARNER J., JOHNSTON R., "A simple demonstration that the global positioning system (GPS) is vulnerable to spoofing", *The Journal of Security Administration*, vol. 25, pp. 19–28, 2002. Available at http://www.ne.anl.gov/capabilities/vat/pdfs/GPS-Spoofing-(2002-2003).pdf.

[WAR 03] WARNER J., JOHNSTON R., "GPS spoofing countermeasures", *Homeland Security Journal*, December 12 2003. Available at http://www.ne.anl.gov/capabilities/vat/pdfs/GPS-Spoofing-CMs-%282003%29.pdf.

Chapter 7

Test and Measurement Aspects of Packet Synchronization Networks

7.1. Introduction

This chapter provides aspects related to the testing and measurement of frequency transfer across a packet network. The chapter focuses on the testing and measurement of two technologies: (1) Synchronous Ethernet (SyncE) and (2) the IEEE 1588 end-to-end telecom profile. Multiple test topologies and scenarios are presented, discussed and the expected results are compared to the requirements and performance objectives specified in the standards. Although the test scenarios are certainly not exhaustive, they provide an understanding of the baseline tests that must be conducted when testing these two technologies.

7.2. Traditional metrics

Various interfering impairments prevent perfect frequency transfer across a packet network. All networks exhibit some form of jitter and wander, which ultimately impacts the user experience of a particular application or service. The use of performance objectives and metrics is one of several ways to quantify the jitter and wander when studying the transfer of frequency across a packet network. This section presents some of the basic definitions and terms. Chapter 2 provided various definitions and concepts related to synchronization and this section adds other definitions and terms when dealing with test and measurements:

Hz is a unit of measure representing the number of cycles in 1 s.

Frequency is the number of cycles that occur in a specified amount of time. For example, a 10 MHz clock is equal to 10 million cycles in 1 s.

Phase is a measure of the difference between the significant instants of a measured signal and a reference signal, usually expressed in units of degrees, radians or unit of time. In telecommunication networks, phase is typically represented in unit of time.

Jitter is defined in the International Telecommunication Union Telecom (ITU-T) standards as the short-term variations of the significant instants of a signal from their ideal points in time. The SyncE standard, for example, defines the jitter region as variations that occur at frequencies greater than 10 Hz and variations less than 10 Hz as the wander region. Jitter is specified both in terms of root mean square (rms) or peak-to-peak measurements and is often expressed in units of time. Note that jitter often refers to the short-term variations found in a physical signal and is not necessarily the same as the definition of packet jitter or also known as packet delay variation (PDV).

Wander is defined in the ITU-T standards as the long-term variations of the significant instants of a signal from their ideal positions in time. Long-term implies that the phase variations are happening for frequencies less than 10 Hz. Wander is usually specified in terms of maximum time interval error (MTIE) and time deviation (TDEV). MTIE and TDEV are two of the most popular metrics for assessing the performance of synchronization networks in telecommunication networks.

Frequency offset is used to indicate the degree to which the frequency of a clock may deviate from a nominal value. Frequency offset is often expressed in parts per million (ppm) as shown below:

$$\frac{f - f_0}{f_0} \times 1 \times 10^6$$

where f is the measured frequency and f_0 is the reference frequency. For example, if the reference frequency is 8 kHz and the measured value is 8000.0020 Hz, then the offset from the reference (in ppm) is:

$$\frac{8000.0020 - 8000}{8000} \times 1 \times 10^6 = 0.25 \text{ ppm}$$

Frequency offset can also be expressed in units of time. For example, a 0.25 ppm means that the measured signal is accumulating 0.25 μs of time for each second, whereas a −0.25 ppm means that the measured signal is losing 0.25 μs of time for each second. After 10 s of real time, the signal would be off by 2.5 µs compared to the reference signal. This is known as the frequency drift and the relation is shown below:

$$X \text{ ppm} = \frac{X \text{ µs}}{\text{sec}}$$

Time interval error (TIE) is defined as the variation in time delay (also called phase difference) of a measured signal compared with a reference signal at a particular point in time, within a measurement period. As shown in Figure 7.1, TIE represents the time delay between two signals (the difference in time between the leading edge of the reference signal and the measured signal) and is measured in the time domain using units of seconds. Each measurement produces a time delay sample. TIE movement can be positive or negative and is due to impairments such as clock noise, reference switching, entry and exit from holdover, temperature variations and mechanical vibration. A continuous increase in TIE means that the measured signal is faster than the reference, while a continuous decrease in TIE means that the measured signal is slower than the reference.

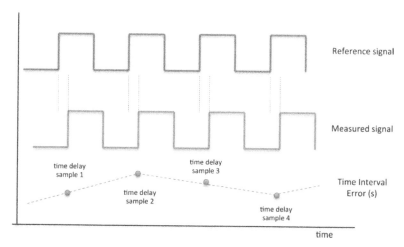

Figure 7.1. *Time interval error*

MTIE is the peak-to-peak time delay (also known as peak-to-peak phase difference) for a given window of time, known as the observation interval. MTIE is the maximum difference between the maximum TIE sample value and the minimum TIE sample value over an observation interval and is always a positive value. MTIE is often used for measuring time delay in the wander region (10 Hz and below). The MTIE either increases or remains constant as the observation interval increases, but can never become smaller as the observation interval increases. MTIE is typically measured for a sufficient amount of time such as 1,000–100,000 s.

Figure 7.2 shows an example of TIE samples versus time. The peak-to-peak phase difference value over the specific observation interval #a (of length 8 s) is calculated as the difference between the maximum TIE value and minimum TIE value observed within that observation period. This corresponds to the difference between "TIE sample $n - 1$" (let us say a maximum value of 3 μs) and "TIE sample 5" (let us say a minimum value of 1 μs). This produces a single MTIE value for the observation interval #a, and is typically defined as *MTIE (8 s) = 2 μs*.

Likewise, the phase difference over each observation interval #1 (each of length 2 s) is calculated as the difference between the maximum TIE value and minimum TIE value observed within each observation interval. Calculating the MTIE (2 s) first consists of sliding a window of length 2 s across the TIE samples and calculating the phase difference for each window. The MTIE then corresponds to the maximum value of all the phase difference that has been calculated. The MTIE (2 s) inFigure 7.2 would likely correspond to the third observation interval and would have a value equal to the difference between "TIE sample $n - 2$" (let us say 2.6 μs) and "TIE sample 5" (let us say 1 μs). Therefore, the *MTIE (2 s) = 1.6 μs*.

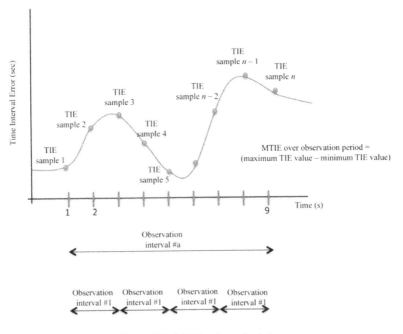

Figure 7.2. *MTIE value calculation*

MTIE essentially consists of the process of sliding a window for intervals of 2, 4, 8, 16, 32 s, etc., and then obtaining the maximum phase difference for each of these intervals. For each observation interval, a value of MTIE (2, 4, 8, 16, 32 s, etc.) would be calculated and plotted against the observation interval.

By definition, a TIE value can be as large, but never larger than an MTIE value or a given observation time. For a sequence of TIE samples x_i, MTIE for a particular observation time is expressed mathematically as follows:

$$\mathrm{MTIE}(n\tau_0) = \max_{j=1}^{N-n}\left[\max_{i=j}^{j+n}(x_i) - \min_{i=k}^{j+n}(x_i)\right] \text{ where } n = 1, 2, \ldots, N-1$$

where N is the number of samples in the sequence, τ_0 = sample period and $n\tau_0$ is the observation interval.

Maximum relative time interval error (MRTIE) is the maximum peak-to-peak time delay of an output signal with respect to a "given input signal". The only difference when compared with MTIE is that the measurement is performed with respect to a "given input signal", where the input signal is not necessarily an ideal signal or directly coming from a primary reference clock (PRC). MRTIE has been found particularly useful in defining the requirements for traffic interfaces (as specified in ITU-T Recommendation G.823 [G.823]) and in cases where an ideal reference might not be available. MRTIE removes any frequency offset that might be present on the given input signal. For example, MRTIE has been useful in measuring the wander of an output signal versus an input signal and sizing the buffer size needed to minimize loss of information.

TDEV is a statistical measure of the amount of time variation (phase noise) measured through a bandpass filter as a function of integration time. TDEV is useful in determining the frequency of the noise (frequency content of the TIE samples) sent to downstream clocks and to verify how much filtering a phase-locked loop (PLL) system might require. The integration time calculates the noise from 10 Hz down to 1/(integration time). TDEV accuracy is dependent on the number of data samples collected and the recommended measurement periods are usually at least 12 times the longest integration period. TDEV is usually expressed in units of nanoseconds.

ITU-T Recommendation G.810 [G.810] defines the formula for computing TDEV from a set of TIE samples x_i as shown below:

$$\text{TDEV}(n\tau_0) \cong \sqrt{\frac{1}{6n^2(N-3n+1)} \sum_{j=1}^{N-3n+1} \left[\sum_{i=j}^{n+j-1} (x_{i+2n} - 2x_{i+n} + x_i) \right]^2}, \quad n = 1, 2, \ldots, \text{integer part}\left(\frac{N}{3}\right)$$

where:

x_i are time error samples;

N is the total number of samples;

τ_0 is the time error sampling interval;

τ is the integration time, the independent variable of TDEV;

n is the number of sampling intervals within the integration time τ.

This formula shows that the term inside the bracket is the double difference of three adjacent TIE samples, for example $(x_2 - x_1) - (x_1 - x_0)$. These are then squared and summed as the double difference values slide along the TIE data. For example, if the TDEV plot shows a peak value of 100 ns at $\tau = 10$ s, this would indicate that the low-frequency component of a measured signal has an approximate frequency of 0.1 Hz with an amplitude of 100 ns.

7.3. Equipment configuration

There are several ways to configure an equipment under test. The most popular and straightforward way is through a command line interface (CLI) or graphical user interface (GUI). The CLI allows a user to change the configuration parameters directly from a serial or through a telnet or secure shell session. The GUI allows a user to change the configuration parameter using a PC-based communication and graphical tool. To access the CLI of the equipment, a serial connection is generally established with the device. This is done by using a tool such as a terminal emulator and configuring the tool with specific parameters such as baud rate: 115,200, data bit: 8, stop bit: 1, parity: none and flow control: none. During power up, the equipment will typically be configured using a default configuration. The user is then free to make changes to the configuration for his particular test scenario. Typical commands used to interrogate the running configuration are in the form of *show-config* or *show-status* or *show clock-status*. Typical commands used to configure the equipment are in the form of *set-parameter* or *set-value*. Each type of equipment will have its own command hierarchy, description and convention. Another way to access the equipment is to assign it an Internet Protocol (IP) address and use tools such as telnet or secure shell to access the CLI of the equipment. Alternatively, some equipment will implement a web server and can be directly accessed and configured using a web browser.

7.4. Reference signals, cables and connectors

This section presents a short summary of the typical signals, cables and connectors used when testing and measuring SyncE and IEEE 1588 equipment. Table 7.1 shows a multitude of reference signals, cables and connectors that are typically found today on several types of equipment: PRCs, measurement equipment, packet-based switches and routers, wireless base stations, etc. These signals are often used to provide a source of synchronization between equipment (e.g. frequency reference or phase/time reference). Some signals typically used for the purpose of providing a frequency reference, such as the 2.048 Mbit/s or the 2.048 MHz synchronization interface, are specified in ITU-T Recommendation G.709 [G.709]. Signals for the purpose of providing a phase/time reference such as the one pulse-per-second (1PPS) are specified in ITU-T Recommendation G.8271 [G.8271]. Sometimes, specific connectors are used to save space on the faceplate of an equipment since the connector might have a small form factor, whereas in other cases certain service providers might require another type of connector to ensure proper connection to an installed base of equipment.

Reference signals	Cables	Connectors
2.048 MHz	Coax	BNC, RJ45, SMA, SMC
10 MHz	Coax	BNC
2.048 Mbit/s (E1)	Coax	BNC, Bantam, RJ45, SMA, SMC
1.544 Mbit/s (DS1)	UTP	Bantam, RJ48c, RJ45, SMA, SMC
1PPS	Coax	SMA, SMC
1PPS + ToD	UTP	DB9 RS422
Ethernet	UTP, Fiber	RJ45, SFP

Table 7.1. *Examples of typical reference signals, cables and connectors for the purpose of synchronization*

7.5. Testing Synchronous Ethernet

As described in Chapter 2, SyncE primarily consists of an Ethernet equipment clock (EEC) with performance defined in ITU-T Recommendation G.8262 [G.8262] and the Ethernet synchronization messaging channel (ESMC) protocol defined ITU-T Recommendation G.8264 [G.8264]. Performance testing of SyncE mainly consists of testing the ability of the SyncE nodes to transfer a frequency reference using the physical layer of Ethernet. The performance can be characterized by measuring the jitter/wander generation, jitter/wander tolerance and wander transfer produced by a single SyncE node. The performance of SyncE can also be tested by measuring the amount of wander accumulation across a chain SyncE nodes, each having an EEC. Protocol testing of SyncE mainly consists of verifying the ability of the SyncE nodes to properly decode and generate ESMC protocol data units (PDUs).

ESMC PDUs are encapsulated in Ethernet frames and are transmitted between each SyncE node. Each ESMC PDU contains various header fields as well as the clock quality level (QL), also known as the synchronization status message (SSM) of the SyncE node system clock. The protocol can be tested by verifying the ability of the SyncE node to select the best clock QL based on the ESMC messages received on ingress ports, as well as verifying the frame format of the ESMC message transmitted on egress ports.

7.5.1. Testing the performance of SyncE EEC

Performance testing mainly consists of verifying the items listed in Table 7.2. The testing of SyncE jitter is omitted since the requirements only apply when interworking a SyncE network with a legacy synchronous optical network/synchronous digital hierarchy (SONET/SDH) network or an optical transport network (OTN). For a network composed of SyncE nodes, testing for jitter is primarily governed by the high-band jitter requirements of the Ethernet interface type found in IEEE802.3 specification [802.3], although ITU-T has also specified wide-band jitter generation and jitter tolerance requirements for 1000BASE-X and 10GBASE-X interfaces. This chapter does not further discuss the aspects of jitter measurements and does not make a distinction between EEC-Option1 and EEC-Option2 as specified in [G.8262], but tries to provide the reader some guidance on a set of standard and baseline test scenarios.

Performance tests	ITU-T Recommendation
Frequency offset	G.8262 section 6
Pull range	G.8262 section 7
Jitter and wander generation	G.8262 section 8
Jitter and wander tolerance	G.8262 section 9
Wander transfer	G.8262 section 10
Short-term transients	G.8262 section 11
Holdover	G.8262 section 11
Wander accumulation	G.8261 section 9.2.1

Table 7.2. *SyncE performance tests*

The reference test bed is shown in Figure 7.3, and will be the baseline test bed for all test scenarios found in this section. The test bed shows a *PRC*, a *tester* and a *SyncE device under test (DUT)*. The role of the *PRC* is to provide an accurate frequency reference signal to the tester and/or to the DUT via the external sync interface (ESI) port, and is also used for the purpose of comparing a measured signal to the reference signal. PRCs used for testing in laboratory are typically cesium clocks or Global Positioning System (GPS) receivers, which can provide an accuracy of at least 1 part in 10^{11}. This is approximately equivalent to gaining or losing 1 s in 3,168 years. The *tester* is an equipment mainly used for measurement purposes. It is capable of performing various functions such as wander measurements and computing performance metrics such as TIE, MTIE and TDEV. The tester is also capable of adding various noise sources to conduct stress testing of the DUT and to generate a set of external synchronization signals that are necessary to connect with various DUTs. The tester also has the capability of acting as a packet traffic generator for the purpose of generating background traffic patterns and/or to generate and measure specific type of protocols such as the SyncE ESMC protocol. Finally, the tester offers Ethernet line interface speeds such as 100BASE-T, 1000BASE-X and 10GBASE-X. The *DUT* is a SyncE node under test that implements the EEC clock and the ESMC protocol.

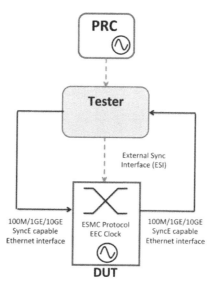

Figure 7.3. *Baseline Synchronous Ethernet test and measurement setup*

7.5.1.1. *Frequency offset measurement*

This test scenario measures the frequency offset of the SyncE node when operating in a free-running condition (the definition of free-running was introduced in Chapter 2). Figure 7.4 describes the test bed. This test is performed when the DUT has not yet been synchronized to a reference signal or has lost its synchronization after a long period of time. The DUT is configured to operate in free-running condition and an egress Ethernet interface that is SyncE capable, is connected to the tester. The tester locks to the frequency of the incoming SyncE signal coming from the SyncE node and computes the frequency offset of the DUT. The tester can also plot the TIE over a specified observation interval such 86,400 s

(1 day). The user can then estimate the frequency offset by calculating the slope of TIE curve. For example, if the TIE value is 153,600 ns after 1,000 s of measurement time, the frequency offset can be estimated to be 153.6 parts per billion.

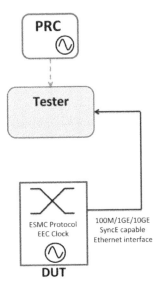

Figure 7.4. *Measurement of the frequency offset*

7.5.1.2. *Pull range measurement*

This test scenario consists of measuring the ability of the SyncE node to synchronize (or reject) to an input reference signal having a predetermined frequency offset modulated on the signal. The definition of pull range was introduced in Chapter 2. Figure 7.5 describes the test bed. The tester and DUT are connected via Ethernet interfaces that are SyncE capable. The tester is configured to modulate a frequency offset onto the SyncE interface connected to the DUT. The frequency offset is controlled and varied and the intention is to verify that the DUT is properly synchronizing or rejecting the input signal depending on the amplitude of the frequency offset.

Reference [G.8262] defines the maximum frequency offset of the EEC clock used in SyncE nodes to be within ±4.6 ppm. That is, the TIE plot cannot deviate more than 4.6 µs/s. Two types of tests can be conducted. The first is when the frequency offset modulated on the input signal is greater than ±4.6 ppm. The objective is to verify that the EEC might synchronize (although it does not have to) to the input signal when the signal is greater than ±4.6 ppm, but becomes synchronized as the modulated frequency offset reaches a value within the range of ±4.6 ppm.

This is also referred as the pull-in capability. The second test consists of applying a zero frequency offset and increasing the offset to and beyond the range of ±4.6 ppm, and to verify that the EEC stays synchronized while the offset is within ±4.6 ppm. This is also referred to as the pull-out capability and has been defined for further study in [G.8262] although it can be tested and measured. The verification can be done by looking at changes in the TIE plot or directly looking at the status of the clock or alarms through CLI access of the SyncE node.

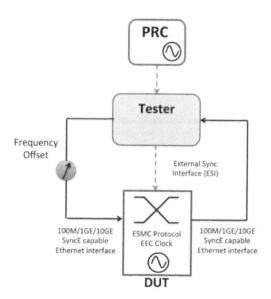

Figure 7.5. *Measurement of pull range*

7.5.1.3. *Wander generation measurement (output wander)*

This test scenario verifies the amount of wander produced by the SyncE node on one of the egress Ethernet interfaces that is SyncE capable. Figure 7.6 describes the test bed. The tester provides an "ideal" reference signal to the DUT and the test is conducted in an environment with controlled constant or variable temperature. The tester is connected to the ESI port of the DUT (to provide the reference signal) and also connected to one of the Ethernet interface of the DUT (to measure the wander generated). [G.8262] defines the requirement for an "ideal" input reference signal as well as the amount of output wander that can be generated by the DUT. The output wander generated by the DUT is measured by the tester (TIE, MTIE and TDEV metrics) and compared to the MTIE and TDEV conformance masks shown in Figure 7.7. For example, the MTIE and TDEV must be within 113 and 6.4 ns, respectively, for an observation interval of 1,000 s.

298 Synchronous Ethernet and IEEE 1588 in Telecommunications

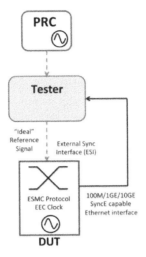

Figure 7.6. *Measurement of output wander*

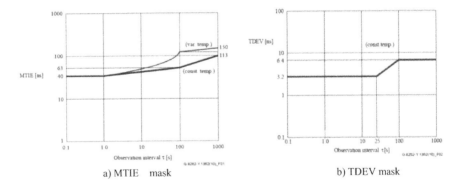

Figure 7.7. *Conformance masks requirement for output wander*

7.5.1.4. *Wander transfer*

This test scenario verifies the ability of the SyncE node to filter some of the noise that is present on an incoming Ethernet interface that is SyncE capable. Figure 7.8 describes the test bed. Although the standard is lacking details on how to properly execute such a test, several test equipment vendors offer noise sources that can be modulated on the input signal. The noise source, for example, consists of specifying the amplitude and frequency of a certain signal type (e.g. sinusoidal pattern) and then measuring the MTIE and TDEV of an egress Ethernet interface that is SyncE capable, or measuring how much of the input noise (amplitude and frequency) is transferred to the output. One way to perform this test is to use the limits for wander tolerance (section 7.5.1.5) defined in [G.8262], section 9.1, Table 9. The test consists of generating various sinusoidal patterns each with a specific amplitude and wander frequency as listed in Table 9. This allows us to determine how much attenuation or amplification is done on the input signal and what the corresponding filter bandwidth of the EEC is.

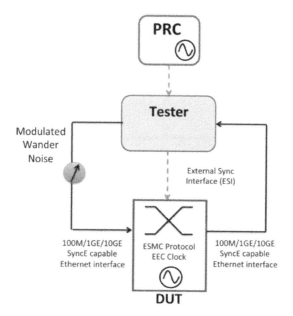

Figure 7.8. *Measurement of wander transfer*

7.5.1.5. Wander tolerance

This test scenario verifies the ability of the SyncE node to accept an incoming reference signal having a minimum amount of wander noise coming from an Ethernet interface that is SyncE capable. Figure 7.9 describes the test bed. [G.8262] defines the amount of wander noise (which corresponds to the SyncE network limits) that the node must tolerate and provides sinusoidal test signals that can be used for the purpose of testing. The modulated wander noise is specified in terms of MTIE and TDEV masks and corresponds to the SyncE network limits. The node should accept the reference signal without causing any alarms, or causing the clock to switch to another reference signal, or causing the clock to go into holdover. The verification is done by accessing the CLI or management interface of the DUT and observing any alarms or related change in the system clock status.

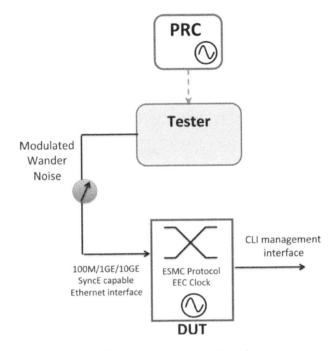

Figure 7.9. *Measurement of wander tolerance*

7.5.1.6. *Short-term transient response*

This test scenario verifies the ability of the SyncE node to limit the amount of phase error on an output SyncE interface when the input SyncE interface is lost and a second reference signal traceable to the same PRC is available. Figure 7.10 describes the test bed. The test scenario shows the tester providing two input reference signals to the DUT; one reference signal coming from an Ethernet interface that is SyncE capable and another via the ESI of the DUT. To measure the amount of phase error, the reference signal currently providing the system clock to the DUT is disconnected. According to [G.8262], the reference could be lost for at most 15 s and the maximum phase error allowed during this time period is shown in Figure 7.11. The time period between the loss of the first reference signal and the lock to a second reference signal is directly related to the time it takes for the synchronization network to rearrange itself with the use of the ESMC protocol. However, the compliance mask shows a maximum phase error of 1 µs, which corresponds to the entire chain of EECs and not to a single SyncE node. When testing for a single SyncE node, two values of the masks are important to measure and verify. The first corresponds to the 120 ns phase jump when entering and leaving holdover, and the second corresponds to the 0.05 ppm slope for the holdover period after the first phase jump.

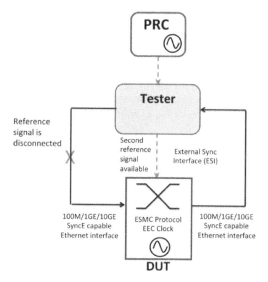

Figure 7.10. *Measurement of short-term transient*

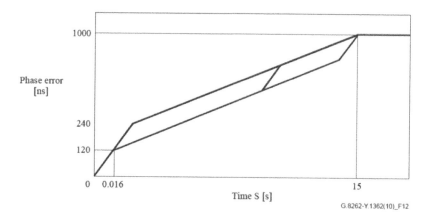

Figure 7.11. *Compliance mask for reference switching*

7.5.1.7. *Holdover*

This test scenario verifies the maximum phase error produced by the SyncE node when it loses its reference signal and another reference signal is not available for an extended period of time. The definition of holdover was introduced in Chapter 2. Figure 7.12 describes the test bed. The test is started by providing a reference signal to the DUT, for example an ingress Ethernet interface that is SyncE capable, and measuring the TIE and MTIE to verify that the DUT is operating in normal operating condition. The Ethernet interface is then disconnected from the DUT, for example after a period of 30 min, and the system clock is expected to change its status from normal locked mode to holdover mode. The change in status can be verified through the CLI or management interface of the DUT. The MTIE is then compared with the mask shown Figure 7.13. It is important to note that several of the devices on the market today can achieve performance that is far below the mask, primarily due to the internal oscillator used in the DUT.

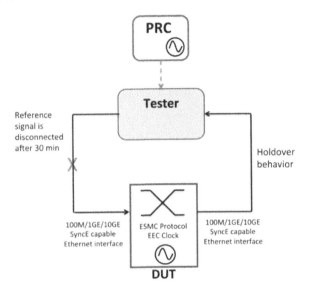

Figure 7.12. *Measurement of holdover*

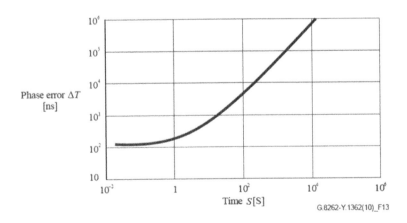

Figure 7.13. *Compliance mask for holdover operation*

7.5.1.8. *Wander accumulation of a chain of SyncE node*

All the previous test scenarios dealt with the performance of a single SyncE node. This test scenario verifies the accumulation of wander as the reference signal propagates through a chain of SyncE nodes. The definition, diagram and details of a reference chain as defined in ITU-T Recommendation G.803 [G.803] was presented in Chapter 2 (section 2.1) and was also discussed in the context of SyncE deployments in Chapter 4 (section 4.3.2). [G.803] specifies the details of the worst-case synchronization chain for the purpose of developing the performance requirements related to wander accumulation. As discussed in the sections mentioned above, the tester measures the TIE at the last SyncE node of the chain. The MTIE and TDEV metrics are also computed and compared with the network limits defined in [G.8261] as shown in Figure 7.14. This test scenario can be lengthy to setup and to verify in a lab environment since it requires configuring a full chain of SyncE nodes or, at a minimum, a chain that is representative of service provider deployments. The measurement time is also conducted for a period of at least 1 day. For practical reasons, the test and measurement of SyncE in lab environments will typically use a chain of SyncE that is smaller than the chain defined in [G.803].

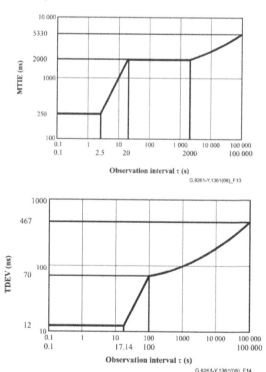

Figure 7.14. *Compliance mask for wander accumulation*

7.5.2. Testing the ESMC protocol

The ESMC protocol, as explained in Chapter 2, section 2.2.1, is the protocol used for distributing clock quality information along the chain of SyncE nodes and to prevent timing loops in arbitrary network topologies. This section provides the reader with a set of standard and baseline test scenarios and is by no means exhaustive. Testing of the ESMC protocol requires verifying the following aspects:

– ESMC frame format as defined in [G.8264].

– Verifying the reception, processing and generation of ESMC messages and SSM codes.

– Reference signal selection and switching.

– Negative testing under the presence of non-conformant ESMC frames.

7.5.2.1. Testing the ESMC frame format

This test scenario verifies the fields of the ESMC frame format. The frame format was presented in Chapter 2, section 2.2.1.1.1. Verifying each field (nibbles and bytes) is essential to ensure conformance and proper interoperability of the protocol between SyncE nodes. The ESMC protocol, although defined by ITU-T, belongs to the family of IEEE slow protocols since it shares the same MAC destination address 01:80:C2:00:00:02 and Ethertype 8809 with other IEEE slow protocols such as Link Operation, Administration, Maintenance (OAM) and Link Aggregation Control Protocol (LACP). The subtype 0A differentiates the ESMC protocol from all other slow protocols.

Figure 7.15 describes the test bed. The tester is connected to the DUT and provides a valid SyncE signal, that is a reference signal that is traceable to PRC as well as valid ESMC messages. The DUT then generates a SyncE signal on one of the Ethernet interfaces, which includes ESMC messages that represent the clock quality of the EEC. The tester captures the ESMC messages and each field is decoded and analyzed. The fields verified are listed in Table 7.3. In addition to the fields, the minimum and maximum message rates and Ethernet frame size can also be verified.

At the time of writing this book, the only specified information carried in the data field of the ESMC message is the QL, specified as a type length value (TLV) format. It is however not prohibited by the standard to add proprietary information to the ESMC message, and verifying the frame size is one way to know if additional information might be added beyond the QL information. The verification of the QL TLV format information involves decoding the QL TLV format and verifying the values as specified in Table 7.4.

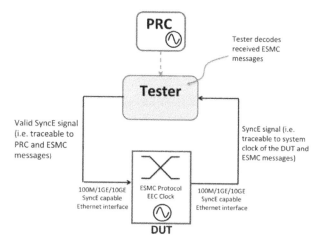

Figure 7.15. *Verification of ESMC frame format*

ESMC field	Expected value
MAC DA	01:80:C2:00:00:02
MAC SA	Valid source address of the transmitting port (i.e. of the tester)
Ethertype	0×8809
Subtype	0×0A
ITU-OUI	0×0019A7
ITU subtype	0×0001
Version	0×1
Flag	0×1 or 0×0
Reserved	
Reserved	0×000000
Data	QL-TLV (verified below)
Frame check sequence (FCS)	Valid 4 byte FCS

Table 7.3. *Ethernet ESMC frame format*

Field	Expected value
Type	0×01
Length	0×0004
Reserved	0×0
SSM code	Valid SSM code defined in ITU-T Recommendation G.781 [G.781]

Table 7.4. *ESMC QL TLV format*

7.5.2.2. Testing the SSM QL codes

This test scenario verifies the SSM QL code sent by a SyncE node. Figure 7.16 describes the test bed. The tester is sending ESMC messages with a specific SSM code to an Ethernet interface of the DUT and the same tester is receiving ESMC messages from another Ethernet interface of the DUT. This test consists of analyzing the SSM code of the ESMC message generated by the DUT when a specific SSM code is present at the input interface of the DUT. Depending on the internal quality of the oscillator (e.g. QL-EEC quality) in the SyncE node, the output SSM code should be the same as the input SSM code, unless the SyncE node is in a state such as holdover. The procedure consists of sending ESMC messages with SSM code QL-PRC and verifying that the ESMC messages transmitted by the DUT to the tester contains the QL-PRC code. The procedure is repeated using QL-SSU code and QL-EEC code. Once the QL value being sent by the tester falls below the QL value of the DUT, the output QL value should stay and reflect the internal quality of the oscillator (e.g. QL-EEC).

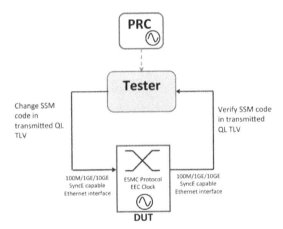

Figure 7.16. *Verification of SSM QL code*

7.5.2.3. Testing the SSM QL reception, generation and reference selection

A SyncE node receiving ESMC frames on more than two ingress Ethernet interfaces will use the SSM QL code and its internal clock quality to select the best reference signal. Figure 7.17(a) describes the first test bed. The tester is connected to the DUT via three Ethernet ports. The tester is transmitting ESMC frames to the DUT with SSM code QL-PRC on the first port and QL-SSU on the second and third port. The tester verifies that the received SSM code from the DUT on the second and third port is equal to QL-PRC (since the SyncE node will select the reference from the first port due to highest clock quality) and equal to QL-DNU on the first port (since that port has been selected as the reference, and such port needs to advertise DNU in order to prevent timing loops). It can also be verified through the clock management interface that the DUT selected the reference signal on the first port. The procedure is then repeated by sending QL-PRC on the second port and QL-SSU on the first and third port, and verifying that the second port will be receiving QL-DNU from the DUT and the first and third port will be receiving QL-PRC from the DUT. Figure 7.17(b) describes the second test bed. Both test scenario provide a minimum verification that the DUT is capable of properly receiving and generating appropriate SSM codes as well as properly selecting the reference signal with the highest quality.

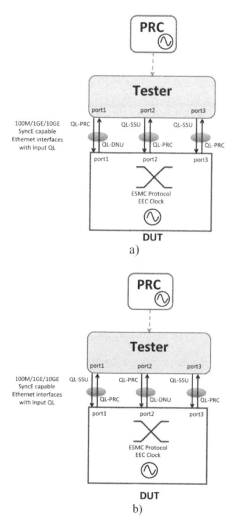

Figure 7.17. *Verification of SSM QL reception and generation*

Another aspect of this test consists of verifying that the DUT will properly select another reference (if one is available) when the current selected reference is lost for at least 5 s, that is when ESMC messages have not been received for at least 5 s. The procedure is based on Figure 7.17(a) and consists of sending QL-PRC on the first port and QL-SSU on the second and third port and verifying that the first port is selected as the reference signal. The first port is then disconnected and after a period of time the DUT should select the reference signal from the second port (assuming that port 2 has higher priority than port 3). The DUT is then expected to send QL-DNU to the tester on the second port and send QL-SSU to the tester on the third port.

7.6. Testing the IEEE 1588 end-to-end telecom profile

The IEEE 1588 end-to-end telecom profile defined in ITU-T Recommendation 8265.1 [G.8265.1] and described in Chapter 2, section 2.5.1, and Chapter 4, section 4.3.3, is only defined for the transfer of frequency. The profile is based on the IEEE 1588 PTP protocol [1588-2008] and all communication as well as the transfer of frequency is done between a packet master and packet slave. In contrast to SyncE, the network that interconnects the packet master to the packet slave are packet switches and routers and do not provide any synchronization assistance (i.e. no boundary clock or transparent clock support). Testing the telecom profile mainly consists of three parts:

– The first is to verify the PTP protocol communication between the packet master and packet slave in order to ensure a high level of interoperability between equipment.

– The second is related to the measurement of the PDV produced by a single packet switch or router and the PDV produced by a chain of packet switches or routers, and verifying that the PDV is within the requirements of PDV network limits. Chapter 4 provides a detailed explanation of PDV network limits.

– The third is to verify the tolerance and performance of the packet slave clock by recovering a frequency under the presence of various impairments and to comply with the performance requirement of a specific application (e.g. base station air interface synchronization).

7.6.1. *Testing the telecom profile – protocol*

The conformance to the PTP protocol is a key element to be verified in order to ensure interoperability between various equipment implementations. The PTP protocol requirements defined in IEEE 1588-2008 must be met as well as all other requirements defined in ITU-T Recommendation G.8265 [G.8265] and [G.8265.1].

Non-conformance to these standards is likely to result in interoperability issues between a packet master and packet slave. This section presents a number of baseline test scenarios and is by no means exhaustive. It provides some of the protocol-related functions that need to be tested:

– one-way and two-way PTP message transfer;

– one-step and two-step clocks message transfer;

– unicast negotiation and message formats;

– PTP mapping (User Datagram Protocol – UDP/IP);

– message transmission rates;

– clock traceability;

– alternate best master clock algorithm (A-BMCA);

– protection and redundancy functions.

Two test beds for testing the telecom profile are shown in Figures 7.18 and 7.19. Figure 7.18 is used to test a packet master and Figure 7.19 is used to test a packet slave. The test beds show the DUT (the packet master or packet slave) and the use of packet master emulators or packet slave emulators. The emulators are telecom profile PTP compliant devices and their primary role is to act as real devices. The emulators are used to stimulate the DUT and are also used to collect and analyze PTP messages sent by the DUT. The emulators are also capable of performing PDV measurements and measuring various clock signals. The switch is a layer 2 Ethernet switch connecting the emulators to the DUT and does not generate any packet network impairments (e.g. packet loss or delay variation). The PRC is used to provide a reference signal to the packet master and is also used as a reference for PDV measurements or output clock wander measurement.

7.6.1.1. *Testing message formats*

This test scenario verifies the format of PTP messages sent by a packet master or packet slave DUT. There are several PTP message types being exchanged between a packet master and packet slave and each message has a large number of fields. Figure 7.20 shows the "SIGNALING Request of Sync" message sent by a packet slave under test and received by the packet master emulator, as well as the "SIGNALING Grant of Sync message" and the "SYNC" messages sent by a packet master under test and received by the packet slave emulator.

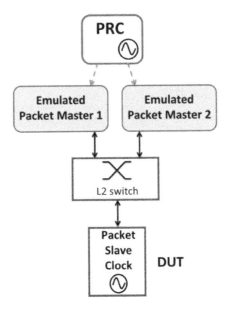

Figure 7.18. *Test bed #1 to verify packet master*

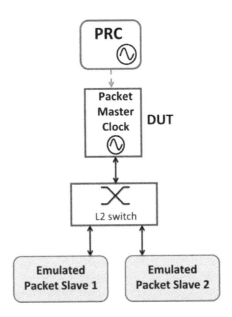

Figure 7.19. *Test bed #2 to verify packet slave*

Each PTP message consists of a common message header part (34 bytes) containing information that can be decoded and verified. In this example, each PTP message fields are defined in [1588-2008], Table 18 (common message header), Table 33 (signaling message fields), Table 73 (REQUEST_UNICAST_ TRANSMISSION TLV format), Table 74 (GRANT_UNICAST_TRANSMISSION TLV format) and Table 26 (Sync and Delay_Req message fields).

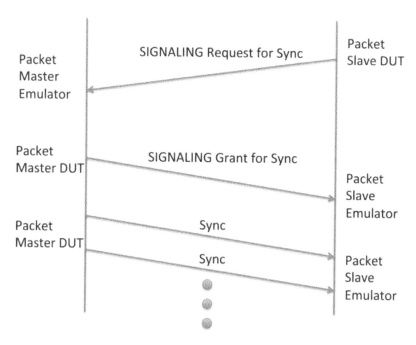

Figure 7.20. *Example of PTP unicast negotiation and message exchange*

Test bed #2 is used to verify the "SIGNALING Request of Sync messages" sent by the packet slave (i.e. the slave is requesting the packet master emulator to provide Sync messages at a certain rate). The packet master emulator receives the message and analyzes each field. Table 7.5 shows each field and the expected value based on the requirements defined in [1588-2008] and [G.8265.1].

PTP message fields	Expected value
transportSpecific	0
messageType	C for SIGNALING message (hex)
reserved	0
versionPTP	2
messageLength	54 (single TLV)
domainNumber	value between 4 and 23
reserved	0
flagField	set to FALSE expect for bit 2 of octet 0
correctionField	0
reserved	0
sourcePortIdentity	clockIdentity of the packet slave, portIdentity set to 1
sequenceID	one greater than the sequenceID of the previous signaling message
controlField	undefined in telecom profile
logMessageInterval	7F (hex)
targetPortIdentity	all 1s or equal to the clockIdentify of the packet master emulator
tlvType	0004 (hex)
lengthField	6
messageType	0 for SYNC message (hex)
reserved	0
logInterMessagePeriod	requested mean period with a value between -7 (128 messages per second) and 4 (one message per 16 s)
durationField	value between 60 and 1,000 s

Table 7.5. *Verification of the "SIGNALING Request for SYNC" message*

Test bed #1 is used to verify the "SIGNALING Grant of Sync" messages as well as the "SYNC" messages (i.e. the master grants the request of the packet slave emulator and starts sending SYNC messages at the negotiated rate). The packet slave emulator receives the GRANT message and the SYNC messages and analyzes each field. Tables 7.6 and 7.7 shows each field and the expected value based on the requirements defined in the [1588-2008] and [G.8265.1].

PTP fields	Expected value
transportSpecific	0
messageType	C for SIGNALING message (hex)
reserved	0
versionPTP	2
messageLength	54 bytes
domainNumber	value between 4 and 23
reserved	0
flagField	set to FALSE expect for bit 2 of octet 0
correctionField	0
reserved	0
sourcePortIdentity	clockIdentity of the packet slave, portIdentity set to 1
sequenceID	one greater than the sequenceID of the previous signaling message
controlField	undefined in telecom profile
logMessageInterval	7F (hex)
targetPortIdentity	all 1s or equal to the clockIdentify of the packet master emulator
tlvType	0005 (hex)
lengthField	8
messageType	0 for SYNC message (hex)
reserved	0
logInterMessagePeriod	granted mean period with a value between –7 (128 messages per second) and 4 (one message per 16 s)
durationField	value between 60 and 1,000 s
Reserved	0

Table 7.6. *Verification of the "SIGNALING GRANT for SYNC" message*

7.6.1.2. *Testing message rates*

This test scenario verifies the rate of SYNC messages requested by a packet slave and also verifies the intermessage time of SYNC messages sent by a packet master. For example, the packet slave is requesting a rate of one message per second, which is within the allowed range of one PTP Sync messages every 16 s up to 128 PTP Sync messages every second defined in [G.8265.1]. Test bed #2 is used to verify the rate requested by the packet slave. This is done by capturing the "SIGNALING Grant of Sync" and verifying that logInterMessagePeriod value is set to 0. Test bed #1 is used to verify the intermessage times of the Sync messages sent by the packet master. The packet slave emulator performs time stamping of each received Sync message and calculates the interarrival between each pair of messages. This is typically done over a long period of time. The packet slave emulator then verifies if 90% of the interarrival times are within ±30% of the requested intermessage period, that is 90% of the values must be between 0.7 and 1.3 s in this example.

PTP fields	Expected value
transportSpecific	0
messageType	0 for SYNC message (hex)
reserved	0
versionPTP	2
messageLength	44 bytes
domainNumber	value between 4 and 23
reserved	0
flagField	set to FALSE expect for bit 2 of octet 0
correctionField	0
reserved	0
sourcePortIdentity	clockIdentity of the packet slave, portIdentity set to 1
sequenceID	one greater than the sequenceID of the previous signaling message
controlField	undefined in telecom profile
logMessageInterval	7F (hex)
originTimestamp	transmission timestamp of SYNC msg

Table 7.7. *Verification of the "SYNC" message*

7.6.1.3. *Testing alternate BMCA – master selection process*

This test scenario verifies the master selection process specified as part of the A-BMCA. Figure 7.21 shows the Slave DUT requesting PTP Announce messages from the two packet master emulators. The emulators grant the request and proceed to send Announce messages to the packet slave. packet master #1 sends Announce messages with a clockClass = 84 (PRC) and packet master #2 sends Announce messages with a clockClass = 90 (SSU-A). Through the use of the CLI or management interface of the DUT, the test verifies that the packet slave has selected packet master #1 (Step 1 in Figure 7.21). The packet master #1 and packet master #2 emulators change their clockClass to 90 and 84, respectively (Step 2 in Figure 7.21). Through the use of the CLI or management interface of the DUT, the test verifies that the Slave DUT has now selected packet master #2 (Step 3 in Figure 7.21) since the clockClass of the packet master #2 emulator is better than packet master #1 emulator. This verifies that the packet slave is capable of master selection based on the clockClass attribute.

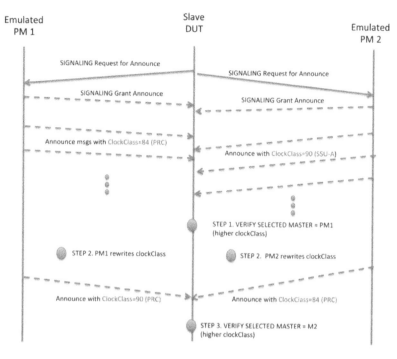

Figure 7.21. *Verification of the Alternate BMCA master selection process*

7.6.1.4. Testing alternate BMCA – packet timing signal fail condition

This test scenario verifies that a packet slave selects an alternate packet master emulator when the current selected packet master emulator becomes unavailable. Figure 7.22 shows the packet slave DUT requesting PTP Announce messages from the two packet master emulators. The packet master emulators grant the request and proceed to send Announce messages to the packet slave. Packet master #1 sends Announce messages with a clockClass = 84 and packet master #2 sends Announce messages with a clockClass = 90. Through the use of the CLI or management interface of the DUT, the test verifies that the packet slave has selected packet master #1 (Step 1 in Figure 7.22) and will synchronize with it. The packet master #1 emulator then stops sending Announce message to the packet slave (Step 2 in Figure 7.22). The packet slave will now lose traceability to the packet master #1 emulator, but is still receiving Announce messages from the packet master #2 emulator. After a certain time period known as the "announceReceiptTimeout", the packet slave will raise a Packet Timing Signal Fail (PTSF) condition called PTSF-lossAnnounce. Once the signal fail condition is raised, the packet slave will now proceed to select packet master #2 emulator (Step 3 in Figure 7.22). The packet slave is now expected to synchronize with this new packet master #2 emulator.

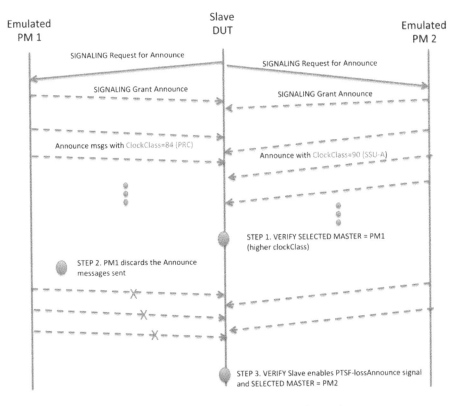

Figure 7.22. *Verification of Alternate BMCA packet timing signal fail condition*

7.6.2. *Testing the telecom profile – performance of packet networks*

This section provides information on testing and measuring packet networks for the purpose of frequency transfer between a packet master and a packet slave. As described in section 4.3.3.3, the performance of a chain of packet network equipment such as switches and routers (for the purpose of frequency transfer and clock recovery) can be defined by the level of PDV the chain generates. A chain of packet network elements producing no PDV can almost be regarded as a "wire" and provides the best condition to perform frequency transfer and clock recovery between a packet master and packet slave.

The characterization of PDV in a packet network can be performed by using what ITU-T defines "PDV metric". The objective of a PDV metric is to identify the PDV characteristics of a network that might impact the ability of a packet slave to generate an appropriate output clock. The Appendix I of [G.8260] contains definitions and properties of two types of "PDV metrics"; one class for the purpose of estimating packet slave clock performance and another class for studying the PDV characteristics of a packet network. Several PDV metrics, such as those defined in Appendix I of [G.8260] have been considered during the study activities of packet clocks but have not been fully agreed as a standard way to define the expected behavior of the packet clocks or the PDV that a packet network can produce. At the time of writing this book the only PDV metric used in a normative way by ITU-T for the purpose of frequency transfer based on the [G.8265.1] telecom profile has been the Floor Packet Percentage (FPP) metric.

The definition of FPP is given in Appendix I of [G.8260]. The FPP metric is used to study the population of packets traversing a network at or very near the observed floor delay of the network. These packets are often referred to as "lucky packets" and are typically those that experience delay close to the propagation delay of the network that connects the packet master and packet slave. Those that experience queuing or congestion are considered "unlucky packets". The FPP metric is a function of two essential parameters. The first is the window interval W representing a certain time period. The second is the fixed cluster range δ representing a value of delay above the minimum observed delay across a set of N samples. Figure 7.23 shows an example of the FPP metric parameters to study the population of packets within a cluster range starting at the observed floor delay.

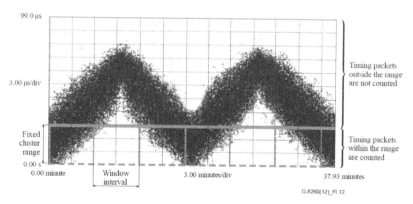

Figure 7.23. *Floor packet percentage parameters*

The FPP in [G.8260] is defined as:

$FPP(n, W, \delta) > p\%$

where the p parameter defines the network acceptance criterion or the proportion of packets that must at least meet the delay criterion δ in any given time period W. As discussed in section 4.3.3.3, ITU-T has specified values of $W = 200$ s, $\delta = 150$ μs and $p = 1\%$ for the purpose of frequency transfer between a packet master and packet slave. These values were obtained by studying the PDV across a chain of switches/routers in a lab environment, and represent the maximum permissible level of PDV expected at the output of a packet network that is consistent with the design rules of hypothetical reference model (HRM)-1 as defined in ITU-T Recommendation G.8261.1 [G.8261.1], and also presented in section 4.3.3.3 in Chapter 4. Any PDV values outside the fixed cluster range are not counted by the metric. This maximum permissible level of PDV is referred to as the "PDV network limit".

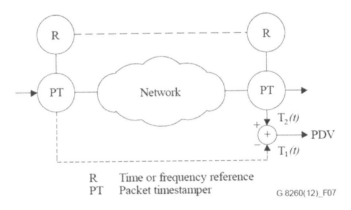

Figure 7.24. *Test bed to measure packet delay variation*

Figure 7.24, extracted from [G.8260], shows a typical test setup to conduct PDV measurement across a packet network. Any PDV measurement requires equipment that is capable of precise time stamping when a packet ingresses the network and egresses the network. These are defined as the packet timestamper (PT) function in the figure. A PT can emulate the sending of a PTP Sync message (or any other type of traffic) and must calculate the delay of a packet, based on the ingress and egress timestamps (T1 and T2 in the figure), as it goes through the network. For the purpose of frequency transfer, the variation in packet delay is sufficient and in such a case the PT device requires either a time of day or frequency reference at both ends of the network. These are defined as reference (R) in the figure. PDV measurements are impacted by the granularity of the time stamp as well as the reference signal used for time stamping, and equipment used for this purpose should minimize these errors in order to obtain a true PDV representation of the network. PDV measurements across a chain of packet networks such as switches and routers can be conducted in a lab environment or in a live environment and the pass or fail criteria is based on the parameters W, δ and p associated with the FPP metric. Test equipment with accurate time stamp, PTP traffic generation, external reference clock and graphical user interface displaying the packet delay distribution and parameterization of the FPP metric (as well as several other PDV metrics) are now commercially available.

7.6.3. Testing the telecom profile – performance of a PTP packet slave clock

Testing a PTP packet slave clock is not straightforward and has been a subject of discussion for almost a decade. The packet slave has to filter the PDV generated by the network (as discussed in the previous section) with the goal of recovering an output clock that meets a certain conformance mask. The ITU-T Recommendation G.8263 [G.8263] describes a functional model of a packet slave with two main functional blocks, as shown in Figure 7.25.

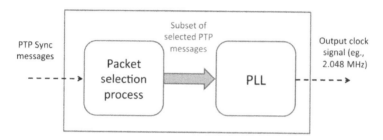

Figure 7.25. *Simplified functional model of a PTP packet slave clock*

In several implementations the packet selection process consists of selecting the fastest packets (those that have traversed the network with minimal delay), while other criteria might be considered in some specific implementations (e.g. selection window around the mean delay). The detection and removal of floor delay steps are also quite common in packet slave clock designs. The PLL block may, in some implementations, not be based on traditional PLL design but rather on a frequency-locked loop (FLL) design. An FLL has the objective of maintaining the frequency error of the output signal as close to zero while the phase error of the signal might not be bounded. This is acceptable for applications such as the air interface synchronization of frequency-based mobile base stations. The testing of tolerance of a PTP packet slave clock under the presence of PDV patterns implies to stress these two main functional blocks and their related features.

7.6.3.1. Testing packet slave clock – noise generation using an ideal input

This test scenario verifies the amount of wander produced at the output of the packet slave when the input is subject to an ideal stream of PTP Sync messages having no PDV (or at a minimum the PDV produced by the packet master itself). Figure 7.26 shows a packet master emulator (tester) and a packet slave connected through an Ethernet interface. As explained in section 7.6.1, the packet slave uses unicast negotiation to request SYNC messages from the packet master emulator. The packet master then generates an ideal input stream of PTP SYNC messages to the packet slave. The packet slave performs clock recovery and generates an output clock signal. After a stabilization period, the tester starts measuring the TIE generated by the packet slave and compares it with the conformance mask of [G.8263] as shown in Figure 7.27. The test is typically conducted in an environment having constant temperature; however, the requirements have also been defined for variable temperature environments.

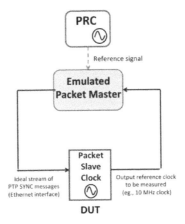

Figure 7.26. *Measurement of packet slave output wander*

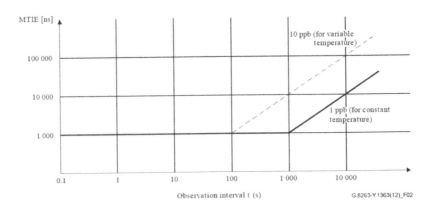

Figure 7.27. *Conformance mask of packet slave output wander*

7.6.3.2. Testing packet slave clock – holdover

This test scenario verifies the holdover capability of the packet slave when the input stream of PTP Sync messages are lost. Figure 7.28 shows the test setup and the lack of SYNC messages at the packet slave (the messages are discarded). As shown in section 7.6.1.4, a packet slave will raise a PTSF if it starts to lose some of the PTP frames. If there is no other available packet master, the packet slave will change its state from normal operating condition to holdover. During a signal fail condition, it is possible for a packet slave that does not provide good holdover capabilities to inform the end-user application (i.e. the application requiring the frequency) that the reference signal has been lost. This can be done by forcing the output signal to a frequency that is out of range or by raising an alarm or by changing the SSM code (e.g. from PRC to DNU) if the output signal supports SSM. These cases can be verified by some means but the holdover performance itself cannot be measured. If the packet slave does however support holdover, then the tester will measure the long-term holdover of the packet slave and will compare with the maximum permissible phase error during holdover defined in [G.8263] based on following, where S is the measurement period.

$$|\Delta x(S)| \leq [(10+1.0)S + 0.5 \times (1.16 \times 10^{-5})S^2 + 150]\text{ns}$$

As an example, the phase error (measured using MTIE) after 1,000 s must be less than 11.1558 µs.

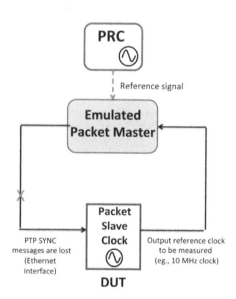

Figure 7.28. *Measurement of holdover*

7.6.3.3. *Testing packet slave clock – tolerance to input PDV patterns*

This test scenario verifies the amount of wander produced on the output clock of the packet slave when the input is subject to a stream of PTP Sync packets having a minimum level of PDV. The minimum level of PDV to be tolerated by the packet slave must at least meet the PDV network limit defined in section 8 of [G.8261.1], and also described in section 4.3.3.3.1 in Chapter 4. Figure 7.29 shows a packet master emulator sending a stream of PTP Sync messages with a predefined PDV pattern that meets the PDV network limit. The packet master emulator measures the output clock wander generated by the packet slave clock.

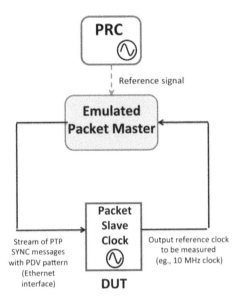

Figure 7.29. *Measurement of output wander in the presence of an input PDV pattern*

A packet slave output clock must meet the conformance mask defined in [G.8263], shown in Figure 7.30, when the PDV network limit has at least 1% of the PTP Sync packets sent by the packet master within a 150 μs fixed cluster range starting at the floor delay in every observation window of 200 s. This conformance mask is typically used for applications such as mobile base stations, where the packet slave is co-located with the base station but provides an output clock to it. In the case where the packet slave is embedded inside the base station, then the air interface synchronization requirement of ±50 parts per billion applies. Chapter 4, section 4.3.3.2.2, provides further details on the location of the packet slave and the applicability of the conformance mask.

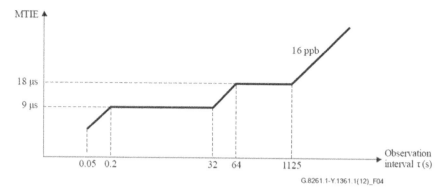

Figure 7.30. *Conformance mask of packet slave clock in normal operating condition*

Many discussions at ITU-T have taken place with regard to the generation of appropriate PDV patterns for the purpose of testing the packet slave clocks. No matter what the type of PDV pattern, it should be mentioned that a PTP packet slave clock is expected normally to tolerate any PDV pattern compliant to the PDV network limits in [G.8261.1], while complying at the same time with the expected performance objective at its output clock interface. The difficulty is in defining suitable test PDV patterns that will generate the PDV network limit, as there are many ways to do so from simple to quite complex. In principle, the delay sequences of the PDV pattern are generated by means of a statistical model, with the parameters chosen to generate a delay distribution with properties similar to the PDV network limits defined earlier. Several methodologies for generating suitable PDV patterns are possible and some commercial devices now offer such patterns. For instance, PDV patterns based on sinusoidal waveforms can be used to generate low-frequency components, where the waveform is characterized by an amplitude parameter and a period parameter. The amplitude parameter can be configured between 0 and 150 μs while the period parameter can be configured from 200 s up to much larger values. These parameters are used to generate PDV patterns that meet the PDV network limit and can be used as input into the packet slave in order to measure the amount of wander produced on the output clock signal and to verify that the conformance mask is respected.

7.6.3.4. Testing packet slave clock – scenarios based on G.8261 Appendix VI

This section briefly discusses the use of the ITU-T Recommendation G.8261 Appendix VI [G.8261] test bed topology for the purpose of evaluating the transfer of frequency across a packet network. It also presents a few test scenarios based on Appendix VI. The test topology was initially defined for the verification of circuit emulation services and adaptive clock recovery techniques, but also later extended to test packet slave clocks that support IEEE 1588 and the telecom profile as defined in [G.8265.1] and [G.8263]. Some of the tests in the appendix have been defined for Slaves using a one-way protocol or two-way protocol. However, it is important to mention that this appendix is not normative and was primarily specified for benchmarking the performance of a packet slave and not testing the tolerance of a packet slave. The appendix provides cautionary words on its use and the statement is given below:

> "Results from the test cases provided in this appendix provide no guarantee that equipment will perform as expected in a complex network situation under a range of complex and changing load conditions. Although test cases in this appendix provide a useful guidance on the performance of Ethernet-based circuit emulation techniques, evaluation in complex network scenarios that mimic the deployment profile is strongly recommended."

The test bed is shown in Figure 7.31 and is primarily composed of 10 Gb Ethernet switches or 9 Gb Ethernet and one fast Ethernet switch. IP routers or MPLS switches can replace the Ethernet switches. Interworking functions (IWF) such as circuit emulation service (CES) devices are placed at both ends of the topology. The topology also shows the use of a traffic generator connected to each switch, where it is configured to send background traffic into each switch and to some extent impair the timing packets sent between the IWF devices. The network traffic models are also defined in Appendix VI. There are however are no guidelines on the type of switches to be used, how to configure the switches or the traffic management mechanisms (e.g. maximum transmission unit (MTU) size, traffic classification, bandwidth profile per priority code points and traffic schedulers), the protection mechanisms or how to configure the services such as Ethernet private lines (EPL) that are carrying the timing packets across the network. These are aspects that are representative of real deployments and ones that can impact the transfer of frequency. One of the disadvantages of such topology is that it does not offer predictability and repeatability when it comes to testing. Alternatively there are devices available that can replace the entire test topology of switches and are often called packet network impairment devices or network PDV emulators. The background traffic generated by such devices are based on PDV patterns obtained from live network measurements or based on PDV patterns that are generated by a statistical model. Various impairments such as loss and delay can be also injected in the traffic. These devices reproduce PDV patterns that would be found at Reference point #2 in the figure and allow greater flexibility and repeatability on how to generate PDV patterns but are certainly not better than making test and measurements on a live commercial network. Finally, the test topology shows equipment that is used for wander and frequency measurements of the output clock signal of the IWF (Reference point #3 in the figure) and the results should be compared with the relevant application requirement and conformance mask. In the case of packet slave clocks based on the telecom profile, the conformance mask shown in section 7.6.3.3 would apply.

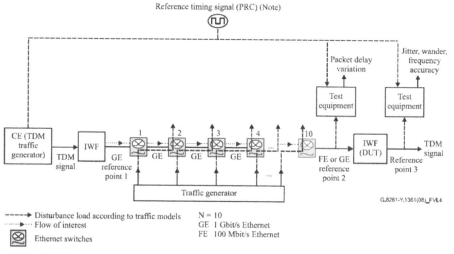

Figure 7.31. *ITU-T G.8261 test topology for packet slave clock recovery*

Appendix VI in [G.8261] has defined several test scenarios that are applicable to the test and measurement of packet slave clock recovery. These scenarios are based on one-way frequency transfer or two-way frequency transfer. The information below shows some of the basic test scenarios and although not exhaustive, the reader should refer to Appendix VI in [G.8261] for more information. The first scenario is shown in Figure 7.32 and is referred in [G.8261] as test case #2. This test consists of generating background traffic across the test topology where the offered traffic load is varied from 80 to 20% of the link speed. This test verifies the ability of the packet slave to handle sudden traffic load changes or sharp changes in the PDV. The output clock is measured and verified against the appropriate conformance mask.

The second scenario is shown in Figure 7.33 and is referred in [G.8261] as test case #3. This test consists of generating background traffic across the test topology where the offered load is slowly varied from 20 to 80% and back down to 20% over an extended period of time such as 24 h. This test verifies the ability of the packet slave to handle low-frequency change in the PDV. The output clock is measured and verified against the appropriate conformance mask.

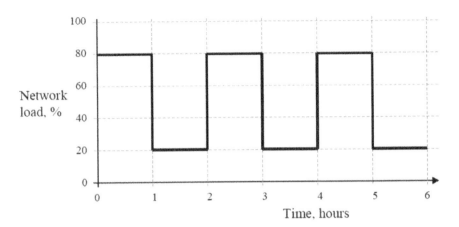

Figure 7.32. *Sudden network disturbance load modulation*

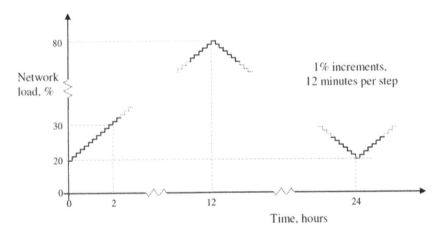

Figure 7.33. *Slow network disturbance load modulation*

7.7. Bibliography

[802.3] IEEE Std 802.3™-2008, Carrier sense multiple access with collision detection (CSMA/CD) access method and physical layer specifications, IEEE Computer Society, December 2008.

[1588-2008] IEEE Std 1588TM-2008, IEEE Standard for a precision clock synchronization protocol for networked measurement and control systems, IEEE Instrumentation and Measurement Society, July 2008.

[G.703] ITU-T Recc. G.703), Physical/electrical characteristics of hierarchical digital interfaces, 2001.

[G.781] ITU-T Recc. G.781, Synchronization layer functions, 2008.

[G.803] ITU-T Recc. G.803, Architecture of transport networks based on the synchronous digital hierarchy (SDH), 2000.

[G.810] ITU-T Recc. G.810, Definitions and terminology for synchronization networks, 1996.

[G.823] ITU-T Recc. G.823, The control of jitter and wander within digital networks which are based on the 2048 kbit/s hierarchy, 2000.

[G.8260] ITU-T Recc. G.8260, Definitions and terminology for synchronization in packet networks, 2012.

[G.8261] ITU-T Recc. G.8261, Timing and synchronization aspects in packet networks, 2008; see also Amd.1 (2010).

[G.8261.1] ITU-T Recc. G.8261.1, Packet delay variation network limits applicable to packet based methods (Frequency synchronization), 2012.

[G.8262] ITU-T Recc. G.8262, Timing characteristics of a synchronous Ethernet equipment slave clock (EEC), 2010; see also Amd.1 (2012) and Amd.2 (2012).

[G.8263] ITU-T Recc. G.8263, Timing characteristics of packet based equipment clocks (PEC) and packet based service clocks (PSC), 2012.

[G.8264] ITU-T Recc. G.8264, Timing distribution through packet networks, 2008; see also Amd.1 (2010) and Amd.2 (2012).

[G.8265] ITU-T Recc. G.8265, Architecture and requirements for packet based frequency delivery, 2010.

[G.8265.1] ITU-T Recc. G.8265.1, Precision time protocol telecom profile for frequency synchronization, 2010; see also Amd.1 (2011) and Amd.2 (2012).

[G.8271] ITU-T Recc. G.8271, Time and phase synchronization aspects of packet networks, 2012.

Appendix 1

Standards in Telecom Packet Networks Using Synchronous Ethernet and/or IEEE 1588

A1.1. Introduction

One of the main purposes of a standard is to provide appropriate description of functional requirements in order to allow implementations to be developed independently. Indeed, a key goal of any standards body is to produce a standard with sufficient detail so that it will guarantee interoperability. This is, of course, the goal, but often a difficult task, depending on the complexity of the issue. In any standard, the document produced represents the contributions of many individuals and often reflects the collected knowledge of that group. When a topic spans the domain of multiple standards development organizations (SDOs), the challenge of interoperability increases. Cooperation between SDOs is critical.

ITU-T Synchronous Ethernet (SyncE) and IEEE 1588 are good examples of cooperation between different standards committees, the Institute of Electrical and Electronics Engineers (IEEE) and International Telecommunication Union – Telecom (ITU-T).

SyncE, specified by ITU-T, added a network layer to Ethernet specifications done by the IEEE.

IEEE 1588-2008 was developed by the IEEE as a precision time protocol. ITU-T has developed an IEEE 1588 telecom profile for the transport of frequency in telecommunication networks. It has also specified the performance aspects and all the related requirement aspects such as network architecture and clock specifications.

ITU-T is currently developing IEEE 1588 telecom profiles for the transport of phase and time in telecommunication networks.

A1.2. General content of ITU-T standards

The standardization of synchronization networks in ITU-T is based on several aspects. All clocks of the network have to deliver a synchronization signal with a quality acceptable by the applications receiving this signal, whatever their position in the synchronization network is.

Standards define the requirements for the network and the equipment levels.

A1.2.1. *Network level*

The network level provides the specification of network limits. It gives the maximum value of noise, jitter, wander, etc., that can be present in any point of a network. This is a very important point, as once the network requirements are defined, it allows us to define the input tolerance on jitter and wander of all types of equipment that can be implemented in the network.

The specification of network architecture is based on the definition of:

– hypothetic reference models (HRMs) that define chains of clocks in tandems (the synchronization network reference chain), number of pieces of equipment, etc.;

– protection and security aspects;

– specification of network performance that meets the requirements of applications, typically jitter and wander specifications.

Figure A1.1, which is based on the G.803 synchronous digital hierarchy (SDH) synchronization reference chain, shows an example of network limits. All equipment is itself compatible with its equipment specification, that is the primary reference clock (PRC) [G.811], synchronization supply unit (SSU) [G.812] and SDH equipment clock/(synchronous) Ethernet equipment clock (SEC/EEC) [G.813]/[G.8262]. The network limit for the synchronization network has been defined according to the application needs and the number of pieces of equipment in the reference chain. They have been defined so that the output of any clock of the network is compatible with the jitter and wander limits defined in [G.825].

Appendix 1 331

G.803 states that N equals 20 for the SEC chain, also applicable to EEC, and that the maximum number of clocks in the entire chain is 60 for the SECs and 10 for the SSUs.

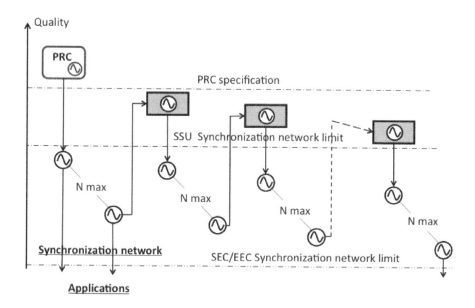

Figure A1.1. *Synchronization network limit*

A1.2.2. *Equipment level*

At the equipment level, the following specifications are relevant:

– the specification of equipment that can be put in telecommunication networks according to the architecture definition with compliance of the network performance;

– the input interface jitter and wander specification;

– the output interface jitter and wander specification;

– the synchronization status message (SSM) algorithm specification;

– the specification of clocks that can be put in tandem according to the network architecture rules;

– the specification of the quality of input signals that can be accepted by a clock (input noise tolerance);

– the specification of the transfer function of a clock (noise transfer);

– the specification of the output of the clock, in the absence of input noise (noise generation).

Figure A1.2 shows the list of characteristics that must be defined to specify a clock in a telecommunication network.

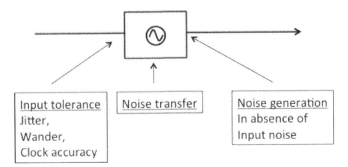

Figure A1.2. *Clock specification*

A1.2.3. *Use of network and equipment specification*

The equipment specification can be used to test a piece of equipment in a laboratory.

In a network, a piece of equipment will present an output level of noise higher than its own equipment specification since its input is degraded by the noise accumulated along the chain of equipment propagating the timing from the reference source. Figure A1.1 shows the three levels of synchronization interface quality that can be present at the output of a PRC, an SSU and an SEC.

It is reminded that the SSM gives only an indication of the quality of the source of the signal; for example, a synchronization interface at the output of an SEC with an SSM transporting a quality level of a PRC does not have to comply with the noise mask of a PRC.

A1.3. Summary of standards

The following sections provide several tables to help the reader find his/her way in the standard literature.

A1.3.1. Standards related to SyncE

The standardization of SyncE by ITU-T is based on the standardization of SDH for EEC-option 1 and synchronous optical network (SONET) for EEC-option 2; Table A1.1 summarizes the applicable recommendations for SyncE.

	ITU-T						
	G.803	G.8260	G.8261	G.8262	G.8264	G.781	O.174
Definitions		X					
General requirements	X		X				
Network architecture	X		X				
Functional model and ESMC					X	X	
Performance			X				
Clock				X			
Test equipment							X

Table A1.1. *Standards for SyncE*

Table A1.2 summarizes the recommendations that are applicable for SyncE architecture.

Recommendation	General description	Main requirements	Testing and network planning considerations
G.803 "Architecture of Transport Networks Based on SDH"	General SDH-based sync network (SyncE is based on the same architecture.)	Requirement on reference network chain	No specific testing. Reference network can be used in the SyncE network dimensioning
G.8261	Overall requirements for timing in packet networks including SyncE	SyncE architecture	No specific testing for SyncE

Table A1.2. *Architectural aspect*

Table A1.3 summarizes the recommendations that are applicable for SyncE network limits.

Recommendation	General description	Main requirements	Testing and network planning considerations
G.8261 "Timing and Synchronization Aspects in Packet Networks"	Overall requirements for timing in packet networks including SyncE (frequency sync)	SyncE network limits (jitter and wander)	

Table A1.3. *Network performance aspects*

Table A1.4 summarizes the clock and equipment specifications that are applicable for SyncE.

Recommendation	General description	Main requirements	Testing and equipment design considerations
G.781	Synchronization layer	Reference selection process SSM Handling	
G.8264	Overall requirements for SSM protocol for SyncE	ESMC	Protocol testing of the ESMC and SSM is described in Chapter 7 of this book.
G.8262	Synchronous Ethernet Equipment clock specification	Free running accuracy, holdover, noise generation, transfer and tolerance (jitter/wander)	EEC test is described in Chapter 7 of this book.

Table A1.4. *Clock and equipment specifications*

A1.3.2. Standards related to IEEE 1588 end-to-end telecom profile for frequency

Table A1.5 summarizes the recommendations/standards applicable for IEEE 1588 in telecommunication networks for frequency synchronization.

	ITU-T						IEEE
	G.8260	G.8261	G.8261.1	G.8263	G.8265	G.8265.1	1588
Definitions	X						
Protocol							X
Network Architecture		X			X		
Functional Model		X					
Performance			X				
Slave clock				X			
Protection					X	X	
Testing		X		X			
PTP Profile						X	X

Table A1.5. *Standards for frequency transport*

Table A1.6 summarizes the recommendations/standards for architectural and packet protocol aspects for IEEE 1588 end-to-end telecom profile for frequency synchronization.

Recommendation/Standard	General description	Main requirements	Testing and network planning considerations
IEEE 1588	IEEE 1588 Protocol and architecture specification	Protocol requirements	Protocol conformance requirements are specified.
G.8265	Packet timing architecture	Frequency distribution	
G.8265.1	Frequency sync profile	PTP modes, mapping Message rates BMCA	

Table A1.6. *Architectural and packet protocol aspects*

Table A1.7 summarizes the recommendations for network limits for the IEEE 1588 end-to-end telecom profile for frequency synchronization.

Recommendation	General description	Main requirements	Testing and network planning considerations
G.8261	Overall requirements for frequency sync over packet networks	Network limits (CES and packet timing)	Packet timing testing (G.8261 Appendix VI) – informative benchmarking
G.8261.1	HRM	PDV Network limits	

Table A1.7. *Network performance aspects*

Table A1.8 summarizes the recommendations for clock and equipment specification for the IEEE 1588 end-to-end telecom profile for frequency synchronization.

Recommendation	General description	Main requirements	Testing and equipment design considerations
IEEE 1588	IEEE 1588 Protocol and architecture specification	PTP clock functional models (from the protocol point of view)	
G.8261	Overall requirements for frequency sync over packet networks	CES requirements	Packet timing testing (G.8261 Appendix VI) – informative benchmarking
G.8263	Packet slave clock	Frequency accuracy, noise generation, PDV noise tolerance, holdover	
G.8265.1	IEEE 1588 profile	- SSM mapping into IEEE 1588 ClockClass - Protocol aspects to be supported by the slave clock	

Table A1.8. *Clock and equipment specification*

A1.3.3. *Standards related to IEEE 1588 full timing support telecom profile for phase and time transport*

Table A1.9 summarizes the recommendations/standards applicable for IEEE 1588 in telecommunication networks for phase and time synchronization.

At the time this book was written, not all of these recommendations were published and work was still in progress.

	ITU-T											IEEE
	G.8260	G.8271	G.8271.1	G.8272	G.8273	G.8273.1	G.8273.2	G.8273.3	G.8275	G.8275.x	G.SUP	1588
Definitions	X											
Protocol												X
Network architecture		X										
Functional model		X										
Performance			X									
PRTC				X								
Telecom Grand Master						X						
Telecom Boundary clock						X	X					
Telecom Transparent clock						X		X				
Slave Clock												
Protection									X	X		X
Testing		X			X							
profile										X		X
Simulation results											X	

Table A1.9. *Phase and time transport standards*

Table A1.10 summarizes the recommendations/standards for architectural and packet protocol aspects for IEEE 1588 full timing support telecom profile for phase and time synchronization.

Recommendation/Standard	General description	Main requirements	Testing and network planning considerations
IEEE 1588	IEEE 1588 architecture and protocol	PTP protocol	
G.8275 (Ongoing work)	Time sync architecture	Time and phase distribution	
G.8275.1 (Ongoing work)	Time sync profile	PTP modes, mapping Message rates BMCA	

Table A1.10. *Architectural and packet protocol aspects*

Table A1.11 summarizes the recommendations for network limits for IEEE 1588 full timing support telecom profile for phase and time synchronization.

Recommendation	General description	Main requirements	Testing and network planning considerations
G.8271	Overall network requirements	End-to-end time accuracy for various applications	Network requirement at the output of the end application
G.8271.1 (Ongoing work)	Network limits for time	Network limits in various points of the reference network	

Table A1.11. *Network performance aspects*

Table A1.12 summarizes the recommendations for clock and equipment specification for IEEE 1588 full timing support telecom profile for phase and time synchronization.

Recommendation	General description	Main requirements	Testing and equipment design considerations
IEEE 1588	IEEE 1588 architecture and protocol	PTP clock functional models (from the protocol point of view)	
G.8271	Overall network requirements	Time sync interface is the only equipment related requirement	
G.8272	PRTC Specification	Free running accuracy Noise generation Holdover	
G.8273 (Ongoing work)	Framework of Phase and Time clocks	Framework for G.8273.1, G.8273.2, and G.8273.3 clock recommendations	Annex of G.8273 describes testing of Time/Phase Clocks
G.8273.1 (Ongoing work)	T-GM Specification	Packet layer performance	Annex of G.8273 describes testing of Time/Phase Clocks
G.8273.2 (Ongoing work)	T-BC Specification	Physical and packet layer performance Time noise generation, time noise tolerance, time noise transfer, holdover	Annex of G.8273 describes testing of Time/Phase Clocks
G.8273.3 (Ongoing work)	T-TC Specification	Packet layer performance	Annex of G.8273 describes testing of Time/Phase Clocks

Table A1.12. *Clock and Equipment specification*

A1.4. Bibliography

[G.781] ITU-T Recommendation. G.781, Synchronization layer functions, 2008.

[G.803] ITU-T Recommendation. G.803, Architecture of transport networks based on the synchronous digital hierarchy (SDH), 2000.

[G.810] ITU-T Recommendation. G.810, Definitions and terminology for synchronization networks, 1996.

[G.811] ITU-T Recommendation. G.811, Timing characteristics of primary reference clocks, 1997.

[G.812] ITU-T Recommendation. G.812, Timing requirements of slave clocks suitable for use as node clocks in synchronization networks, 2004.

[G.813] ITU-T Recommendation. G.813, Timing characteristics of SDH equipment slave clocks (SEC), 2003.

[G.823] ITU-T Recommendation. G.823, The control of jitter and wander within digital networks which are based on the 2048 kbit/s hierarchy, 2000.

[G.824] ITU-T Recommendation. G.824, The control of jitter and wander within digital networks which are based on the 1544 kbit/s hierarchy, 2000.

[G.825] ITU-T Recommendation. G.825, The control of jitter and wander within digital networks which are based on the synchronous digital hierarchy (SDH), 2000.

[G.8260] ITU-T Recommendation. G.8260, Definitions and terminology for synchronization in packet networks, 2012.

[G.8261] ITU-T Recommendation. G.8261, Timing and Synchronization aspects in Packet Networks, 2008.

[G.8261.1] ITU-T Recommendation. G.8261.1, Packet Delay Variation Network Limits applicable to Packet Based Methods (Frequency Synchronization), 2012.

[G.8262] ITU-T Recommendation. G.8262, Timing characteristics of a synchronous Ethernet equipment slave clock (EEC), 2010.

[G.8263] ITU-T Recommendation. G.8263, Timing characteristics of packet based equipment clocks (PEC) and packet based service clocks (PSC), 2012.

[G.8264] ITU-T Recommendation. G.8264, Timing distribution through packet networks, 2008.

[G.8265] ITU-T Recommendation. G.8265, Architecture and requirements for packet based frequency delivery, 2010.

[G.8265.1] ITU-T Recommendation. G.8265.1, Precision time protocol telecom profile for frequency synchronization, 2010.

[G.8271] ITU-T Recommendation, G.8271, Time and Phase Synchronization Aspects in Packet Networks, 2012.

[G.8272] ITU-T Recommendation G.8272, Primary reference time clock, 2012.

[IO.174] ITU-T Recommendation ITU-T O.174, Jitter and wander measuring equipment for digital systems which are based on synchronous Ethernet technology, 2009.

[1588™-2008] IEEE Std 1588™-2008, Standard for a Precision Clock Synchronization Protocol for Networked Measurement and Control Systems.

Appendix 2

Jitter Estimation by Statistical Study (JESS) Metric Definition

A2.1. Mathematical definition of JESS

Jitter Estimation by Statistical Study (JESS) metric has been defined to analyze the Packet Delay Variation (PDV) generated by a piece of network equipment and to determine the number of nodes that can be cascaded between a Precision Time Protocol (PTP) packet master clock and a PTP packet slave clock.

The usage of this metric is introduced in Chapter 4 of this book, section 4.3.3.3.1. The mathematical definitions of this metric (JESS and JESS-w) are provided hereafter.

JESS metric is a function having three arguments (h, w, p) and resulting in a value n:

– h: the Probability Density Function (PDF) of a random variable defined over an interval [0, d_{max}], for example *estimated using the normalized histogram of a "stationary" PDV measurement (here, "stationary" means, for instance, that the level of load applied on the equipment is fixed)*;

– w: a positive number, for example *corresponding to the size of the floor delay window considered for the jitter estimation*;

– p: a probability, for example *corresponding to the minimum value considered for the jitter estimation of the probability that the relative delay suffered by a timing packet lies within the floor delay window of interest (or equally, the minimum percentage of packets to be considered in the floor delay window of interest)*;

– n: a positive integer, for example *this value provides an estimation of the maximum number of network nodes that can be cascaded between a PTP packet master clock and a PTP packet slave clock.*

The mathematical definition of JESS metric is as follows:

Let us define: $h_n = h \otimes h \otimes ... \otimes h$ (n times) [A2.1]

where $\otimes$ represents the convolution operator

Let us also define: $P_n = \int_0^w h_n(x)dx$ [A2.2]

which is the probability that the relative delay suffered by a timing packet after n nodes lies within the floor delay window.

JESS (h, w, p) is defined as the *maximum value of n such that* $P_n \geq p$ [A2.3]

An alternative way of defining the JESS metric is to use Fourier transforms. Let us define:

– $F(h)$ the Fourier transform of h;

– $F^{-1}(h)$ the inverse Fourier transform of h.

Since it is known that:

$$h_n = F^{-1}(F^n(h))$$ [A2.4]

P_n previously defined in [A2.2] can alternately be defined as follows:

$$P_n = \int_0^w F^{-1}(F^n(h))(x)dx$$ [A2.5]

equation [A2.3] of the JESS metric definition remaining the same.

A2.2. Mathematical definition of JESS-w

It is sometimes useful to estimate for a given histogram the minimum size of a floor delay window required to meet a percentage objective of timing packets within this floor delay window. In particular, this can be useful for chains mixing different types of nodes or links. JESS-w metric is defined for this purpose.

JESS-w is a function having two arguments (h, p) and resulting in a value w_{min}:

– h: the PDF of a random variable defined over an interval [0, d_{max}], for example *estimated using a normalized histogram corresponding to the mixed chain*;

– p: a probability, for example *corresponding to the minimum value considered for the jitter estimation of the probability that the relative delay suffered by a timing packet lies within the floor delay window of interest (or equally, the minimum percentage of packets to be considered in the floor delay window of interest)*;

– w_{min}: a positive number, for example *this value provides the minimum size of the floor delay window required for the PDF h to meet the objective of p% of timing packets within the floor delay window*.

The mathematical definition of the JESS-w is as follows:

$$JESS\text{-}w(h, p) = w_{min} \text{ such that } \int_0^{w_{min}} h(x)dx = p \qquad [A2.6]$$

Permissions and Credits

The authors would like to thank the following organizations for granting them permission to use or quote their publications.

European Telecommunication Standards Institute (ETSI)/Third-Generation Partnership Project (3GPP)

The following text on GSM requirements is reproduced from ETSI TS 145 010 with permission from ETSI:

> The BTS shall use a single frequency source of absolute accuracy better than 0.05 ppm for both RF frequency generation and clocking the time base. The same source shall be used for all carriers of the BTS. For the pico BTS class, the absolute accuracy requirement is relaxed to 0.1ppm.

Note, the following applies with respect to this copyright license:

© European Telecommunications Standards Institute 2007. Further use, modification, copy and/or distribution are strictly prohibited. ETSI standards are available at http://pda.etsi.org/pda/.

The following text on WCDMA requirement is reproduced from 3GPP TS 25.104 with permission from ETSI:

> The modulated carrier frequency of the BS shall be accurate to within the *following* accuracy range [...] observed over a period of one timeslot.

The following text on UMTS TDD requirements is reproduced from 3GPP TS 25.123 with permission from ETSI:

> maximum deviation in frame start times between any pair of cells on the same frequency that have overlapping coverage areas.

The following text on LTE TDD requirements is reproduced from 3GPP TS 36.133 with permission from ETSI.

> maximum absolute deviation in frame start timing between any pair of cells on the same frequency that have overlapping coverage areas.

Note, the following applies with respect to the ETSI copyright license on the above 3GPP Technical Specifications:

© 2013. 3GPP™ TSs and TRs are the property of ARIB, ATIS, CCSA, ETSI, TTA and TTC who jointly own the copyright in them. They are subject to further modifications and are therefore provided to you "as is" for information purposes only. Further use, is strictly prohibited.

Institute of Electrical and Electronics Engineers (IEEE)

The following text is reprinted with permission from IEEE 1588-2008 [1588-2008]: "Profile: The set of allowed Precision Time Protocol (PTP) features applicable to a device". Copyright IEEE. All rights reserved.

The following text is reprinted with permission from IEEE 1588-2008 [1588-2008]: "A logical grouping of clocks that synchronize to each other using the PTP protocol, but that are not necessarily synchronized to clocks in another domain." Copyright IEEE. All rights reserved.

Figure 2.20. PTP flow with Delay messages: adapted from IEEE 1588-2008 Figure 12 – Basic synchronization message exchange. Copyright IEEE. All rights reserved.

Figure 2.21. Offset and delay calculations with PTP timestamps: adapted from IEEE 1588-2008 Figure 34 – Delay request-response path length measurement. Copyright IEEE. All rights reserved.

Figure 2.25. Pdelay message flow: adapted from IEEE 1588-2008 Figure 35 – Peer delay link measurement. Copyright IEEE. All rights reserved.

Permissions and Credits   347

The following text is reprinted with permission from IEEE 1588-2008 [1588-2008]: "the node, system and communication properties necessary to support PTP". Copyright IEEE. All rights reserved.

The following text is reprinted with permission from IEEE 1588-2008 Interpretation Committee [1588-2008 IC]: "PTP does not attempt to change the behavior of the transport protocol". Copyright IEEE. All rights reserved.

The following text is reprinted with permission from a liaison from IEEE 802.1 Working Group to ITU-T Q13/15: "Rigorous adherence to layering principles is fundamental to the continued growth of networking". Copyright IEEE. All rights reserved.

International Telecommunication Union (ITU)

For Figure 1.3 – Source: Recommendation ITU-T G.823 (2000), The control of jitter and wander within digital networks which are based on the 2048 kbit/s hierarchy (Fig. 1)

For Figure 1.4 – Source: Recommendation ITU-T G.823 (2000), The control of jitter and wander within digital networks which are based on the 2048 kbit/s hierarchy (Fig. 10)

For Figure 1.8 – Based on Recommendation ITU-T G.8261 (2008), Timing and synchronization aspects in packet networks (Fig. 11)

For Figure 7.7 – Source: Recommendation ITU-T G.8262 (2010), Timing characteristics of a synchronous Ethernet equipment slave clock (Figs. 1 and 2)

For Figure 7.11 – Source: Recommendation ITU-T G.8262 (2010), Timing characteristics of a synchronous Ethernet equipment slave clock (Fig. 12)

For Figure 7.13 – Source: Recommendation ITU-T G.8262 (2010), Timing characteristics of a synchronous Ethernet equipment slave clock (Fig. 12)

For Figure 7.14 – Source: Recommendation ITU-T G.8261 (2008), Timing and synchronization aspects in packet networks (Figs. 13 and 14)

For Figure 7.23 – Source: Recommendation ITU-T G.8260 (2012), Definitions and terminology for synchronization in packet networks (Fig. I.12)

For Figure 7.24 – Source: Recommendation ITU-T G.8260 (2012), Definitions and terminology for synchronization in packet networks (Fig. 8)

For Figure 7.27 – Source: Recommendation ITU-T G.8263 (2012), Timing characteristics of packet-based equipment clocks (Fig. 2)

For Figure 7.30 – Source: Recommendation ITU-T G.8261.1 (2012), Packet delay variation network limits applicable to packet-based methods (Frequency synchronization) (Fig. 4)

For Figure 7.31 – Source: Recommendation ITU-T G.8261 (2008), Timing and synchronization aspects in packet networks (Fig.VI.4)

For Figure 7.32 – Source: Recommendation ITU-T G.8261 (2008), Timing and synchronization aspects in packet networks (Fig. VI.6)

For Figure 7.33 – Source: Recommendation ITU-T G.8261 (2008), Timing and synchronization aspects in packet networks (Fig. VI.7)

The following text sourced from Recommendation ITU-T G.8261 (2008), Timing and synchronization aspects in packet networks (Appendix VI) is reprinted with permission from ITU-T:

> Results from the test cases provided in this appendix provide no guarantee that equipment will perform as expected in a complex network situation under a range of complex and changing load conditions. Although test cases in this appendix provide a useful guidance on the performance of Ethernet-based circuit emulation techniques, evaluation in complex network scenarios that mimic the deployment profile is strongly recommended.

Others

The authors would like to thank Marc Weiss (NIST) for writing the Foreword, providing his expertise on reviewing Chapters 2 and 6 of the book, and for allowing us to use his text in the Introduction.

The authors would like to thank Tommy Cook (Calnex Solutions) for kindly providing the basic files for creating Figure 2.16.

Biography

Jean-Loup Ferrant

Jean-Loup Ferrant, after graduating from INPG Grenoble (France), joined Alcatel in 1975 and worked on analog systems, PCM and digital cross-connects. He has been working on SDH synchronization since 1990 and on SDH and OTN standardization for more than 20 years in ETSI TM1, TM3 and ITU-T SG13 and SG15. He has been the rapporteur of SG15 Q13 on network synchronization since 2001. He was one of the Alcatel-Lucent experts on synchronization in transport networks until he retired in March 2009. He is still the rapporteur of SG15 Q13, sponsored by Calnex Solutions, and an active member of the ITSF.

Mike Gilson

Mike Gilson is a technical specialist with BT Technology, Service and Operations (TSO) based at Adastral Park and is part of the team developing the strategy, architecture and design of BT's future time and frequency synchronization. He has a BA (honors) degree in business studies from the University of East Anglia and is a Member of the Institute of Engineering and Technology (MIET). He joined BT in 1983 and has worked on synchronization aspects since 1988. He has actively contributed to many standards bodies and groups on the subject of synchronization, most recently in ITU-T SG15.

Sébastien Jobert

Sébastien Jobert is one of the experts in Orange, participating in the definition of the strategy, architectures, and engineering related to time and frequency synchronization, QoS and performance of telecommunication networks. He received a masters degree from "Pierre et Marie Curie" University in Paris and joined France

Télécom Orange in 2006. He takes an active part in standardization: in ITU-T SG15 Q13 and SG12, where he is the editor, and in IETF on various working groups.

Michael Mayer

Michael Mayer has been actively contributing to ITU-T and COAST-SYNC standards since 1996 and is the editor of several ITU-T Recommendations related to synchronous Ethernet and IEEE1588. He began his career at Bell-Northern Research and Nortel Networks where he was involved with systems design and standardization of Nortel optical network products. Most recently, he was Ciena's expert in timing and synchronization. He is currently providing consulting services to carriers and network equipment vendors in the area of network synchronization and standardization. He holds a degree in electrical engineering from Queen's University, Kingston, Canada.

Laurent Montini

With an electrical engineering degree and 8 years of networking experience, Laurent Montini joined Cisco in January 1997 as a systems engineer working on early technologies such as ATM, IPv6 and IP multicast. In 2000, as consulting engineer, he introduced MPLS and pseudo-wires in mobile operator networks, prompting him in 2005 to actively participate in synchronization development in IEEE-1588 and ITU-T Q13/15 workgroups. As technical leader in Corporate Consulting Team since 2009, he leads synchronization over NGN expertise in Cisco.

Michel Ouellette

Michel joined Iometrix as vice president of engineering in 2011. He is responsible for equipment vendor and service provider conformance testing for Carrier Ethernet 2.0 services, mobile backhaul, IEEE1588 and SyncE. He worked at Huawei Technologies and Nortel Networks as a systems architect and project manager for 13 years. He attends ITU-T, IEEE and MEF standards. He is a senior member of the IEEE and has published more than 25 journal papers, granted 18 patents and received best paper award at ITSF and OPNETWORK. He holds an MASc and a BASc degree in electrical and computer engineering from the University of Ottawa and attended l'Ecole Nationale Supérieure des Télécommunications.

Silvana Rodrigues

Silvana Rodrigues is a director of system engineering at Integrated Device Technology (IDT). She graduated in electrical engineering from Campinas

University, Sao Paulo, Brazil. She started her career at the Telecommunication Research Center (CPqD, Telebras) in Brazil. She has been working with Network Synchronization for several years. She actively participates in several standards groups; she is the secretary of the IEEE-1588 standards committee and the editor of several ITU-T Recommendations (ITU-T G.8262, G.8263, G.8272 and G.8273.x).

Stefano Ruffini

Graduated in telecommunication engineering from the University of Rome "La Sapienza", Italy, Stefano Ruffini joined Ericsson in 1993. He is one of the Ericsson experts working on the mobile backhaul aspects and is involved in the definition of the equipment and network synchronization solutions. He has been working on synchronization aspects for approximately 20 years. He is currently actively contributing to ITU-T SG15 Q13 (serving as an associate rapporteur and editor) and to other relevant synchronization standardization bodies and forums.

Index

A

accuracy, 23
adaptive, 51, 92–94
   clock recovery (ACR), 92–94
   clocking, 68
   method, 51
alarm reporting, 246
algorithm, 90, 314–316
asymmetry, 228–230
attack, 258
augmentation system (LBAS, SBAS), 94, 95
authentication, 258, 270, 273
availability, 82, 102, 258, 273

B

base station, 159, 161, 175, 208–210
BMCA, 57, 78, 314–316
boundary clock, 59, 119
   distributed boundary clock, 234
   Telecom Boundary Clock (T-BC), 279

C

carrier operator, 165
CDMA, 26
CES, 82–86,
cesium, 97
clock, 94–105, 319–326
   master clock, 38, 203–207
   ordinary clock, 58
   packet-based clock, 54
   packet master clock, 203–207
   packet slave clock, 207–210
   reference clock, 174
   selection, 314, 315
   slave clock, 35, 38, 207–210
   Synchronous Equipment Clock, 9, 42, 43
   Synchronous Ethernet Equipment Clock, 121, 187, 294–303
combination, 139
COMPASS, 94, 95
congruency, congruent, 145

D

DCN, 251
differential, 50
   clock recovery, 86
   clocking, 84
   method, 84
DSL, 3, 210

E

EEC, 294–303
eLORAN, 102, 103
eMBMS, 26
encryption, 283
ESMC 44, 304–308
ETSI, 7, 17, 345, 346
external timing interface, 279

F

FDD, 24, 182
filtering, 53, 90
FCAPS, 241
FPP, 217, 317
frame format, 294, 304, 305
free running, 35, 37

G

Galileo, 94, 95, 164
GLONASS, 94, 95
GNSS 94–105, 221–225, 263–270
GPS, 94, 95, 164
Grandmaster, 144
GSM, 17

H

hold-in, 36
holdover, 194, 223, 224, 301, 302, 321

I

IEEE 1588 (end-to-end, node-by-node), 55–75, 196–220, 308–326, 335, 336
    end-to-end, 55–75, 196–220, 308–326, 335, 336
    node-to-node, 197, 198
integrity, 270, 273
ITU-T, 81, 82, 330–332

J

jamming, 222, 265–267
JESS, 341–343
jitter, 341–343
justification, 8

L

leased line, 185
linear chain, 124–127
LORAN (LORAN-C, eLORAN), 102, 103
LTE 159

M

make or buy, 161–182
    management (channel, network, plane), 233, 234, 240–245
    management channel, 233, 234
    management network, 240–245
    management plane, 240
mask, 298, 301, 302, 320, 323
master-slave, 35, 57
MBMS, 26
MBSFN, 26, 27
measurement, 40, 41
microwave system, 21
mitigation (mitigating), 270–274
mobile, 17–27, 158–161, 170–174
    mobile backhaul, 17–27
    Mobile Network Operator (MNO), 171–173
    mobile networks, 17–27, 158–161, 170–174
MRTIE, 291
MTIE, 289–291
MTOSI, 251, 252

Index 355

N

network, 3–27, 33–41, 111–117, 120–122, 130, 131, 152–182, 240–245, 250–252, 259–261, 316–319, 330–332
  network asymmetry, 249
  network limit, 331, 334, 336, 338
  network management interface, 246, 247
  network operator, 171–173
NGN, 12–17
NTP, 87–91

O

offset (time), 36, 62
one-step (clock), 61, 69, 226, 309
operating systems, 88, 240
ordinary clock, 58
OTN, 3–12

P

packet timing, 47–55
packet-based, 82–94, 141, 142
  packet-based method, 52–55, 141, 142
  packet-based synchronization, 92, 93
  packet-based timing, 82–86, 277–282
path delay, 54, 71, 76
PDH, 3–12
PDV, 318
performance, 294–303, 316–326
physical layer, 42–46
  physical layer synchronization, 42
plesiochronous, 3–12
PLL, 35, 36
plug and play, 64
pointer adjustment, 8
PON, 173
PRC, 38, 126

primary reference clock, 38, 126
priority, 80
proactive maintenance, 249
probes, 249
profile, 75–82, 197, 217, 236, 308–326, 335–339
protection, 242–245
provisioning, 245
PRTC, 140, 141
pseudo-wire, 350,
PSTN, 3–7
PTP, 55–75
  Precise Time Protocol, 55–75
  PTP communication path, 59
  PTP master, 59
  PTP message, 60
  PTP slave, 59, 319–326
PTSF, 80
  PTSF-unusable, 80
pull-in, 36
pull-out, 36

Q

Q adapter, 241
Q interface, 241
QL, 306–308

R

receivers, 225, 227
redundancy, 121, 122, 142–146
reference, 140, 141, 293
  primary reference (clock), 140, 141
  signal, 293
  timing signal, 35, 47
revertive, 243
ring, 127–129
rubidium, 97

S

SDH, 7–11
SEC, 9
Selection, 306–308
service quality, 246
SLA, 178
slip, 5, 6
SNMP, 250–252
spoofing, 265, 268–270
SRTS, 48
SSM, 43–46
SSU, 130, 131
stability, 99
STM-N, 9
SyncE, 42–46, 119–133, 184–196, 210–212, 294–303, 333, 334
synchronization, 3–12, 14–17, 22–29, 33–41, 43, 44, 111–117, 119–146, 161–236, 242–245, 255–261, 274–283,
   synchronization network design, 133
   synchronization network, 242–245
   synchronization plan, 242–245
synchronous (clocking, network, method), 28–39, 33–41
   synchronous method, 28–39
   synchronous network, 33–41
synchronous Ethernet (SyncE), 42–46, 119–133, 184–196, 210–212, 294–303, 333, 334
syntonization, 149

T

TDD, 174–181
TDEV, 291
TDM 86, 87
TDM pseudo-wire, 83
test, testing, 293–326
threat (security), 255–282
TICTOC, 91
TIE, 289
time constant, 37
timing, 42–76, 82–87, 94–102, 182–236, 261–282, 315, 316
   timing loop, 120, 123, 124
   timing service, 68, 86
TLV, 46
ToD, 47, 48
transparency, 166
transparent clock, 134, 140, 141
   distributed transparent clock, 140, 141
   Telecom Transparent Clock (T-TC), 279
two-step (clock), 61, 89
TWTT, 61

U

UTC, 94

V

vulnerability, 263–270

W

wander, 10, 11, 16
WCDMA, 21, 26

Printed and bound by CPI Group (UK) Ltd, Croydon, CR0 4YY
08/02/2023

03189970-0005